RIVULINS
OF THE OLD WORLD

by

Jorgen Scheel

Photos by the author

t.f.h. Publications

ISBN 0-87666-137-1

Distributed in the U.S.A. by T.F.H. Publications, Inc., 211 West Sylvania Avenue, P.O. Box 27, Neptune City, N.J. 07753; in England by T.F.H. (Gt. Britain) Ltd., 13 Nutley Lane, Reigate, Surrey; in Canada to the book store and library trade by Clarke, Irwin & Company, Clarwin House, 791 St. Clair Avenue West, Toronto 10, Ontario; in Canada to the pet trade by Rolf C. Hagen Ltd., 3225 Sartelon Street, Montreal 382, Quebec; in Southeast Asia by Y.W. Ong, 9 Lorong 36 Geylang, Singapore 14; in Australia and the south Pacific by Pet Imports Pty. Ltd., P.O. Box 149, Brookvale 2100, N.S.W., Australia. Published by T.F.H. Publications, Inc. Ltd., The British Crown Colony of Hong Kong.

TABLE OF CONTENTS

ACKNOWLEDGMENTS

I am greatly indebted to Dr. Herbert R. Axelrod for having had the opportunity to write the present review of rivulins of the Old World.

I also wish to thank the directors of the Institute of Genetics and the Institute of Zoology for advice and support during the cytological study, the Carlsberg Foundation for grants which made the zoological expeditions to Africa possible, the Danish State Research Foundation for microscopic equipment, and the Ministry of Defense and the Army Command for leave during the African expeditions.

I dedicate this book to

JOH. PAUL ARNOLD, Aquarist, of Hamburg, Germany

and

GEORGE STANLEY MYERS, Professor of Zoology, Stanford University,

two leading authorities on rivulins

HOW TO USE THIS BOOK

Though this book presents for the first time in book form the revolutionary work of Jorgen Scheel, it also has been written in a language that most serious hobbyists can understand. A Glossary on page 472 will assist in the few technical words which are not fully explained within the context of their use.

It is necessary to read through the first 79 pages in order to familiarize yourself with the great work done by the author and to understand what makes one rivulin different from the other.

To identify a living fish would require that you go through the book page by page and find the fish which looks most like the one you are trying to identify, then refer to pages 7 through 11 and ascertain its current scientific name. If the specimen you are trying to identify has been preserved so you can count the fins rays and scales, you can refer to Table 1 on pages 457 and 458. This table is your best place to begin in order to determine the scientific name of any unknown rivulin for it is really a résumé of everything in this book.

THE PUBLISHER

Tentative list of valid names for African and Asian rivuline species

AHL:	*Aphyosemion ahli* Myers 1933	Coastal Cameroon
ANN:	*Epiplatys annulatus* (Boulenger 1915)	Guinea-Liberia
ARN:	*Aphyosemion arnoldi* (Boulenger 1908)	Niger Delta
AUS:	*Aphyosemion australe* (Rachow 1921)	Gaboon?
BAR:	*Epiplatys barmoiensis* Scheel 1968	Sierra Leone/Liberia
BAT:	*Aphyosemion batesii* (Boulenger 1911)	Cameroon
BER:	*Roloffia bertholdi* (Roloff 1965)	Sierra Leone/Liberia
BIF:	*Epiplatys bifasciatus* (Stdr. 1881)	West Africa/Nile
BIV:	*Aphyosemion bivittatum* (Lönnberg 1895)	Togo/Sp. Guinea
BLO:	*Aplocheilus blockii* (Arnold 1911)	S. India
BUA:	*Aphyosemion bualanum* (Ahl 1924)	E.Cameroon/C.Afr.Rep.
CAL:	*Aphyosemion calliurum* (Blgr. 1911)	S. Nigeria
CAM:	*Aphyosemion cameronense* (Blgr. 1903)	S. Cameroon
CHA:	*Epiplatys chaperi* (Sauvage 1882)	Ivory Coast/Ghana
CHE:	*Epiplatys chevalieri* (Pellegrin 1904)	C. Congo
CHR:	*Aphyosemion christyi* (Blgr. 1915)	C. Congo
CIN:	*Aphyosemion cinnamomeum* Clausen 1963	W. Cameroon
DAG:	*Epiplatys dageti* Poll 1953	Liberia/Ghana
DAY:	*Aplocheilus dayi* (Steindachner 1892)	Ceylon
DUB:	*Aphyoplatys duboisi* (Poll 1953)	C. Congo
ELE:	*Aphyosemion elegans* (Blgr. 1899)	C. Congo
ESE:	*Epiplatys esekanus* Scheel 1968	E. Cameroon
EXI:	*Aphyosemion exiguum* (Blgr. 1911)	E. Cameroon/N. Gaboon
FAS:	*Epiplatys fasciolatus* (Günther 1866)	Guinée/Liberia
FER:	*Aphyosemion ferranti* (Blgr. 1910)	Kasai, C. Congo
FIL:	*Aphyosemion filamentosum* (Meinken 1933)	S. Nigeria
FLA:	"*Aphyosemion*" *flavipinnis* (Meinken 1932)	S. Nigeria
GAM:	*Fundulosoma gambiense* (Svensson 1934)	Gambia
GAR:	*Aphyosemion gardneri* (Blgr. 1911)	Nigeria/W. Cameroon
GER:	*Roloffia geryi* (Lambert 1958)	Guinée/Sierra Leone
GRA:	*Epiplatys grahami* (Blgr. 1911)	S. Nigeria/Cameroon
GUE:	*Nothobranchius guentheri* (Pfeffer 1893)	Zanzibar
GUI:	*Roloffia guineensis* (Daget 1954)	Guinée/Liberia
GUL:	*Aphyosemion gulare* (Blgr. 1901)	S. Nigeria
HOM:	*Pachypanchax homalonotus* (Dumeril 1861)	Madagascar
LAB:	*Aphyosemion labarrei* Poll 1952	Lower Congo
LAM:	*Epiplatys lamottei* Daget 1954	High Guinée
LIB:	*Roloffia liberiensis* (Blgr. 1908)	S. Liberia
LIN:	*Aplocheilus lineatus* (C&V 1846)	India
LON:	*Epiplatys longiventralis* (Blgr. 1911)	Niger Delta
LOU:	*Aphyosemion louessense* (Pellegrin 1931)	Brazzaville-Congo

MAE:	*Roloffia maeseni* (Poll 1941)	W. Ivory Coast/Guinée
MIL:	*Nothobranchius microlepis* (Vinciguerra 1897)	Somali
MUF:	*Epiplatys multifasciatus* (Blgr. 1913)	C. Congo-L. Congo?
NDI:	*Aphyosemion ndianum* Scheel 1968	W. Cameroon
NIC:	*Epiplatys nigricans* (Blgr. 1913)	C. Congo
OCC:	*Roloffia occidentalis* (Clausen 1965)	Sierra Leone
ORT:	*Nothobranchius orthonotus* (Peters 1844)	Mozambique
PAN:	*Aplocheilus panchax* (Hamilton-Buchanan 1822)	India/Indonesia
PET:	*Roloffia petersii* (Sauvage 1882)	Ivory Coast/Ghana
PLA:	*Pachypanchax playfairi* (Günther 1866)	Seychelles/Zanzibar
RAC:	*Nothobranchius rachovii* Ahl 1926	Mozambique
ROL:	*Roloffia roloffi* Roloff 1936	Sierra Leone
SEX:	*Epiplatys sexfasciatus* Gill 1862	Ghana/Gaboon
SIN:	*Epiplatys singa* (Blgr. 1899)	Lower Congo
SJO:	*Aphyosemion sjöstedti* (Lönnberg 1895)	S. Nigeria/Cameroon
SPI:	*Epiplatys spilargyreius* (Dumeril 1861)	Trop. Africa
SPL:	*Aphyosemion splendidum* (Pellegrin 1930)	N. Gaboon
STR:	*Aphyosemion striatum* (Blgr. 1911)	N. Gaboon
TAP:	*Nothobranchius taeniopygus* (Hilgend 1888)	Tanganyika
THI:	*Fundulosoma thierryi* Ahl 1924	W. Africa
WAL:	*Aphyosemion walkeri* (Blgr. 1911)	Ivory Coast/Ghana
WER:	*Aplocheilus werneri* Meinken 1966	Asia

Tentative list of valid subspecies of African and Asian Rivulinae

DOM: *Aplocheilus panchax dorsomarginatus* Klausewitz 1957
MON *Epiplatys dageti monroviae* Daget & Arnoult 1964
SHE: *Epiplatys chaperi sheljuzhkoi* Poll 1953
TOD: *Roloffia occidentalis toddi* (Clausen 1965)

Tentative list of ill-defined names, which might prove to be valid *(Nomen dubium)*

AND:	*Aplocheilus andamanicus* (Köhler 1906)	Andaman Islands
ANS:	*Epiplatys ansorgei* (Blgr. 1911)	Gaboon
BAU:	*Epiplatys baudoni* (Pellegrin 1922)	Cent. Afr. Republ.
BEA:	*Aphyosemion beauforti* (Ahl 1924)	S. Cameroon
BRI:	*Nothobranchius brieni* Poll 1938	Katanga, Congo
COG:	*Aphyosemion cognatum* Meinken 1951	C. Congo
DEC:	*Aphyosemion decorsei* (Pellegrin 1904)	C. Congo
FAL:	*Aphyosemion fallax* Ahl 1935	Nigeria??
KIY:	"*Fundulosoma*" *kiyawense* (Ahl 1928)	N. Nigeria

LUJ:	*Aphyosemion lujae* (Blgr. 1911)	S. Congo
MAC:	*Epiplatys macrostigma* (Blgr. 1911)	Lower Congo Area
MAG:	*Aphyosemion margaretae* Fowler 1936	E. Congo area
MAY:	*Nothobranchius mayeri* Ahl 1935	Mozambique
MEI:	*Aphyosemion meinkeni* Myers 1933	Unknown
MEL:	*Roloffia melanteron* (Fowler 1950)	Liberia
MEP:	*Nothobranchius melanospilus* (Pfeffer 1896)	Tanganyika
NEU:	*Nothobranchius neumanni* (Hilgendorf 1905)	Tanganyika?
NIM:	*Epiplatys nigromarginatus* (Schultze 1937)	S. Nigeria?
OBS:	*Aphyosemion obscurum* (Ahl 1924)	E. Cameroon
OGO:	*Aphyosemion ogoensis* (Pellegrin 1930)	Gaboon
OLB:	*Epiplatys chaperi olbrechtsi* Poll 1941	W. Ivory Coast
ORN:	*Epiplatys ornatus* (Ahl 1928)	Unknown
PAL:	*Nothobranchius palmquisti* (Lönnberg 1907)	Tanganyika
PAT:	*Nothobranchius patrizii* (Vinciguerra 1927)	Somali
RUR:	*Nothobranchius rubroreticulatus* B&M 1960	Chad
SAN:	*Epiplatys sangmelinensis* (Ahl 1928)	E. Cameroon
SCH:	*Aphyosemion schoutedeni* (Blgr. 1920)	C. Congo
SEY:	*Nothobranchius seychellensis* Ahl 1935	Seychelles Isl.
SIA:	*Aplocheilus panchax siamensis* Scheel 1968	Thailand?
ZIM:	*Aphyosemion zimmeri* (Ahl 1924)	Unknown

Tentative list of synonyms

ACU:	*Epiplatys acuticaudatus* (Pellegrin 1913)	= SPI
BEL:	*Aphyosemion bellicauda* (Ahl 1924)	= OBS?
BIT:	*Aphyosemion bitaeniatum* (Ahl 1924)	= BIV
BOU:	*Epiplatys boulengeri* (Pellegrin 1926)	= MUF
BRU:	*Aphyosemion brucei* (Boulenger 1911)	= GAR?
CAB:	*Roloffia calabaricus* (Ahl 1935)	= LIB
CAR:	*Aphyosemion carnapi* (Ahl 1924)	= OBS
CAS:	*Aphyosemion castaneum* Myers 1924	= SCH
COE:	*Aphyosemion coeruleum* (Boulenger 1915)	= SJO
CON:	*Aphyosemion congicum* (Ahl 1924)	= CHR?
DEF:	*Epiplatys grahami decemfasciatus* (Pellegrin)	= SPI
DOR:	*Epiplatys dorsalis* Mayer 1936	= FAS
ELB:	*Aphyosemion elberti* (Ahl 1924)	= BUA
ESC:	*Aphyosemion escherichi* (Ahl 1924)	= CAM?
GUS:	*Aphyosemion gustavi* (Ahl 1924)	= BEA
HOL:	*Aphyosemion bivittatum hollyi* Myers 1933	= BIV
INF:	*Epiplatys infrafasciatus* (Günther 1866)	= SEX
JAC:	*Aphyosemion jacobi* Ahl 1928	= EXI
JAU:	*Aphyosemion jaundense* (Ahl 1924)	= EXI

KUH:	*Nothobranchius kuhntae* (Ahl 1926)	= ORT?
LEO:	*Epiplatys sexfasciatus leonensis* Ahl 1937	= FAS
LOB:	*Aphyosemion loboanum* (Ahl 1924)	= EXI
LOE:	*Aphyosemion loennbergii* (Boulenger 1903)	= BIV
LOL:	*Aphyosemion loloense* (Ahl 1928)	= EXI
MAR:	*Epiplatys marnoi* (Steindachner 1881)	= SPI
MAT:	*Epiplatys matlocki* Fowler 1950	= FAS
MIC:	*Aphyosemion microstomum* (Ahl 1924)	= CAM
MUC:	*Aphyosemion multicolor* (Meinken 1930)	= BIV
NDE:	*Epiplatys ndelensis* Fowler 1949	= BIF
NIG:	*Aphyosemion nigerianum* Clausen 1963	= GAR
NII:	*Aphyosemion nigri* Ahl 1935	= BIV?
NOR:	*Aphyosemion normani* (Ahl 1928)	= OBS
NUC:	*Pachypanchax nuchimaculatus* (Guichen 1866)	= HOM
PAP:	*Aphyosemion pappenheimi* (Ahl 1924)	= BIV
PAR:	*Aplocheilus parvus* (Sundaja Raj)	= BLO
PAS:	*Aphyosemion pascheni* (Ahl 1928)	= AHL?
POL:	*Aphyosemion polychromum* (Ahl 1924)	= AUS
PRE:	*Aphyosemion preussi* (Ahl 1924)	= OBS
RIG:	*Aphyosemion riggenbachi* (Ahl 1924)	= BIV
RUF:	*Aphyosemion rubrifascium* Clausen 1963	= BUA
RUM:	*Aplocheilus rubrostigmus* (Jerdon 1849)	= LIN
RUS:	*Aphyosemion rubrostictum* (Ahl 1924)	= BIV
SAK:	*Pachypanchax sakaramyi* (Holly 1928)	= HOM?
SRE:	*Aphyosemion schreineri* Ahl 1935	= BEA
SEN:	*Epiplatys senegalensis* (Steindachner 1870)	= SPI
SPM:	*Epiplatys spillmanni* Arnoult 1960	= SHE
SPP:	*Aphyosemion splendopleure* (Meinken 1930)	= BIV
SPU:	*Aphyosemion spurrelli* (Boulenger 1913)	= WAL
STE:	*Epiplatys steindachneri* (Svensson 1934)	= BIF
TAE:	*Epiplatys taeniatus* (Pfaff 1933)	= BIF
TES:	*Aphyosemion tessmanni* (Ahl 1924)	= BUA
TRO:	*Nothobranchius troemneri* (Myers 1926)	= ORT
UNC:	*Aphyosemion unicolor* (Ahl 1924)	= BIV?
UNS:	*Aphyosemion unistrigatum* Ahl 1935	= BIV
VEX:	*Aphyosemion vexillifer* (Meinken 1929)	= CAL

List of probably invalid names for African and Asian rivuline species *(Nomen nudum)*

AMS: *Nothobranchius amsinki* Ahl 1928
BUR: *Aphyosemion burundi*
HJE: *Aphyosemion australe hjerreseni* Meinken 1953
OES: *Aphyosemion oeseri* (Ahl 1928)
PIC: *Aphyosemion pictum* (Meinken 1932)
RUO: *Aphyosemion rubrofasciatum* (Brünning 1930)
SEF: *Epiplatys sexfasciatus* (Peters 1863)

List of names attached to very ill-defined phenotypes *(Nomen vanem)*

CHI: *Epiplatys chinchoxcanus* (Ahl 1924)
EIM: *Nothobranchius eiminki* Ahl 1935
FLV: *Epiplatys flavus* (Ahl 1924)
NYO: *Epiplatys nyongensis* (Ahl 1928)
ROB: *Nothobranchius robustus* Ahl 1935
SUP: *Epiplatys superbus* (Ahl 1924)
VOS: *Nothobranchius vosseleri* Ahl 1924
ZEN: *Epiplatys zenkeri* (Ahl 1928)

List of names which may or may not be valid *(Nomen dubium)*

LUT: *Aplocheilus panchax lutescens* (Köhler 1906)
MAI: *Aplocheilus panchax mattei* (Köhler 1906)
MAN: *Aplocheilus panchax marginatus* (Köhler 1907)
MKU: *Nothobranchius mkuziensis* Fowler 1934 (description not seen)

INTRODUCTION

Our knowledge of all aspects of *Rivulinae* is rapidly increasing and certainly it is not premature to introduce a book on this theme at the present time. The reliable information concerning these interesting fishes is, however, to be found in many popular, technical and scientific periodicals and books, and also much has previously been published concerning rivulins which is not in agreement with our present knowledge. Thus, a review of rivulins probably is needed just now.

In the study of rivuline species there is a tradition of zoologists and aquarists having joined hands, because it has become increasingly clear that the study of live material in the aquarium and in the laboratory is indispensable in the taxonomic study of these fishes.

This book represents a review of information given by amateurs and zoologists and often it is possible on the basis of the information given by both groups of people to unravel certain taxonomic problems.

As will be seen, I attach much importance to the publications of certain aquarists who studied newly introduced strains of aquarium fishes which were later described as new species. J. P. Arnold of Hamburg who studied almost all newly introduced rivuline strains in Germany was a very keen observer and his many publications in German aquarium magazines have been an invaluable source of reliable information. Arnold very often was not in agreement with the leading rivuline-zoologist of the time, G. A. Boulenger of the British Museum. But, as we shall see, Arnold was generally right and Boulenger was wrong, usually because of the incomplete knowledge of the nature of the rivuline morpho-species and because he generally ignored colors and color patterns in the systematics of these fishes.

I am greatly indebted to many aquarists and zoologists for their kind help through the years.

The habitat (biotope) of EXI located east of Yaunde. (See facing page.)

The habitat of the type of RUF (=BUA). This area is east of Ndop.

I

THE RIVULINS

The *Rivulinae* (or *Aplocheilinae*) represents the major group of *Cyprinodontidae* making up more than half the known number of nominal species of the family. About 250 different nominal species of rivulins are known at the present time, and it is likely, that in the future, many new species of rivulins will be described from Africa and South America.

The rivulins live in tropical climates in Asia, Africa and America and only a few species extend into the warm temperate climates in the New World and in southeastern Africa. No other subfamily of cyprinodonts have a similar world-wide distribution. Most species are restricted to the tropical forest, but some species groups and genera extend into the savannah, generally being restricted to temporary pools, swamps and the like.

Although most species (Africa and Asia) are found in the coastal lowlands, very few species have been found in brackish water or in the sea. Generally speaking, the rivulins are strictly freshwater fishes, inhabiting freshwater which often is so pure that it corresponds to distilled water of average quality.

Rivuline species are small fishes. The average species does not exceed two inches. A few species reach four inches. This subfamily also contains very small fishes (Africa: FLA, ANN, DUB and Asia: BLO) although *Procatopodinae* (*Aplocheilichthys myersi* Poll) and *Oryziatinae* (*O. minutillus* Smith) contain forms which are even smaller.

This large group of cyprinodonts have developed many different phenotypes adapted to different ecological conditions. There are pike-like, surface forms (*Aplocheilus* and *Epiplatys*) with dark crossbars on the sides, and deep-bodied, bottom-living forms with large dorsal fins.

The rivulins also contain a number of species highly adapted to the extremely difficult ecological conditions in temporary pools. These are the sole true annual fishes, which live for less than one year and which survive as embryos in the soil during the dry season.

Because of these qualifications, the annual rivulins have gained increasing importance in the biological anti-malaria control and probably in future, these fishes will prove to be the superior tool in anti-mosquito control. Many of the non-annual forms are very effective in anti-mosquito control (*Aphyosemion* and *Rivulus*).

Since the introduction of these fishes as aquarium pets at the beginning of the century, these fishes have been popular among aquarists because of their small size and gorgeous colors in the males, but it was not until quite recently that the maintainance and reproduction of the most handsome forms has been fully mastered.

To the museum taxonomist, these fishes always have been very difficult to identify, because of the extreme individual variation of all morphological characters, and much taxonomic research is needed in the future before the true relationships between various populations is fully understood.

Because of the difficult maintainance of rivuline species very little biological research has been prepared on the basis of rivuline species. At the present time it is possible, however, to maintain constant laboratory strains which can be used for various biological research including cytological studies of these strains and of hybrids. Because of the quick growth of the individuals and of some of the hybrids, individuals might be ready for cytological research on gonads three or four months after spawning. To the cytologist, these fishes are highly interesting because of the extreme variation of chromosome numbers ($n = 9$ to 27) and of karyotypes even within the morpho-species.

The transfer of live rivuline material is very easy because eggs can be sent all over the world by mail. During one year I introduced fifteen rivuline species to New Zealand, where rivulins could not be had before that time.

There are groups of experienced aquarists specialized in the keeping and breeding of rivuline species in many countries.

The *American Killifish Association* is an international organization with members in many countries. This organizatjon (AKA) has its own printed magazine (The Journal of the American Killifish Association) and also the AKA distributes Newsletters and lists of eggs for trade.

The *British Killifish Association* is a similar, but younger society, also with members in different countries.

The *ZAG Eierlegende Zahnkarpfen* is a similar organization in DDR and in many larger cities there are local groups of cooperating killifish fans.

II

KEEPING RIVULINE SPECIES

Generally, individuals of most rivuline species are not suited for the community tank. These fishes do not exhibit any social behavior and even the juveniles do not shoal. On the contrary, adult rivuline individuals often are rather aggressive towards other fishes and in particular towards individuals of their own kind. Also most rivuline individuals do not thrive in the community tank which does not offer the proper milieu for these fishes.

Killifish fanciers usually have a large number of rather small tanks and in each tank they place a single male and some females. If kept with a single female only, the male usually will dominate, and, at times, the female will be killed if there are not sufficient hiding places. Tanks containing from two to four gallons of water usually will be sufficient for species of average size.

The heating of a number of tanks is too complicated if each tank is supplied with an electrical heater and a thermostat. For that reason, the fancier usually keeps all his tanks in a single room in which the air temperature remains sufficiently constant (20–25° C or 68–77° F) all year round. He also will have small containers for the development of eggs in moist peat in that room. If such a room is not available, the tanks and peat might be stored in a cabinet. Transparent plastic on frames has proven to be very useful for the construction of such cabinets, in which only one (air) heater and one thermostat are placed. For most tropical species the temperature should not drop below 18° C (64° F) and not exceed 30° C (86° F). Adult individuals should not be kept too warm, otherwise they will not live for a long time.

Usually, no aeration is needed for tanks containing killifish, but pumps are very useful for the production of Daphnia using the "dirty-water/sugar-years" system (Scheel 1966c: 40–41).

All tanks should be completely covered by glass or other means. Most rivuline individuals are excellent jumpers and will escape from the tank if they find a hole sufficiently large to climb through.

The natural food for most killifish is terrestrial insects. Since ants, the usual natural food, are not easily cultured, they are replaced by fruit flies (*Drosophila melanogaster, D. funebris* and other small species) which are very easily cultured and handled (Scheel 1963c: 7–9). During summertime

Larvae and pupae of *Culex* in the aquarium.

various sorts of mosquito larvae might also be cultured (Scheel 1963b: 9–11). All rivuline species of the Old World can be trained to take various sorts of dry food and they thrive well on such food, if they get plenty of live food once a week. It seems that juveniles which are fed on dry food plus live food grow more quickly than juveniles raised on live food only.

Plants are not a must if other hiding places can be offered for the protection of females, but in tanks containing quick-growing plants the ecological conditions can be kept rather constant by the methods described below. If plants are not used, 10–20% of the water should be replaced each week in order to keep down the concentration of waste products. My own tanks are all artificially lighted and have hard, alkaline water. For such tanks I found *Symnea triflorum* to be an almost ideal plant, but each week the stems have to be cut down (2–3 inches below the surface of the water) to keep a good and dense growth. *Riccia* is often used as a surface layer in tanks housing small forms which are not too aggressive (ANN, DUB and the like) but *Riccia* does not grow well in hard water.

Most rivuline individuals take very much food and it is often not too easy to keep down the concentration of waste-products. Also, these fishes prefer food rich in proteins and their waste-products simply cannot be taken up by the plants because of the deficiency of carbon dioxide (Scheel 1962: 9–12) and an artificial supply of this gas should be given. The usual system, with fermenting sugar-solution in a separate bottle for each tank, is not very practical for the killifish fancier. When much care is taken, another system might be used. Prepare a 5% solution of sugar in water (about 6 ounces of sugar to each gallon of water) and add 16 cc. of this solution to each four gallons of aquarium water (1 cc. to each liter). Best results are had when the sugar solution is stirred into the water and bad results are usually had when solid sugar is added directly to the aquarium. This method is very useful when the possible bad effects are understood and counteracted. First the sugar is converted into carbon dioxide and alcohol, which do no harm. Then the alcohol is converted into acetic acid and the oxygen needed for this process is taken from sulphates, resulting in a production of the poisonous hydrogen sulphide gas, which is very harmful for fishes. I have used this system in fifty tanks for more than two years without killing a single fish. The sugar solution usually is added to the tank every two weeks. It has been possible to maintain many tanks without renewal of the plants, peat and water for more than one year.

In nature, most rivuline species are used to a very soft and acid water containing very little dissolved mineral matter and often rather much organic matter. If rivuline individuals are kept in this type of water in the aquarium, various diseases often become very serious, and the individuals are easily killed. The "pest of killifishes," the mycobacteria, seems to thrive

Colonies of *Mycobacteria* in live gonad tissues of AUS/AHL hybrid male. Phase-contrast photo.

very well in such natural waters, whereas in the unnatural hard, alkaline water, the attacks from this disease are less serious, and the fishes survive. Also the attacks from "velvet" or *Oodinium* are less serious in hard, alkaline water. The mycobacteria prefer to live inside the cells containing fat, and the attacks from this particular disease are very serious when the juveniles are about to mature sexually, because at that phase of the development the gonads contain much fat. These bacteria attack other tissues also.

At the present time we have no good cure for mycobacteria and the only thing which we can do is to improve the conditions for the fishes. The first thing to do is to keep the fishes in hard, alkaline water. Secondly, it appears to be important to keep down the concentration of waste products in the tanks by renewing 10–20% of the aquarium water each week or by a good plant culture.

If the tapwater of your district is not hard and/or alkaline, add solid lime.

Compared to other aquarists, the killifish fancier is fortunate because he is able to get most of his supply of new species, not as adult individuals, but as eggs which usually do not carry any disease. Mycobacteria and spores of velvet probably cannot be kept out of the tanks. The real nature of the attacks from velvet is not well understood. For some periods this disease might be a serious problem to the individual aquarist and at other times he might not notice any attack for years. When peat is used as a bottom layer, no medicine is useful because the drug will be absorbed very quickly by the peat. In such tanks I use solid copper, and I found the thin bars which are used in workshops for hard-soldering very useful. One foot of this material has been used with good results in two to three gallons of hard, alkaline water. If used in soft and slightly acid water one should observe the fish (and snails) very carefully. Plants usually are not hurt, but the snails (and Hydra, if present) usually are killed. In hard, alkaline water the bars may remain in the tank for weeks. My aquarium water measures about 15 German degrees of temporary hardness and almost no permanent hardness.

I have used peat as a bottom layer in my tanks for almost twenty years. Coarse peat usually is preferred because this material is more easily handled, permits better circulation of the water within the bottom layer and gives better protection for the fishes which are chased by larger individuals. I prefer not to discard the old peat and the peat used in my tanks are more like fine mud than like common peat. The old peat, which does not produce brown water, appears to be just as useful for the protection of the fishes against disease as the new peat.

The layer of peat at the bottom of the tank should normally measure one to two inches.

CARE OF PEAT

Aquarists who are used to sand or gravel as a bottom layer of their tanks, usually resist using peat because they consider this material to be very troublesome and so it is. When keeping so called "difficult" tropical fishes from the forest, the use of peat instead of sand renders certain major advantages, not only for reproductjon but also in the keeping of feeble fishes.

Peat is an active material. It absorbs various chemical compounds and this is one of the reasons why peat is able to maintain a very clear water even in tanks with many fishes. Also the peat gives a number of chemical compounds to the water which are useful during the reproduction of difficult fishes and to prevent attack from bacteria. Generally, it seems that the many advantages compensate for the disadvantages produced by the more difficult handling of the material.

First, the peat should be soaked in water and all the air in the material should be forced out by keeping the material in water for some weeks or by boiling it for some hours. You may add one handful of bicarbonate of soda to each three gallons of water used for the soaking or boiling of the fresh peat. The soda will extract brown compounds (and also some useful hormones) and the aquarium water will be less brown, but when soda has been used the peat should be washed in water until the water does not become dark brown.

Peat should never be used together with sand or gravel. Simply add sufficient amounts of soaked peat to the tank and soon the material will settle on the bottom. Plants are kept down by small stones placed on their roots.

By the absorbtion of various waste products the peat may become so saturated by these compounds, that a regeneration of the material might be needed. This is done simply by a treatment with soda as above for at least one week and washing of the material, but often it is even better to place all used peat in a large vat and to add sugar. If the vat is large enough, there might be much water above the peat-layer and Daphnia are produced with very fine results in this water, particularly if an air-stone is placed above the surface of the peat. In this way the waste products are converted into live food and no light is needed. The water above the peat-layer should remain rather cloudy, and if it becomes clear more sugar should be added.

Spawning behavior of *Pachypanchax*. Photos by Rudolf Zukal.

III

BREEDING NON-ANNUAL RIVULINE SPECIES

It has already been pointed out that rivuline individuals should be kept in hard and alkaline water to reduce attacks from mycobacteria and other diseases.

Some rivuline species reproduce sufficiently well in hard, alkaline water, and for these species, this sort of water should be used. Many species do not, however, reproduce sufficiently well in hard, alkaline water. The reason for this is not fully understood, and often different aquarists have different results under apparently, similar breeding conditions. Sometimes, the percentage of fertilized eggs is much reduced or even nil. It might be, that with these species, the sperm does not survive a sufficiently long time or is not able to swim, but there might be other reasons also. During my cytological study of rivuline males, which I had directly from other breeders or from the trade, I often observed, that the gonads were rudimentary, although the males had reached almost full size and had developed all the colors of a sexually mature male. In 1966, I brought back from West Cameroon live individuals of GAR and the sympatric *Procatopus nigromarginatus* from pools in a river bed west of Mamfe. These individuals were taken at the end of the dry season. Males of both cyprinodont species were almost fully grown and had developed the full colors of the males of their species. During the first cytological investigations I found that the males of both species had very rudimentary gonads and these contained almost no spermatozoa. Also, these males did not exhibit sexual behavior in the aquarium. Six weeks after their arrival in Copenhagen, these males still had rather rudimentary gonads, but they all exhibited sexual behavior. Three months after their arrival, the gonads of males from both species were normal, although these males had not increased much in size. Similar conditions are well-known from other groups of fishes, and in particular among fish species which aquarists have called "the problem fishes": *Acanthophthalmus kuhli*, *Nannostomus trifasciatus* and *Sphaerichthys osphromenoides*, just to mention three species representing three different orders of bony fishes. Stolk (1957: 35, 1958: 12 and 1960: 12 and 264) studied the pituitary glands and the gonads of aquarium kept individuals of the three species above mentioned (and several other species of tropical fishes) and proved that sexual maturation did not take place in the aquarium unless urine from pregnant women had been added to the aquarium water (0.5–1.0 cc./L.). The gonatropine hormones from the urine act on the pituitary glands, which start growing and after some time sexual hormones are produced, which force the gonads to develop, and after three or four weeks spawning takes place.

In nature a similar system works. During the dry season, the young fishes often have plenty of food (insects) and grow to about full size without any development of the gonads. Instead the individuals store large amounts of fat within and around the tissues which later on will mature as gonads. When the rain starts, the dead vegetation which has accumulated on the forest floor and which has in part decomposed, imparts to the rainwater various organic compounds and among these compounds there are some which are very similar to the gonatropine hormones and which act on the pituitary glands of the immature tropical fishes of the forest. About one month after the rains have started, the individuals reach sexual maturity and spawning begins. At that time the rain probably has flooded large areas near the streams, plenty of microscopic live food is available and everything is ready for the new generation of fish. If the reproduction had started just after the rains, the conditions certainly would have been less favorable.

Thus, it is likely that some of the rivuline species which we have from other breeders, have not matured sexually, although they appear to be adults. If these individuals have not been raised in the "right way" (which we do not know), they may not reproduce, or their reproduction is not too satisfactory. Peat is known to contain various hormones, and some of these appear to act as gonatropine hormones. My own *Acanthophthalmus* reproduced in my tanks with a bottom-layer of peat many years ago. In my tanks with very old and mudlike peat (in steady use for ten or more years) all my rivuline species mature well and on time. Probably, the many different microscopic organisms which live in this peat produce the hormones needed.

With other species, we are able to get a sufficient percentage of fertilized eggs in hard, alkaline water, but the eggs are ruined before the embryo is ready to hatch out. It seems that the egg-membrane is not able to withstand the attacks from various bacteria and fungi and decompose. Probably, the membrane reacts with the calcium and magnesium ions of the hard water and becomes less resistant. In 1962, I received many samples of live cyprinodont species from Ghana and southern Nigeria, together with these fishes I probably also received a new and strong "strain" of egg-destroying bacteria, for since that time I had to reproduce most species in soft, acid water. I am informed that several friends in other countries have not similar problems, but probably they have not yet received these bacteria.

Thus, some of us have to use two different sorts of water for our killifishes, but this does not represent any serious problem.

If you move rivuline individuals from a water containing much dissolved mineral matter to water containing little such matter, these individuals probably will soon be swimming at the surface and they soon die. If you inspect the gills under a microscope you will realize the reason for this. You will find the gills completely ruined. The many brittle cells of the gills burst shortly after the individuals had been placed in the water of low salinity, because of the large differences in osmotic pressures. If fishes are moved from water containing little dissolved mineral matter to water containing much such matter nothing usually will happen.

Most fishes will tolerate a decrease of salinity corresponding to half the amount of dissolved mineral matter, than that which they are used to.

If two different types of water are used for the keeping and the breeding of rivuline species, it is useful if these two sorts of water have similar salinity (amounts of dissolved mineral matter).

The simplest way to produce soft, slightly acid water from hard, alkaline water is first to remove all calcium and magnesium ions of the hard water with sodium ions (the common permutit ion-exchange material is useful), and when these ions have been removed, the water is treated with hydrochloric acid until the pH reaches about 3.6–4.0 (methylorange indicator change from brownish to violet). When the liberated carbon dioxide has been removed by aeration, the water is stored in a tank with peat, which will supply organic compounds to the water. When the carbon dioxide has been removed from the water, the pH-value usually will stabilize between 6.0 and 7.0. I raise plants in the tank where I store my soft, slightly acid water.

If you prefer to use rainwater for the reproduction of your killifishes it is recommended to add 300 to 500 ppm of common rock-salt to the rainwater, otherwise this water probably will contain too little dissolved mineral matter and your breeding pairs cannot be moved into the rainwater without risk. Common salt does not do any harm, and often this material renders an additional protection for the eggs and the individuals.

The non-annual rivuline species usually place their eggs near the surface of the water, within roots of plants and the like. The eggs are provided with long, slimey filaments usually concentrated at one pole and these filaments anchor the eggs to the roots. In the aquarium mops are often used instead of plants. These mops are prepared from nylon-yarn. The mop is anchored to a piece of cork or the like, and killifishes like to spawn within these mops. There should be sufficient yarn that the mop is sufficiently deep, otherwise the individuals will eat the eggs.

Eggs from rivulins cannot stand rough handling just after spawning. The eggs need a few hours to harden. After that time they may be taken out of the mop, with your fingers gently. The eggs are easily found if the mop is not too moist. Usually the eggs adhere so strongly to the mop, that even a violent shaking of the mop to remove water will not dislodge the eggs.

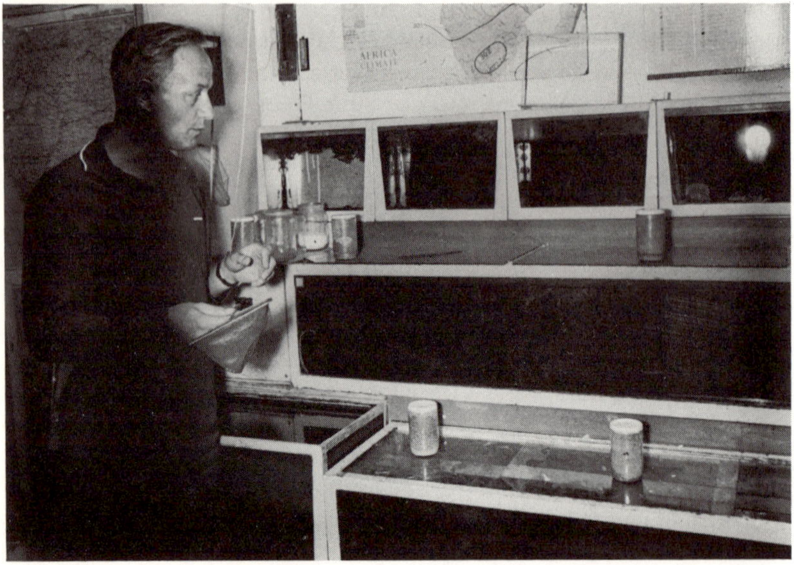

The author in a section of his fish room. The primary food fed to the adult rivulins is *Drosophila* fruit flies which the author raises in a special room where the flies are allowed complete freedom and are trapped in their feeding jars when required as food. Photo by Dr. Martin R. Brittan.

I spawn my non-annual species in rather small tanks (about one gallon of water) and the tanks contain nothing but a mop and the water. Females are separated from males for at least one week, while both sexes are heavily fed. Only one female is placed with the male in the tank, because one female may eat eggs when the other is spawning. As my two types of water have similar salinities I can move the individuals right in and right out of one type of water without any harm. Differences in pH values, even from 5.0 to 8.0 have not produced any harm when individuals suddenly are moved from one type of water to the other.

The individuals usually start spawning a few moments after having been brought together, and within one hour most eggs are spawned, and the individuals are moved to their ordinary water. Eggs are left untouched for six hours at least. Then they are taken out of the mop and placed in small glasses, with airtight lids, half filled with water from the breeding tank. Usually methylene blue (tetra-methylthionine chloride) is added (1–2 ppm) to protect the eggs against fungi and bacteria, and in case the egg-membrane is known to be sensitive (ANN) similar amounts of neutral trypaflavine (or acriflavine, euflavine, or rivanol) is added. This green mixture has proved to be excellent in case the ovaries of the females are infected by certain bacteria, and the eggs are thus infected, too.

If the eggs take up the dye and become strongly colored, the membrane is not in good condition, and often the eggs will not be able to develop normally. The interior parts of the egg should take up no dye or be slightly colored only. Eggs which become very dark, usually are not fertilized and should be removed.

The eggs are inspected twice a day during the first week, and all dead eggs are removed.

After two weeks, the embryos normally should have become ripe, but sometimes they do not hatch out naturally. They might be forced to hatch simply by breathing deeply into the glass and replacing the lid or by adding a small amount of fine mud (which will cover the eggs and reduce the amount of oxygen) or a *little* sugar (be careful) or dry food. As Jacob Scheidnass once said: "The hungry fry catch sight of the good food and come out to take it." Jack was joking of course but he knew that any method which will change the oxygen and carbon dioxide will be useful to induce the ripe embryo to hatch.

If the fry hatch out too early they should be kept in the glass until able to swim normally, but all normally swimming fry should be transferred to a separate tank containing hard, alkaline water and in this tank they are raised.

The fry of most rivuline species stay near the surface of the water during their first few days or weeks. They do not move much, and usually try to hide in corners or between plants. Those of ANN are more pelagic, than those of other rivuline species.

The fry feed on all small live organisms and most fry also take some powdered dry fry food. If brine shrimp is used, a slight aeration placed near the bottom in the lightest part of the tank will be useful to bring the newly hatched shrimp to the surface. Microworms, newly hatched larvae from various sorts of mosquitos and other sorts of small natural food are also taken.

The rate of growth is fastest among the annual species, which may mature sexually after three or four weeks, and the rate is slowest for the typical non-annual species although exceptions occur. Also, within a single brood, the rate of growth often is highly variable and usually the females grow rather slowly. If the differences in size become too large, the larger individuals may eat the smallest, and thus it is useful to sort the juveniles by removing all large males.

Spawning behavior of *Nothobranchius*. Photos by Rudolf Zukal.

IV

BREEDING ANNUAL RIVULINE SPECIES

The development of the egg from annual rivuline species has been described. The biological characteristics of the rivuline egg should be taken into consideration when reproducing annual rivulins.

The annual species are usually characterized by large dorsal fins (*Pterolebias* is an exception) and during spawning the male holds the female using the large dorsal which might be provided with ctenoid spines (*Nothobranchius*). The anterior parts of the anal fins of both sexes often are folded into tubes which serve to bury the egg within the upper layer of the bottom or to insure proper fertilization. The anal fin of males often have prominent fleshy ctenoid sensory papillae (African species) which probably also have certain functions during the spawning act, and which probably also are useful in pre-mating play. During the pre-mating play the male "rides" on the back of the female pressing his throat against the neck of the female and leading the female with his pectorals, anal and caudal fin. The filamentous pectorals found in some species and the sensory papillae which develop on the upper ray of this fin probably function during this part of the reproduction procedure. In this manner the female is lead towards the bottom and, having reached the bottom, the male comes to the side of the female and to grasp her by the dorsal and anal fin. South American annual species often dive deep into the soft bottom and place their eggs at some distance from the upper layers, whereas African annuals generally do not dive deeply into the mud.

Annual rivuline species usually are very fertile compared to the non-annual species and spawning usually takes place every day until the individuals die. The annual species are not sensitive to environmental conditions and if males and females are brought together they often will spawn within the transportation can or even in a glass which you hold in your hand.

Eggs usually are spawned into or onto the bottom layer but if males and females have been separated annual species might also spawn at the surface or any place where the individuals encounter.

If mud is used as a bottom layer the eggs usually will not get sufficient oxygen to develop, and these eggs will remain transparent until they are brought into aerobic conditions or die. In nature these eggs have to remain without visible development until the pools dry up during the dry season.

The habitat (biotope) of EXI located west of Sangmelima.

In the open (*Nothobranchius* and *Fundulosoma*) the upper layer of the dry bottom of the pool will be exposed to large fluctuations of temperature. During daytime the upper layers of the soil will be very warm, whereas during nights the radiation will cool down these layers of soil. The air stored in the porous soil will expand and contract according to the changes of temperature and by this way fresh air is brought down to the eggs hidden in the soil. Some of the eggs (true annuals) will remain transparent under these new conditions whereas others will develop. Some of the eggs which develop an embryo probably will hibernate as "resting embryos" (diapause II) for some weeks or months whereas others will develop into a ripe embryo, which will hibernate as "resting fry" (diapause V+). When the pool has been dried up for a certain time, the soil will contain a number of eggs representing all three diapauses.

When the rain starts, the supply of oxygen to the eggs suddenly is cut off and the eggs which contain ripe embryos have to hatch at once or die, but these conditions probably will not hurt eggs which hibernate at the earlier diapauses. The fry which hatch out in the soil escape tail first. Having reached the water the fry usually swim to the surface and take in air (?) to fill the swimbladder.

If the first rain was untimely, the pool soon will dry up and all fry which had hatched will die. But soon other eggs will develop into mature eggs which are ready to hatch when the next shower fills the pool. This complicated system (see OCC) guarantees that even in case a large number of showers temporarily fill the pool, there still will be eggs ready to hatch when the pool becomes permanent. The bottom of the pool is not level and for this reason conditions at the center of the pool will differ from those near the banks and an "annual reproduction" thus might take place even in case the pool does not dry out completely during the dry season. As my experiments with the South American *Cynolebias nigripinnis* has shown, eggs from this species may remain viable even if they remain under anaerobic conditions (in the soil below water) for three years.

Aquarists usually do not use fine mud, because this material is rather troublesome when reproducing annual rivuline species. After drying up of the bottom layer, the half-dry fine mud is not easily crumbled. For this reason coarse peat is often used as a bottom layer in tanks for annual species.

The breeding individuals usually are heavily fed for two or more weeks and then the bottom layer is taken out and placed in a strainer. When the surplus of water has drained off, the peat is placed on old newspapers or the like to drain off more water and a "cake" is formed to secure an even drying of the surface. After some days the surface of the "cake" becomes dry in part and is crumbled. If not sufficiently dry, the material is left to dry for some hours. Now the bottom-layer is placed in a plastic bag or a glass completely closed and may be safely stored for several weeks or months. Eggs from most African species might be ready to hatch after four weeks, but usually not all eggs will be ready. Eggs from these species usually will not stand storage of one year or more, except for OCC.

For hatching, the moist peat is placed in a tank and water is added and stirred in order to get all material wet. The amount of water used is so measured that the peat floats. Hatching is induced when cool water (18° C/ 64° F) is used. In case the eggs from a particular species are difficult to hatch, I add a little dry food or sugar.

During storage the peat should neither be too moist nor too dry. A consistency of "smoking tobacco" is recommended. If kept too dry the eggs may remain in diapauses Ib or II for too long a time and if kept too moist some of the eggs probably will not get sufficient oxygen because the material does not crumble. The peat is stored at 20–25° C (68–77° F).

After three or four days all fry should have hatched out and when they have been transferred to their tanks, the peat usually is re-dried and restored for another four weeks. The results of the second or even the third hatching attempt may exceed that of the first.

Eggs from all rivuline species can be sent to friends all over the world by common mail (annual species) or by air-mail (non-annual species). If weather conditions are cold, it is recommended to send the eggs before an embryo has developed otherwise they probably will die. Eggs from South American rivuline species hibernating in the first diapause stand cold weather very well. All eggs in which embryos had developed will always arrive dead.

THE BIOLOGY OF THE RIVULINE EGG

The development of rivuline eggs have been studied by many aquarists, but it was not until the late fifties that the different phases of development in annual rivulins were understood by some aquarists. Foersch (1956: 8–13, 39–45, 113–117 and 177–184) studied the development of the eggs from many rivuline species. Generally, when the development took place in moist peat, he found that the eggs from many annual species did not exhibit any indications of development, even months after the spawning, and also that the ripe embryo from these species often did not hatch out when the development within the egg had finished. He used the terms "resting egg" and "resting fry" to indicate the two periods of hibernation (or diapauses).

Scheel (1962b: 9–15) summed up the evidence concerning the development of the rivuline eggs provided by modern killifish fanciers and pointed out that "resting eggs" (*Cynolebias nigripinnis* Regan) under certain anaerobic conditions may remain fully viable for three years.

Peters (1963a: 257–313) carried out an extensive scientific study of the factors governing the development of the eggs from non-annual and annual rivuline species. He found that within African annual rivulins there are usually three phases of hibernation or diapauses.

A few hours after spawning and fertilization, eggs from all rivuline species will start development. First a rounded body (blastodisk or blastula) will appear at the animal pole of the yolk. Gradually this body will flatten out into the multidisk, which will extend onto the yolk-membrane. When the equator of the egg has been reached, the yolk usually will form a deep groove all round, but this groove soon will disappear and the cells of the embryo will cover the whole surface of the yolk membrane.

If the egg has been spawned by a true annual rivuline species, no further development will take place and the egg may remain unchanged for months. Eggs from less annual rivuline species might be forced into this phase of hibernation by a reduction of the tension of oxygen around the egg. This situation occurs when the egg has been spawned into dirty mud which will prevent the oxygen from reaching the egg. The development of such eggs usually will continue if the eggs are washed out from the mud or, if the mud dries out. Eggs from such almost-not-annual species as ARN, GUL and WAL and from most species of *Nothobranchius* may or may not continue their development when gastrulation has finished, or, some eggs may develop whereas others hibernate. Eggs from such forms and from non-annual species should not be kept under anaerobic conditions because they may be killed.

When the development continues beyond gastrulation, a short and rather thin body soon will appear on the yolk membrane. This body, the embryo, will usually be situated in a shallow groove on the yolk. Within one or two days, the embryo measures half the way round the yolk, or 180°. At this phase, embryos from African annual and semi-annual species again may hibernate ("resting embryo" or "Diapause II" of Peters) for months, whereas among American annual rivuline species this phase does not seem to occur. Also this diapause appears to be governed by the amount of oxygen available to the egg.

When the embryo continues the development beyond the "180° phase," the first indications of a working blood system occurs, and from then on it seems that we are not able to arrest the development of the embryo without killing it. When the embryo has consumed the yolk, the third phase of hibernation or diapause may occur. In annual rivuline species, this diapause ("resting fry" or Diapause V+ of Peters) is characterized by a complete stop of the circulation of the blood. This stop may last for weeks or months, and if the ripe embryo or fry is not disturbed by sudden changes of the illumination and/or temperature (?), no movements of the heart appear to occur during this phase. If the sleeping embryo is exposed to sudden strong light (or the heat from this light?) we may observe slight movements of the heart a few seconds or minutes after the light has been turned on. Soon the heartbeats become stronger and the blood starts moving back and forth in the veins until suddenly, the circulating blood breaks through, and the embryo starts moving in the egg. If the egg is now brought back to the dark and is inspected a few hours later, the circulation of blood has again stopped completely. If the ripe embryo is stimulated many times in this way, it soon will become exhausted because of the increased consumption of energy, and probably it will not hatch out as a fully viable fry.

It is remarkable that the ripe embryo from several rivuline species which are not considered to represent annual species very often develop this phase of hibernation and even within FAS, a nominal species of *Epiplatys*, I have observed a complete stop of the circulation of blood when the embryo has finished development within the egg. In FAS and similar non-annual species of *Aphyosemion*, the movements of the heart usually start a few seconds after the exposure of the egg to a strong light, whereas within the group of "true" annual species, no movements of the heart are often characteristic of the first few minutes.

Eggs from species which do not (?) develop diapause V+ might be forced into this phase by a reduction of the amount of water within the egg, to the extent that the blood is not able to circulate (*Rivulus*: see Scheel 1962b: 15).

Eggs from non-annual rivuline species which do not exhibit diapause V+ often do not hatch when the embryo has finished the development within the egg. Such eggs might be forced to hatch by any increase of the amount of carbon dioxide around the egg. Often, it is sufficient to breathe deeply into the glass within which the ripe eggs are stored and to replace the lid. Hatching will take place a few minutes later on. Eggs which develop within green plants will hatch at night when the amount of carbon dioxide is a maximum and the oxygen at a minimum.

The author at work in his laboratory studying material he collected during numerous expeditions in Africa. Photo by Dr. Martin R. Brittan.

V

THE RIVULINE MORPHO-SPECIES

The major section of the present review of the *Rivulinae* of the Old World deals with the different zoological names which have been created and which have been attached to the phenotype of a certain preserved individual (the holotype) which was thought to represent a distinct phenotype different from all previously described phenotypes of similar fishes. These names represent the *nominal rivuline species* of the Old World. It will be shown that some of the phenotypes described do not differ sufficiently from previously described phenotypes and thus these nominal species or names become synonyms for the previously described forms. Some of the nominal species appear to represent distinct new phenotypes or *morpho-species* distinctly separate from all other similar morpho-species by details of the external morphology, including color patterns.

The modern definition of the zoological term "species" does not, however, separate distinct and valid species on the basis of differences in morphology, but by their reproductive isolation. Mayr's (1963: 19) definition of the word "species" has been used by the present author during the preparation of this review of the African and Asian rivuline species. Mayr defined "*species as groups of actually or potentially interbreeding natural populations which are reproductively isolated from other such groups.*"

Except for BER, no rivuline species of the Old World have been based on Mayr's definition. This means that their status as zoological species is questionable until the reproductive isolation of these morpho-species or phenotypes from all previously described similar phenotypes have been proved.

On the other hand, my crossing experiments with individuals from different populations of a single morpho-species (BIV, GAR, BUA and others) strongly indicate that some populations are reproductively isolated from other populations without exhibiting corresponding differences in external morphology, and thus these populations deserve the status of distinct species. Mayr used the term *sibling species* for such groups of apparently identical populations. I prefer to use the word *bio-species* for such species which cannot be defined by details of the external morphology.

As it will be seen, no zoological name is attached to such bio-species. The zoological names are attached to holotypes representing major phenotypes or morpho-species and thus it appears that no new zoological name can be attached to a holotype (preserved fish specimen) unless the new holotype is different from all previously deposited holotypes, otherwise museum taxonomists are not able to identify individuals of the new species and to separate these individuals from individuals of similar species.

For that reason all bio-species are referred to the name of the morpho-species to which they belong and the different bio-species are characterized by their locality in nature. GAR-PH (Port Harcourt, SE Nigeria) is reproductively isolated from GAR-AK (Akure, SW Nigeria) and thus these two populations appear to belong to different bio-species within the GAR morpho-species or superspecies (morpho-species containing more than one bio-species are cryptic superspecies). The term *superspecies* was used by Mayr (1963: 672) for groups of allopatric species originating from a common ancestor.

The BUA-EXI-RUF species complex undoubtedly represents a superspecies, the EXI superspecies, composed of two morpho-species (EXI and BUA) but one of these (BUA) again is composed of two bio-species or sibling species, because RUF is reproductively isolated from BUA, whereas by the phenotype it does not differ from BUA. Thus BUA represents a cryptic superspecies and EXI represents a superspecies of the usual type.

Most nominal morpho-species have been based on a very limited material. About 70% of the species from Atlantic Africa have been based on less than five individuals (27% on one individual, 18% on two individuals and 11% on three individuals). The individual variation in color-patterns, in counts for fin rays and for scales, and in proportions, generally were not known to the authors. Even in cases where these characters could be studied and evaluated on the basis of large samples the opportunities were not used.

This is the reason why various authors of new names, separated their new nominal species from previously described species on the basis of differences in morphology which often did not exceed the individual variation found within a single local population of a rivuline morpho-species. On the basis of descriptions, reliable re-descriptions and my own study of several thousand preserved individuals I tried (unpublished) to evaluate the average individual meristic variation of the 120 nominal rivuline species of Atlantic Africa. On the basis of these figures and the total variation of each meristic characteristic within the whole group, the distinguishing value of each morphological character generally used in descriptions of rivuline-phenotypes could be worked out. It was found that proportions (relative length of head, snout in eye, eye in head, etc.) were almost worthless because within the best known species the individual variation almost covered the range of variation for the whole group. The juxtaposition of the dorsal and the anal fins was found to represent the most valuable meristic character as was also the number of dorsal fin rays, the ratio of the caudal peduncle, and the number of scales along the sides were much more useful for separation of morpho-species than the usual proportions and the number of anal fin rays and of scales round the body in front of ventrals.

It appears that the individual variation of the number of dorsal or of anal fin rays is five fin rays and of the number of scales in a longitudinal series a variation of five such scales is common.

On the basis of this study each nominal species was compared to all other (119) nominal species of Atlantic Africa using all published information concerning meristic characters and it was found that one species only (RUR, the sole species of *Nothobranchius* in Atlantic Africa) could be separated from all other species by at least a single meristic character. For INF, TES and PAS I found that each of these species could not be separated from 70 or more of the 119 species. The mean number of species from which the "average nominal species" could not be separated by meristic characters was found to be about 30.

To make things more complicated, some of the original descriptions are unreliable. Holly's (1930: 195–282) redescription of Ahl's Cameroon rivuline species disclosed gross inconsistency for many nominal species but not for all, and for that reason the differences in measurements and counts are not caused by different methods. See BEA, ELB, ESC, NOR, NYO, OBS, PAS, RIG, SAN and TES. Many "species descriptions" prepared by Ahl have not, however been controlled by Holly or other zoologists and some of these descriptions might be just as unreliable as the ten descriptions above mentioned.

Different systems have also been used to count fin rays and scales and to measure proportions. Boulenger and others undoubtedly did not count all fin rays visible at the base of the fin and his figures usually are lower than those of other authors.

Also, many "new species" have been described on the basis of aquarium kept individuals and the origin in nature of these individuals usually is not known. Forty-six of the 170 names reviewed in this book were based on aquarium kept individuals of unknown origin.

Thus the rivuline morpho-species appears to represent a highly variable phenotype which is very difficult to define and very difficult to separate from other rivuline morpho-species. The original descriptions are not sufficient to define the different species and much biological and morphological study has to be carried out before a satisfactory definition of each morpho-species can be prepared.

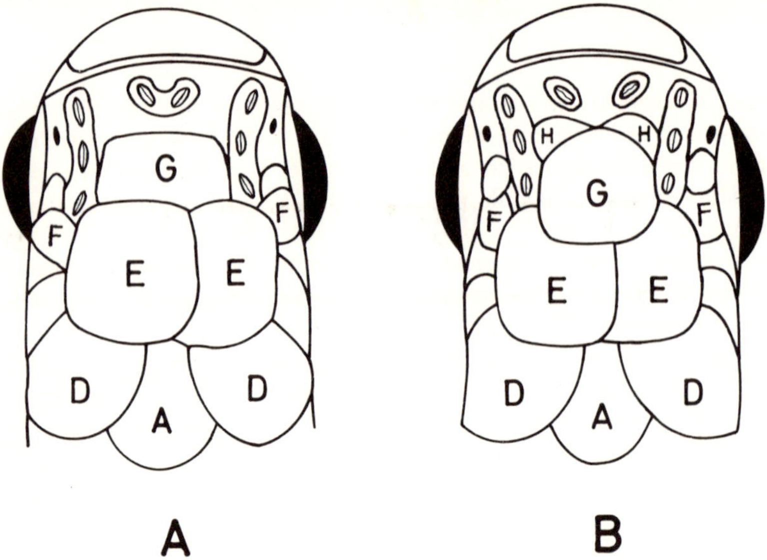

Pattern of frontal scales and of frontal neuromasts in two species of Rivulinae from Atlantic Africa:

A. GUI: E-type without H-scales. The F-series of scales above the eyes have not reversed. Closed neuromast pattern.

B. GUL: G-type with two H-scales. The F-series of scales above the eyes have reversed. Open neuromast pattern.

Front of preserved individual of GUI. No. 35.227 in the Paris Museum, from Kissidougou. Identified as CAM by Pellegrin 1934. One of the syntypes for GUI. Notice the pattern of frontal scales (E-type). The F scales above the eyes have reversed on the left side, not on the right side.

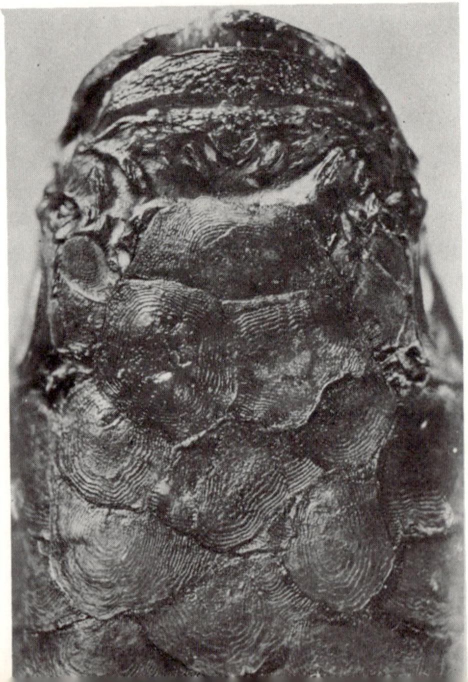

VI

THE FRONTAL SCALATION

The pattern of frontal scales of rivuline species first was used as a systematic character by Hoedeman (1956: 199–202), who on the basis of this character managed to split up the genus *Rivulus* into three major, apparently natural, groups. In 1957 (:294–96) he tried to use this character in the systematics of rivuline species of the Old World, but he failed to find major differences in the scalation of these species. In 1958 (:25–28) he studied this character in many species of *Cyprinodontidae* and in some species of *Goodeidae* and *Poeciliidae*. This study confirmed the finding that this character is very constant within Asian and African rivuline species, which all develop the G-type of scalation, except for *Pachypanchax* which develop the E-type. Whitehead (1962: 122–126) also used this character in his study of *Pantodontinae*.

Hoedeman (1961: 82–92) tried to prepare a theory for the evolution of frontal scale patterns in cyprinodontiform fishes, assuming that the basic and most ancient pattern was represented by the G-type pattern found among Asian and African rivuline species. This basic assumption is difficult to understand, and the present author developed the reverse idea and considered the scalation found in most Asian and African rivuline species to represent the final step in the evolution of the frontal scalation of cyprinodontiform fishes.

On the sides of a fish, the exposed edges of the scales are generally directed backwards. This seems to represent the ideal pattern and the scales are protected from being torn off if the fish hits some obstacle in the water. It also seems that this type of pattern produces the lowest resistance when the fish moves through the water. If this assumption is correct, it seems that the ideal scalation of the head should be similar, and in most rivuline species of the Old World this is true not only for the frontal scales, but also for the scales of the preopercle. In *Fundulus heteroclitus* (L.), just to mention one of the *Fundulus* species which exhibits an apparently primitive scalation, the scales of the preopercle are inserted in a way that the exposed edges are directed forward and the frontal scalation is of the A-type and thus all scales of the dorsal part of the head are inserted as those of the preopercle. This means that a large number of scales on the head of this species are unprotected, and if also, as in *Cyprinodon variegatus* (Lacépède 1803), these scales develop many ctenoid spines in males, such scalation will not be practical if the fish has to live in water containing many obstacles.

In all species of *Rivulinae* (and *Procatopodinae*) which I have studied, the scales of the preopercle have reversed, i.e. the exposed edges are directed backwards.

The frontal scalation of South American rivuline species, generally appear to be more ancient and primitive than that of the Old World species, except for *Cynolebias* and *Cynopoecilus* which develop patterns rather similar to those of the Old World species. Because of the very constant character of the frontal scalation of most Asian and African rivuline species, this character is not very useful in the systematics of this group of rivulins.

Pachypanchax is, however, distinctly separate from other Old World genera by the very constant E-pattern with prominent H-scales which together with the full development of the series of F-scales above the eyes indicate that this genus is ancient in respect to scalation. A rather similar pattern occurs in GUI, a nominal species of *Roloffia*, but in this species, living in mountain brooks generally, the G-scale is more prominent and there are no H-scales and the posterior part of the F-series above the eyes is not present, indicating that the development of the frontal scalation is a bit more advanced in GUI than in *Pachypanchax*, where HOM appears to develop the most ancient pattern. Also in BLO, of *Aplocheilus* the E-pattern with or without H-scales occurs in some populations, whereas in other populations the usual G-type pattern without H-scales occurs, indicating an unstable condition which also exist in the pattern of lateral line organs.

Hoedeman often pointed out that the large G- and E-scales probably originated by fusion of a number of small frontal scales. In species in which no H-scales occur, the anterior part of the G-scale usually is almost square, whereas in the species which have H-scales, the two foremost corners of the G-scale are absent and instead there are two H-scales usually situated below the anterior part of the G-scale, but in some species (GAR) these H-scales may also be situated on the anterior part of the G-scale. Prominent H-scales occur in GUL and SJO, indicating a scalation which has not yet reached the final evolutionary stage.

In *Nothobranchius* the frontal scales are much smaller than in other genera and these scales have receded from the anteriormost part of the front; also, the foremost scales are covered by a rather thick layer of spongy skin.

VII

THE CTENOIDY OF MALES

Ctenoid structures occur in all subfamilies of *Cyprinodontidae* (except perhaps *Oryziatinae* and *Pantanodontinae*) and also in some of the related families (I found such structures in *Anableps*).

In all subfamilies but *Rivulinae*, these ctenoid structures are of the usual spine-type, with a number of short spines situated on the edges of scales (more rarely on the surface of the scale). Ctenoid spines also often develop at the junctions of fin ray segments, but as these segments are considered to represent transformed scales, such spines are likely also to occur in fins.

The spines of scales and fin rays are often covered by tissues which extend far beyond the end of the spine, producing slender soft filaments or "sensory-papillae." This character is well developed in *Fundulopanchax* and *Nothobranchius* and GUI and OCC. Usually it appears that if ctenoid scales occur close to the root of fins, these fins develop ctenoid spines in the fin rays.

Whereas in *Aphyosemion* (*Fundulopanchax*) the ctenoid scales are usually found anteriorly and low on the sides and very rarely on the head, these scales frequently occur on the opercle and preopercle in *Nothobranchius*, and in this genus the ctenoid scales generally have a less restricted distribution, resembling the situation found in other subfamilies where ctenoid scales often are more numerous on the head than on the body. In all cyprinodonts the ctenoid spines generally are situated at a right angle to the surface of the scale.

In *Roloffia* (and *Cynopoecilus* in Brazil and Uruguay) a quite different type of ctenoidy occur in males. In these males the ctenoid scales have a number of radiating ridges and the surface of the scale becomes "wrinkled" (Scheel 1966: 141). The structure of this type of ctenoidy is described in detail in Clausen (1967: 16.) In these species of *Roloffia* (and in *Cynopoecilus*) sensory papillae may occur in fin ray segments, but the supporting bony element is not a spine, but a bud-like, often rather concave element, quite different from that of the fin ray segments of other genera. The supporting element of the sensory papillae of the pectoral fin rays in RUF appears to be of the *Roloffia* type.

The presence or absence of ctenoidy (spine type) appear to represent a rather distinct diagnostic character separating the subgenera *Fundulopanchax* and *Aphyosemion.*

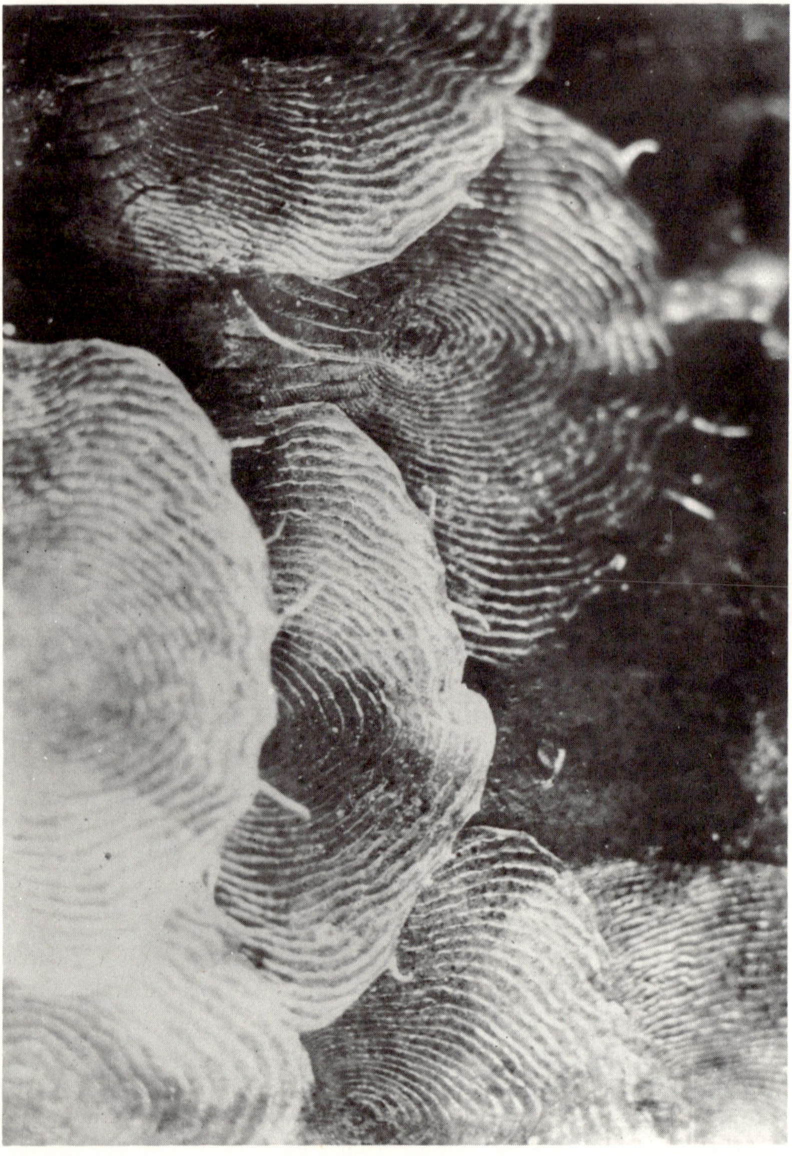

Ctenoid spines on scales above the root of anal in holotype of SJO (male) of the Stockholm Museum.

VIII

THE LATERAL LINE OF THE HEAD

In *Cyprinodontidae* the lateral line organs generally are restricted to the head of the fish, although some neuromasts might be present on the anterior part of the body. The structures found on the edges of the scales of the body in the BUR male may or may not represent sensory organs.

Within *Rivulinae* of the New World the lateral line appears to be much more variable than among the species of the Old World. On the dorsal part of the head, the lateral line organs are always situated in open grooves, usually with some fleshy lobes on the edges of the grooves. In some South American rivuline species the neuromasts (sense organs) are numerous and produce rather regular rows in shallow grooves, whereas in other species the neuromasts are much larger and are arranged in a regular pattern. The species of the Old World resemble the group last mentioned, except for some species (or individuals) of *Nothobranchius* which appear to be closest to the group first mentioned and very similar to the species of *Cynolebias* and *Cynopoecilus* which have very many and very small neuromasts, resembling the situation in *Orestias.*

In the species of the Old World there usually are three neuromasts in the supraorbital groove, which is separated from the upper and open part of the postorbital groove with three neuromasts by the bridge of scales which are produced by the F-series. These two grooves are confluent in ANN and most individuals of *Nothobranchius,* but in the latter this character is often difficult to judge because the grooves are very shallow. There is no bridge of F-scales in *Nothobranchius* in which this series of scales is completely absent above the eyes.

Within these species there is usually a pair of neuromasts situated on the center of the dorsal part of the snout. From a study of the lateral line of American rivuline species, it seems that these neuromasts are part of the supraorbital series. In some Asian species in particular (LIN and BLO) there is more than one pair of neuromasts on the snout, but usually the neuromasts of one of the pairs are considerably larger than the neuromasts of other pairs. Many years ago Clausen drew my attention to the pattern produced by the pair of neuromasts of the dorsal part of the snout pointing out the differences between GRA and SEX and between the western and eastern group of *Aphyosemion* in the Myers 1933-sense. Clausen based

(1966: 331–41 and 1967: 1–64) the new subgenus of *Epiplatys* (named *Parepiplatys*) and the new genus *Roloffia* generally on this character. He used the pattern of fleshy lobes edging the depressions for the neuromasts ("trapezoid" versus "trianguloid") whereas the present author used the distance between the corresponding neuromasts ("open" versus "closed"), because in some forms (CAL, DUB and BLO are examples) the lobes are absent or almost vestigial. In species having more than one pair of neuromasts on the snout, this character sometimes is difficult to judge (BLO). When the two neuromasts are "close," the anteriormost lobes converge into a single undivided lobe. My hybrids from the SEX/MON crossing, representing the "open" and the "closed" pattern, developed a somewhat intermediate (central lobe with furrow or divided lobe) pattern or the "closed" pattern, but in all individuals the pattern was closer to MON than to SEX.

The open upper part of the postorbital part of the lateral line usually is connected to the lower part by a canal with two pores and below the lower pore the lateral line becomes almost vestigial. A series of small neuromasts continue, however, below the eye and above the preopercle to the preorbital area, where again the neuromasts become more prominent and usually are housed in a short canal with two pores. The two canals are not present in *Nothobranchius* and in ANN and HOM. In subfamilies in which the preorbital space is broad and scaled, the anterior part of this part of the lateral line is not so close to the eyes and there might be a number of scales between this line and the lower rim of the eye, in front and below the eye.

There is another part of the lateral line along the lower and the posterior part of the preopercle. Usually the neuromasts of this series are housed in canals with many pores, except for *Nothobranchius*, FLA, ANN and HOM. In the latter species the condition is somewhat variable and in some individuals there is not a continuous long groove around the preopercle, but a number of isolated depressions housing one or more neuromasts.

There is one more series of very small neuromasts just behind the lower lip. The red pigments of the anterior part of the throat usually are associated with these neuromasts. Other series of neuromasts on the opercle and on the back near the B–C series of frontal scales are of no interest in this subfamily.

IX

PIGMENTATIONS

Cells containing black pigments (melanophores), rather evenly distributed, are parts of the basic color of rivuline individuals, but in many species, these pigment-cells produce distinct dark patterns of a more or less temporary nature.

Within *Aplocheilus*, *Epiplatys* and *Roloffia* regular black crossbars often develop on the sides of juveniles and females, whereas in the adult male of these species, this marking usually is absent except for certain situations (fighting). Usually these bars are rather regular and broad and they generally are distributed rather regularly on the sides with one such bar just behind the root of pectorals, another above the root of ventrals and four above and behind the root of anal, the last band usually on the root of the caudal. In SPI, this bar-pattern continues into the caudal fin. In *Roloffia*, these bars are more irregular and irregularly distributed. The crossbars might bc rather vertical (SEX and allied) or oblique (FAS and allied). Temporary black crossbars often develop in between the more permanent bars in females. These bars usually develop posteriorly on the sides. In SJO, RUF and EXI of *Aphyosemion* rather similar, but more narrow, bars may develop in the female, whereas in the males a corresponding pattern of vertical red lines occur. Such bars are not considered to represent "true" black crossbars.

Many species of the above mentioned genera also may develop a broad dark longitudinal band from the root of pectorals to the root of the caudal (extending into this fin in LIN). Often this band might be traced right to the lower lip along the gill cover and below or through the eye. This marking is median or submedian. In most species this marking is temporary and it develops only during the pre-mating display (females) and when the individuals (both sexes) are afraid. When this band develops, the dark crossbars usually disappear or are reduced. This marking is very common in *Epiplatys* and occurs in almost all species, whereas in *Aplocheilus* it is less common and occur usually in DAY only, although in LIN it may develop under certain conditions (see LIN). It is also rare in *Roloffia* and a regular broad dark band occurs in OCC only, whereas in GER and some populations of ROL, this band, when present is much more irregular and it is not used as in the genera first mentioned. Traces of a similar temporary marking occurs in *Aphyoplatys*. It appears that EXI of *Aphyosemion* may develop a similar marking but the nature of the longitudinal band in this species is not fully understood. In BIF of *Epiplatys* and BIV of *Aphyosemion* this band is very marked and of a more permanent nature, although in the species last mentioned, eastern and southern populations usually are less constant in this respect. Also in these species the lateral band is usually lost in the adult male, but in the males of western populations of BIV a short segment of this band remains just behind the root of the pectorals, this particular marking resembling the "wound-like" red-and-black marking which occurs in males of all species of *Aphyosemion* and *Roloffia*.

A similar but narrow dark longitudinal band occurs low on the side in some species of *Aplocheilus* (DAY), *Epiplatys* (BIF) and *Aphyosemion* (BIV). In the two species last mentioned this marking is rather permanent (females and juveniles) whereas in the species first mentioned this marking is temporary. In WER this marking is present, whereas the upper black lateral band is absent.

It has been mentioned under ANN, BLO, FLA and PAN, that in the juveniles of these species (also in adults of FLA) similarly situated but much more narrow bands or lines develop and that these black lines correspond to those which are present in almost all species of *Procatopodinae*.

In the females of some species (DAG, CHA, GRA, MAC) a thin black line is present just below the root of the anal, and in the males of some of these species (GRA–MAC group) the male develops a dark marbled pattern just above the root of that fin.

The lower dark band of the side continues to the throat in BIF and BIV producing the posterior dark band of the throat pattern, whereas the upper band produces the anterior dark band of the throat, this band being present in the lower lip or just behind that lip. A similar two-band-pattern occurs in many species of *Epiplatys* and more or less in some species of *Aplocheilus* quite independent of the presence or absence of the dark longitudinal band on the sides. In the females of these species, the lightly colored area between these two bands may develop a dark color (the two bands merge) during some phases of the sexual play. This alteration of the throat pattern is very marked in CHA, and more or less in FAS. In the males of these species and many more species the lower surfaces of the head develop a deep dark color which does not generally extend into the belly. The darkening of the throat of males develops during some phases of the sexual play and during fighting.

In *Pachypanchax* (PLA) and in GUI of *Roloffia* the throat pattern of *Epiplatys* appears to be present, but much reduced. There is a single dark band anteriorly on the throat, but this band is visible in the activated female only. In other species of *Roloffia*, except for GER, this pattern is also visible but even more reduced and the dark marking is reduced to a round dark spot anteriorly on the throat. Similar markings also occur in a few species of *Aphyosemion* (CAL, AHL and OBS).

In all the species which develop a more or less temporary dark marking on the throat of the active female, the males usually develop the overall dark color on the lower part of the head, and generally, it seems, that the male of all rivuline species have more melanophores on the lower part of the head than have females.

Some females of BUA and EXI develop a dark throat pattern rather similar to that of *Epiplatys*, but in the males the corresponding pattern is red.

Black pigments also produce patterns in fins In males of some of the species which have dark crossbars, the bars above the anal fin may extend into that fin and the last bar on the side may produce a black marginal band, particularly in the lower caudal fin. Black marginal bands also occur frequently in most genera, but often these bands also contain red pigments and are more red than black. Females and juveniles of *Aplocheilus* and PLA in *Pachypanchax* develop a round or crescent shaped black marking in the lower part of the dorsal. Irregularly distributed rounded dark markings occur in females of some species of *Nothobranchius*.

Pattern of dark pigments of the throat of preserved individual of DAG DAG from Port Bouet, Ivory Coast.

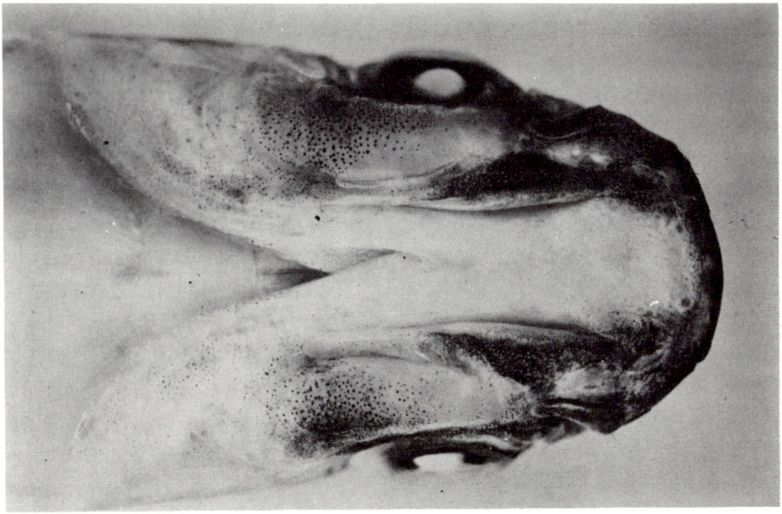

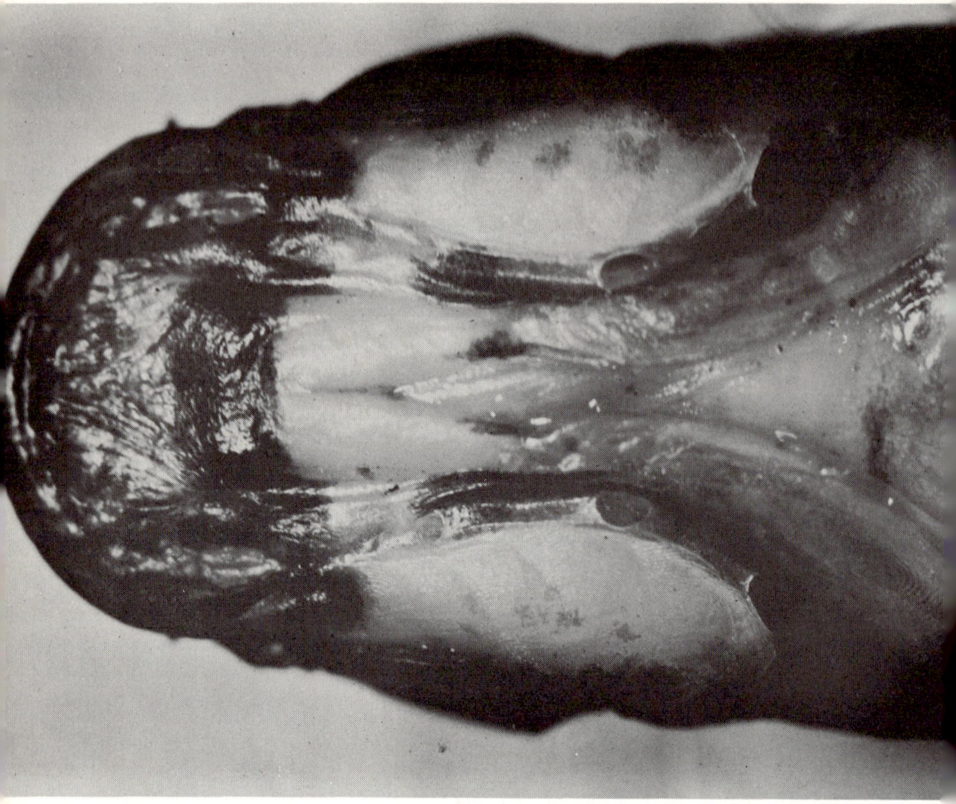

Throat pattern of preserved CHE male from Stanley Pool, Congo. Notice the dark marking just below the center of eye, resembling the pattern seen in BIF.

The dark markings and patterns undoubtedly play an important part in the behavior of the rivuline individuals and these markings are used as signals between individuals of different sexes. Generally it appears to be much more difficult to cross individuals which develop dark patterns than to cross individuals which do not develop such markings.

THE RED PIGMENTATIONS

The red patterns of the rivuline individual are more constant than are the corresponding black patterns, but the red patterns do change, because the red patterns usually contain black pigments also. Larger red markings (spots) often are edged with black and when the black color is present the marking appears more conspicuous. This is particularly true for the "wound-like" assemblance of red spots just behind the root of pectorals. These spots often are less narrowly edged with black than other red spots and during fighting, this marking is very conspicuous because of these black pigments, whereas around other spots the dark color fades.

The basic red pattern of the rivuline male probably is the regular red reticulation produced by the red scale edges. This pattern is present in the females of most species as a regular brown reticulation, which, without black pigments, would appear red.

The regular red reticulation is not useful for the development of variations and thus for species recognition the red pattern should be transformed. It seems that the first step on the evolution of the red pattern of the sides of males is a certain concentration of red pigments towards the junctions of the reticulation, thus producing oblong or crescent shaped red spots on all scales. The oblong red spots may merge producing thin red lines as in some species of *Nothobranchius*. The next step appears to be a rounding of the red spots and then a reduction of the number of red spots. Many different patterns may develop by horizontal or vertical merging of the red spots or by an increase or decrease of the number of spots within certain areas of the bodyside (NDI and CAM are examples).

The supposed more primitive phases of the evolution of the red pattern of the sides occur in *Nothobranchius*, in some species of *Epiplatys* and in a few species of *Roloffia*. The more advanced red patterns occur in *Aphyosemion*, less marked in some species of *Roloffia*, *Epiplatys*, *Pachypanchax* and *Aplocheilus*.

The red pattern of the sides also may disappear completely or almost completely (BLO, PAN, GUI, MAE, CIN and HOM).

The red pigmentation of fins of males is also important in identification. Males with a dark basic color on the sides often develop bright orange or red color in the pectorals. In these males the color pattern of the throat usually contains very little red color and the brightly colored pectorals probably compensate for that. Males of a few species which develop very brilliant colors on the sides and a prominent red pattern on the throat may possess bright orange or red pectorals (BUA). In the GAR/CIN hybrid males the very conspicuous orange-red color of the CIN male is lost and these males develop an almost uncolored pectoral fin with a blue marginal band, but in some males the margin is golden and the blue band becomes submarginal. Within BIV the whole scale of pectoral fin colors develop. As in western males, these fins are orange or lemon, whereas in males from the Niger Delta the fins are almost colorless.

Other fins may develop an irregular pattern of small red spots, streaks or fine red lines on the fin membrane between the rays. These patterns are not well suited for identification. In other species regular rows of red spots occur in unpaired fins and often these rows are attached to junctions of fin ray segments. These spots may merge producing red submarginal or marginal bands which are useful for identification. In *Nothobranchius*, *Aplocheilus* and *Pachypanchax* regular red submarginal bands do not develop in unpaired fins, but in some species of *Nothobranchius* a number of irregular red bands may occur in the anal fin. In most species of *Epiplatys* no submarginal bands develop in the fins, except for ANN, which have a fine lyre-pattern in the caudal and FAS which may develop thin submarginal bands in the anal and the lower part of the caudal. In *Aphyoplatys* the unpaired fins of the males may develop thin red submarginal bands. In *Fundulosoma* the caudal fin may develop an upper and lower submarginal band, very close to the fin edge. In *Aphyosemion* and in *Roloffia* conspicuous red submarginal bands often occur in the caudal fin ("lyre-pattern"), less often also in the anal fin and more rarely in the dorsal fin. These bands are often considered to be useful for identification, but as we shall see in such well-known forms as FIL, GUL and GAR these prominent submarginal bands may be absent in some males of the local population. It seems, that generally these submarginal bands occur in males which develop golden or yellow marginal bands in these fins, separating the golden distant part of the fin from the green or blue basal part.

In some males of *Aphyosemion* (BUA and EXI) the red spots of the caudal fin merge producing broad (or thin) vertical bars. It has been assumed under BIF and SPI (SEN) that this pattern characterzies a "rivuline species of the savannah." In other species of *Aphyosemion* the red pigments of the caudal fin produce thin red lines between the fin rays. These "flame-tailed" species represent one of the most difficult systematic problems in this genus and at the present time we do not know if this particular pattern indicates a natural group or not.

The red color may cover the caudal fin all over or almost all over. Most species of *Nothobranchius* are characterized by their red caudal fin, which may or may not develop a black, marginal band. The red (or orange-red) central (or subcentral) flame in the caudal fin of ANN and SJO may or may not represent a variation of the *Nothobranchius* pattern.

THE YELLOW PIGMENTATION

Yellow pigments do not generally produce any conspicuous pattern on the sides or the head of the individual rivuline. When the males of some species are preserved in alcohol, a very conspicuous lemon color may develop on the anterior part of the throat. The area so colored usually is well separated from the posterior part of the throat which remains white or grey. In some individuals this color also extends to the upper part of the snout. Microscopic examinations of the lemon area did not disclose the usual yellow pigmentation cells and the skin appears to be uniformly colored. Also this lemon color disappears after a few days.

The orange-red color of the anterior part of the MON male appears to be of the same nature and if males from the Ghana strain of DAG are preserved in alcohol, the area which is orange-red in the MON male develop the lemon color, whereas on other parts of the throat this color is absent. I have observed the development of this lemon color in males of

BLO, DUB, PAN, PLA, ROL, SPI and SEX and other species undoubtedly may develop this particular marking. In SEX this marking seems to occur in a few males of certain Niger Delta populations only. In ROL the marking also develops in females.

Yellow fin color is a very common character in both sexes of African and Asian rivulins. Usually this color is more evenly distributed in females than in males, although in most females the basal part of fins remain uncolored. In males the yellow pigmentation of fins usually increases markedly from the basal to the edge of the fin and often the color changes from yellow to golden. Red (or dark) markings very rarely occur within areas which develop a golden color.

In a number of species of *Aphyosemion*, in SEX in *Epiplatys* and PAL in *Nothobranchius* some males develop a bright yellow or golden color in their fins, whereas in other males, no trace of this color is seen. From my crossing experiments it appears that the yellow color usually is recessive (genetically) and the lack of yellow color is dominant, but probably the genetic background for the development of yellow fin color is more complicated (in GAR at least).

THE METALLIC BRILLIANCE

The brilliant green-blue-violet shine on the sides and sometimes also the fins of the male rivuline are produced by the guanine crystals and by the refraction of the light within narrow slits (Tundall's effect). As Foersch (1958: 258) pointed out for *Cynopoecilus ladigesi* Foersch, males which were raised in tanks with a strong light develop more metallic brilliance than males which were raised in dark tanks. In the dark tanks the development of pigments on the sides of the male is more intensive than in light tanks. This appears to be true for all rivuline males.

The intensity of the metallic brilliance also seems to be highly influenced by other aquarium conditions which probably govern the amount of slime on the sides of the fish. In males of the lampeye, *Procatopus gracilis* Clausen, I observed (Scheel 1960: 602) an increase of the brilliance when the males were kept in water containing an abundance of oxygen and found this to be in agreement with the observation of this species in nature by Birket-Smith. When the tanks were treated with carbon dioxide (Scheel 1962: 9–12, 78–79) these males produced a maximum of metallic brilliance, probably because of the very high concentration of free oxygen in these tanks. When males of rivuline species are moved from one tank to another they may lose their metallic brilliance for some time, probably because the layer of slime on the sides increases.

The exact color of the metallic brilliance is difficult to describe, mainly because it changes according to the angle of the light. When the light comes from above, the reflection is often greenish-blue; when it comes from below it is more violet-blue.

The metallic brilliance interacts with the black, red and yellow pigments producing mixed colors, but it also seems that the concentration of guanine increases markedly in areas containing much dark pigments (as in the guppy). True blue pigments probably do not occur in fishes, but the bright blue color of the throat of the OCC male needs more study, because in the young male the color of the throat may change from red to blue and back to red many times during the maturing phase. I wonder how this change comes about?

In rivuline species the active female usually looks for the male and the brilliant colors of the male probably makes him more conspicuous to the female. The different color patterns probably serve a purpose in species recognition in particular in geographical areas where closely related species are sympatric. I was very successful in my crossing experiments with *Aphyosemion* by using very dim light or by using light from below.

Dugout canoes at shore of Lake Barombi. pH 7.6; 3 German degrees of hardness. Habitat for SEX near shore. West Cameroon (near Kumba).

X

THE RIVULINE BIO-SPECIES

All distinct morpho-species comprise one or more bio-species and those which comprise two or more bio-species are called *cryptic superspecies* in this presentation. All these bio-species are reproductively isolated from all other bio-species. Within *Rivulinae* the reproductive isolation usually is based on post-mating factors and rarely on pre-mating factors, although both systems may occur, the pre-mating factors preventing individuals of different species from mating. The sole example of apparently active pre-mating isolating mechanisms, without the presence of post-mating similar mechanisms, is the interaction of LON and SEX within the Niger Delta area.

Compared to other groups of animals and plants which have been studied in crossings, the post-mating isolating mechanisms are highly developed within *Rivulinae* and generally speaking it seems (from more than a hundred crossings) that individuals which differ sufficiently by morphology or by color pattern to be considered to belong to different species, are not able to produce fertile offspring. (The RUF-EXI crossing is an exception, but the BUA-EXI crossing is not.) In these crossings, the hybrids are often not even viable and if they survive they usually are completely sterile. Similar results might be had from the crossing of sibling species but generally it seems that these crossings produce fully viable F_1 hybrids and that individuals of both sexes often are able to produce gametes, but the F_2-individuals, or backcross individuals, are not viable.

The reason for the occurrence of strong post-mating isolating mechanisms is not known, but it is likely that very often they are the results of re-organization of the chromosome material during the evolution of these fishes. Also, the highly variable chromosome number and the probably even more variable karyotypes of rivuline species appear to be unique among fishes and even among animals.

The frequent occurrence of annual fishes among rivuline species is another interesting feature. These annual species occur in both major groups of rivulins (Old World species and New World species) and probably these biological characters had developed to a certain extent before these two species-groups were separated.

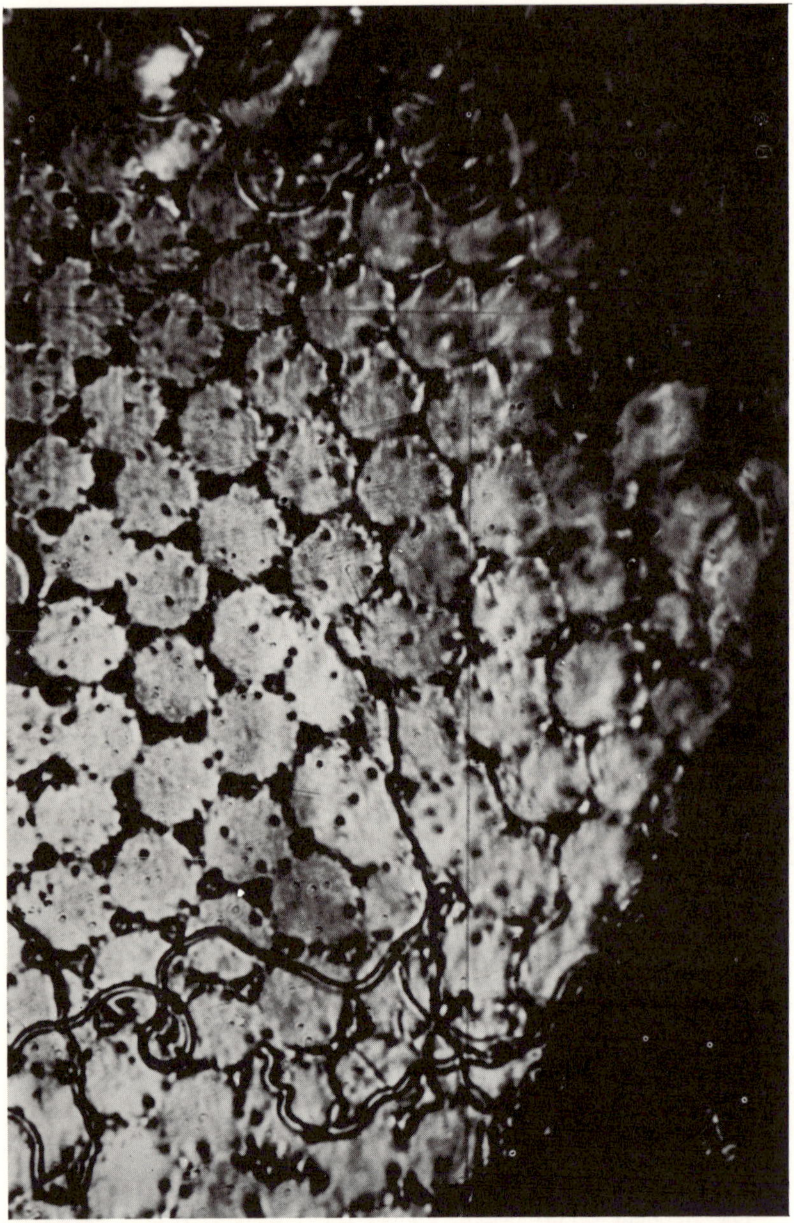

Pattern of egg membrane in FIL.

XI

THE MORPHOLOGY OF THE RIVULINE EGG

The eggs of African and Asian rivuline species measure from 0.9 to 1.9 mm. in diameter. Usually the size of the egg is very constant within different populations of a given species and even within the species-group. Exceptions from this rule occur, but usually constant differences in egg size between two apparently similar populations indicate more profound genetic differences and often such populations represent sibling species (BIV, GAR), although sibling species may have similar eggs (CHR, ROL-BER). In *Nothobranchius*, it seems that large species have large eggs and small species have small eggs. Large differences in egg size represent a problem in some crossings, in particular in those with LAB being one of the components. There seems to be no general rule for an increase or decrease in egg size along the path of evolution of these species.

The color of the egg depends generally on the food which is offered to females, but it seems that eggs from species of *Aphyosemion* and *Roloffia* are more colored (yellow-orange) than are eggs from *Epiplatys* and *Aplocheilus*. The color of the egg clearly separates PET from the species of *Epiplatys*, independent of the feeding. When peat is treated with hydrochloric acid to destroy eggs from annual species, these eggs become almost brown, at pH 3.6.

The shape of the egg varies in different taxonomic units. Those from *Nothobranchius* and *Fundulosoma* are oval; those of other genera are spherical. During decomposition these two sorts of eggs differ and in the group first mentioned, the yolk releases oil drops and becomes granular, before the yolk membranes decompose.

The pattern of the egg-membrane is very useful in the systematics of these species.

All species of nominal *Roloffia* (except for PET), most species of *Aplocheilus*, some species of *Epiplatys*, and *Aphyosemion* develop a regular hexagonal pattern on the membrane, this particular pattern being absent in all American species except for *Cynopoecilus*. According to Peters (1963b: 202), who studied the development of eggs from various rivuline species, the hexagonal pattern of the membrane represents a reinforcement of the membrane.

The development of the hexagonal pattern varies within different species which are not true annual rivulins. Even within a brood from a single female, some eggs may exhibit a perfect hexagonal pattern, whereas in other eggs, the reticulation is obscure or the pattern becomes so indistinct, that it is almost impossible to recognize. The species of *Roloffia* (except for PET) all exhibit annual characters, whereas within *Aplocheilus* no species is known to reproduce as annual rivulins. If, for this reason, the hexagonal membrane pattern indicates a reinforcement of the egg and is associated with reproduction of the annuals, the species of *Aplocheilus* may represent derivatives of some ancient annual rivuline species.

The hexagonal membrane pattern appears to be absent within other subfamilies of cyprinodonts, and within American rivuline species this pattern is very rare. It occurs in *Cynopoecilus* only.

The decomposition of the hexagonal membrane pattern, which is seen on eggs from various species of *Epiplatys* and *Aphyosemion*, results in an almost plain membrane, as in PET within *Roloffia*.

Eggs from *Nothobranchius* and *Fundulosoma* do not develop any hexagonal pattern. The membrane of these eggs have a large number of evenly distributed rather short, stiff filaments. Those of *Fundulosoma* (THI) also have a few long filaments concentrated in one pole, and these eggs resemble those of *Oryzias* and *Horaichthys setnai* Kulkarni 1941. Eggs from ARN are rather similar, although the number of evenly distributed short filaments appears to be lower than in *Nothobranchius* and *Fundulosoma*, and also eggs from ARN develop a perfect hexagonal reticulation. During the cytological study of females of *Aphyosemion* and *Roloffia* I have noticed that during the early phases of the development of the egg-membrane, there are evenly distributed, rather long filaments and no trace of any hexagonal pattern on the membrane, indicating that the *Nothobranchius-Oryzias* pattern represents an ancient pattern.

Eggs from *Nothobranchius* and *Fundulosoma* differ from eggs of most species of cyprinodonts by their non-sticky surface. Eggs from most cyprinodont species adhere particles to the membrane and to the filaments (when present). Eggs from ARN and FIL are not sticky, and the egg from ARN appears to represent a type somewhat intermediate between *Fundulopanchax* and *Fundulosoma*. Among American rivulins, eggs from *Cynopoecilus* (*C. ladigesi* Foersch and *C. melanotaenia* Regan) have non-sticky surfaces. Recently I have studied the eggs from *Austrofundulus dolichopterus* Weitzman and Wourms and was much surprised to see that these eggs were similar to those of *Nothobranchius* (non-sticky surface, evenly distributed short filaments) except for shape and to be quite different from eggs from all other American rivuline species, except perhaps for *Cynopoecilus*. In the genus *Cynopoecilus*, the hexagonal pattern is present and the filaments coalesce six by six producing highly specialized palm-like protuberances on the surface of the membrane. Probably the structure of the egg-membrane represents an important systematic character in *Cyprinodontidae*.

Eggs from rivuline species are very resistant to pressure. According to Peters (1963a: 269), the egg from African species are able to withstand a pressure up to 100–300 grams.

The Koubou River: Bertoua-Betare Oya in East Cameroon, near the rain forest. Current strong. $21\frac{1}{2}$° Centigrade; pH 6.4; German hardness 1.3 degree. Habitat of BUA.

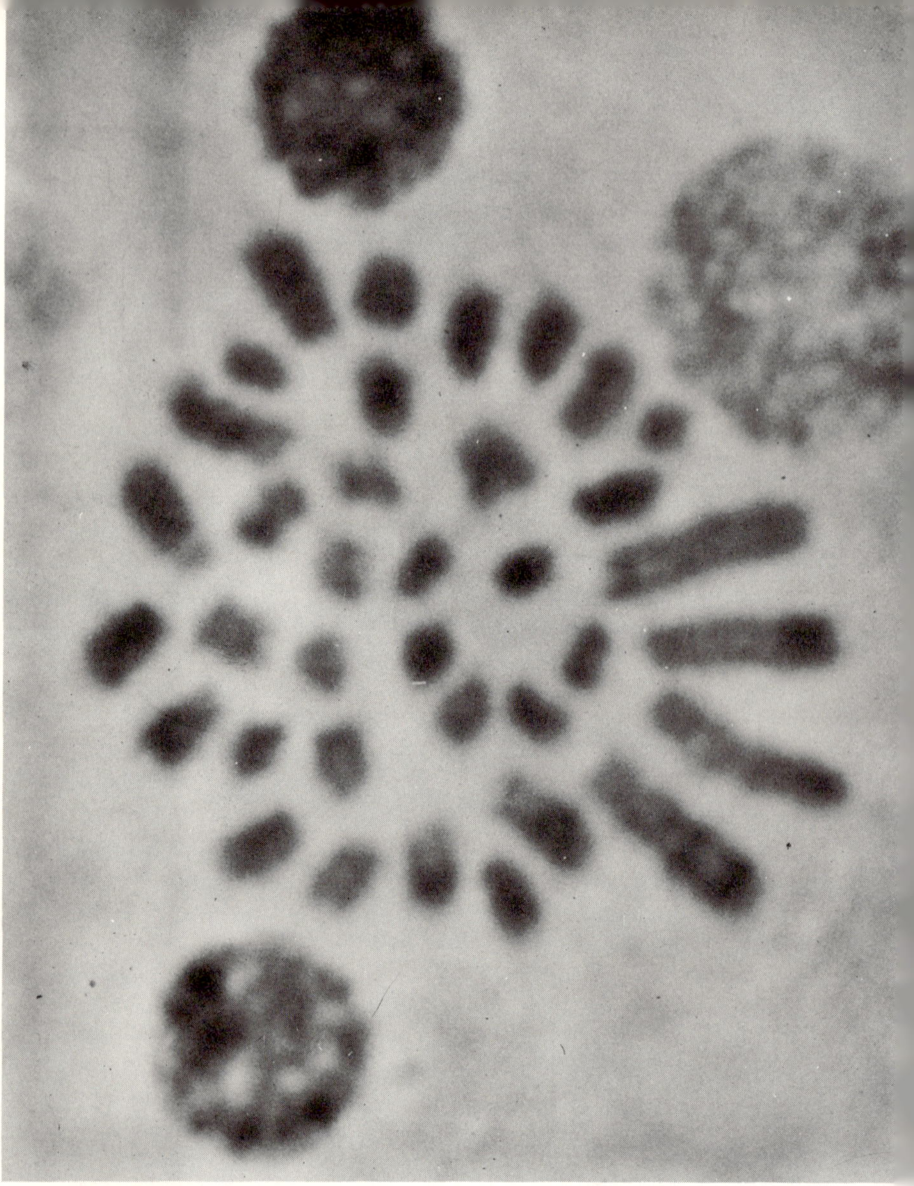

Mitosis of FIL. $2n = 36$.

XII

THE CHROMOSOME CONDITIONS

The study of the chromosomes of animals is still far behind the corresponding study within plants, probably because until recently there was no easy and exact technique for the handling of these small units in animals. This situation has changed markedly since the introduction of the modified squash-technique prepared by Hsu, T. C. and Pomerate, C. M. (1953: 23) and Makino, S. and Nishimura (1952: 1), who expanded the cell nucleus in hypertonic solutions in advance of the fixing and staining of the chromosomes. Karbe (1961: 73–104) modified this method for the study of the chromosomes of fishes and the present author (Scheel 1966: 123–148) again modified Karbe's method for his study of the chromosomes of rivuline species.

In future research in *Rivulinae*, the study of the chromosome number and, in particular, the morphology of the individual chromosome, probably will prove to be the superior tool in the reconstruction of the evolution and systematics of this difficult group of fishes.

The number of haploid chromosomes of rivuline species of the Old World was first studied by Post (1965: 47–93) who counted the number of chromosomes in 103 different species of fish, including 12 rivuline species of the Old World (four *Aphyosemion*, two *Epiplatys*, five *Nothobranchius* and one *Pachypanchax*). Some of the results published by Post agree with my own findings; others do not, probably because Post used different names (ARN for FIL, AHL for CAL, SJO for OCC, HOL for BIV, CHA for DAG). Also, some of the strains which he used might possess slightly different chromosome numbers than those used in my study.

From my own study it seems that the haploid chromosome number varies between n = 9 (RAC and CHR) and n = 25 (ANN, DAG, CHA, LIN) among the species of the Old World, whereas in the New World n = 27 is reached (*Pterolebias peruensis* Myers). Most of the numbers between 9 and 25 have been found in one or more species: 9, 10, 14, 15, 16, 17, 18, 19, 20, 21, 23, 24 and 25 have been found so far.

The haploid chromosome number by itself is of little use, when the morphology of the individual chromosomes is not known.

From his extensive study of the chromosome number for many species of bony fishes, Post concluded that n = 24 was the basic and ancient haploid chromosome number for the bony fishes. My own study of cyprinodont species supports this idea and it seems that n = 24 is the basic chromosome number for *Cyprinodontidae* and related families. In *Rivulinae* there are strong indications that n = 25 is the basic number for the species from the Atlantic drainages of Africa and possibly for all rivuline species of the Old World, except for a few more ancient forms.

The basic or ancient karyotype of *Cyprinodontidae* seems to be characterized by a smoothly graded series of rather small chromosomes, which are generally acrocentric. In acrocentric chromosomes the "center" of the chromosome (the centromere) during movements is situated very close to one end of the chromosome. The smallest of these acrocentric elements measures about half the length of the largest. This karyotype was present in three nominal species of *Procatopus* (*P. cf. similis* Ahl, *P. roseipinnis* Clausen and *P. nigromarginatus* Clausen, the individuals of the two species last mentioned originated from the type localities for these species), and also in *Xiphophorus helleri* and *Xiphophorus* hybrids. The karyotypes for most of the rivuline species having n = 24 or n = 25 were similar.

There are no indications of any karyotype having more than 24 haploid chromosomes in other subfamilies of *Cyprinodontidae* and thus the n = 25 karyotype of some rivuline species probably originated from aneuploidy rather early in the evolution of African rivuline species. Aneuploidy is a sort of miss-division during mitosis or meiosis during which the chromosome complement is not divided into two similar groups and one (or more) extra chromosome occurs in one of the cells produced by the division. There are some slight indications that aneuploidy occurred more than once during the evolution of these rivuline species.

Whereas in birds and reptiles very small chromosomes (*microchromosomes*) occur, such chromosomes usually are, but not always, absent in *Cyprinodontidae* and this fact makes the study of the evolution of these species easier.

The reduction of the chromosome number in plants and in animals usually takes place when two acrocentrics break during the division of the cell and the broken parts fuse in a new way. One large metacentric with long arms and one small metacentric with very short arms may be the result of the fusions. In the metacentric chromosome the centromere is situated close to the middle of the chromosome. These breaks must occur rather close to the centromere, otherwise the fertility of the individuals in the next generation will be seriously affected. In GAR from Eyomojok of West Cameroon there are 20 haploid chromosomes, and in GAR from Owo of West Nigeria there are 18 such chromosomes. In the latter there are two metacentric elements which correspond to four acrocentrics of the former. During the meiosis of the hybrid males these six chromosomes produce two trivalents, each containing one metacentric and the two corresponding acrocentrics.

The F_1 x F_1 spawnings of these hybrids disclosed that they were almost fully fertile.

Within *Rivulinae* of the Old World the small metacentric element usually is lost, probably soon after its occurrence in the karyotype. The reason for this disappearance of the small element may come from the fact that the short arms of the chromosomes of these species do not contain any active genetic material (genes); thus this element will not be able to pair with the corresponding element in meiotic divisions. In *Pterolebias peruensis* from

Primary meiotic division in hybrid male GAR-Owo/Eyomojok. 18 separate bodies are visible. Two of these are trivalents containing one (metacentric) chromosome from the Owo strain and two (acrocentric) chromosomes from the Eumojok strain. This male was quite fertile.

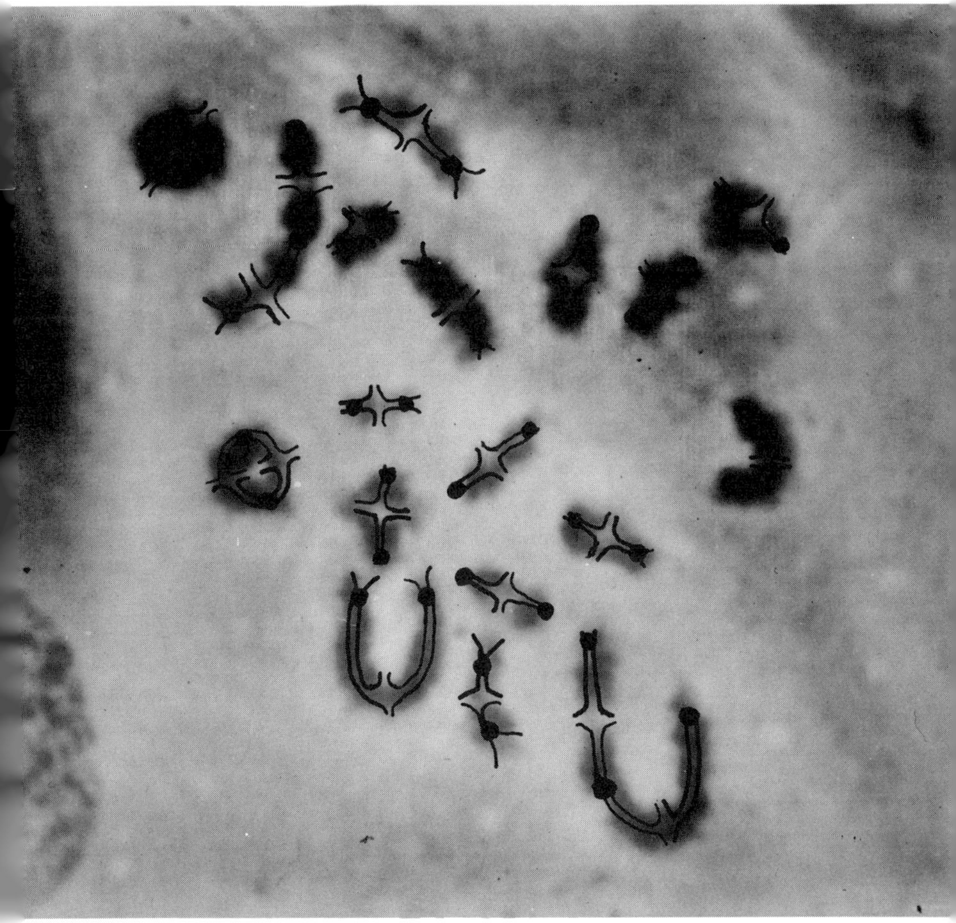

the Upper Amazon River of South America there are 27 haploid chromosomes. Four of these are large metacentrics and another four are very small metacentrics which take Orcein stain and thus contain active genetic material. The basic chromosome number for this species therefore is $n = 27$ and not $n = 27 + 4 = 31$. This finding indicates that in the rivuline species of the New World the structure of the acrocentrics might be somewhat different.

Together with the small metacentric element, the corresponding centromere is lost. This means that if the large metacentric element breaks, the corresponding acrocentrics cannot be reconstructed. Thus an increase of the haploid number is much more difficult than a decrease. On the basis of this theory we may assume that in a series of closely related forms the form with the lowest chromosome number probably is also the youngest form, being a derivate of the form which has the highest number.

In *Epiplatys* and *Aplocheilus* the reduction of the chromosome number appears to take place by a production of large metacentric elements, the small elements being lost. This means that in these species the metacentric elements range between the longest acrocentric of the basic complement and double that size. The nucleus of the individual cell should be able to handle these large chromosomes because during cell division they bend at the centromere and the active lengths thus correspond to that of the acrocentrics from which they originated. Within these two genera the chromosome number has been reduced to 17 (BAR and SPI) within *Epiplatys* and 18 (SIA) in *Aplocheilus*, but *theoretically* the number might be reduced to 12 or 13, by the process above described.

A quite different type of reduction of the number of chromosomes occurs in *Aphyosemion, Roloffia* and *Nothobranchius*, but in these species the production of the large metacentric elements as described above also play an important part within the species exhibiting very low chromosome numbers. The initial or basic type of reduction appears to be of a different type. Two acrocentric break, one near the centromere as when metacentrics are produced, but in the other acrocentric the break occurs near the far end of the long arm. The large broken parts fuse producing a "double-sized acrocentric element" and the small broken parts produce a small metacentric element which usually soon is lost. This type of translocation of the chromosome material was considered to be rare in most groups of animals according to White (1954: 188, 192) who observed such super-acrocentric elements in *Drosophila spinofemora* and a few more species of fruit flies, but he concluded that the superacrocentrics in these species arose from pericentric inversions in similarly sized metacentric elements. There are no indications that the superacrocentric elements in *Rivulinae* arose from metacentric elements.

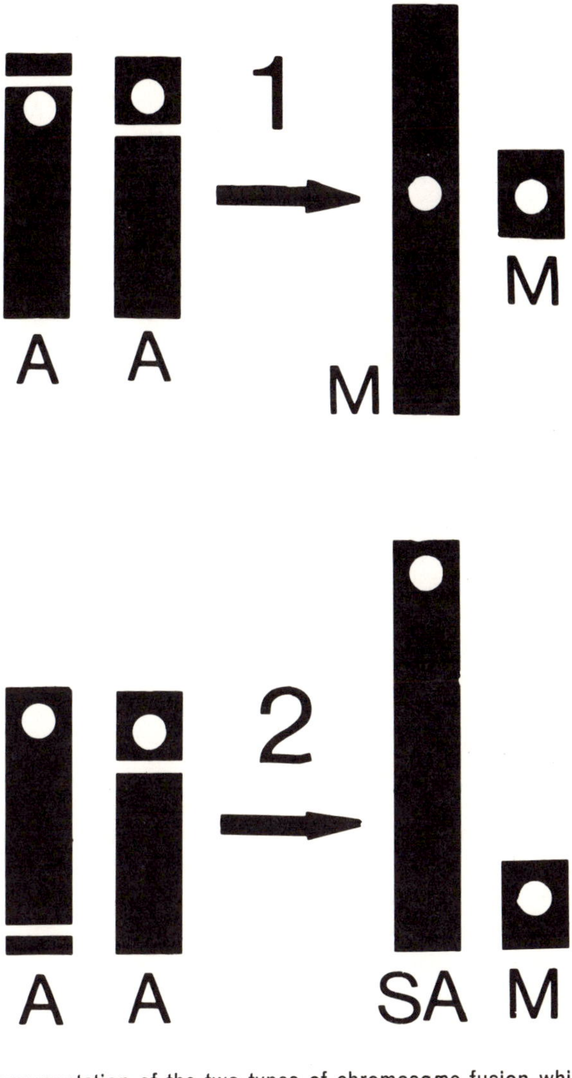

Diagram representation of the two types of chromosome fusion which occur in Rivulinae.

1: (*Above*) the fusion of two basic acrocentric elements (A) producing one large and one small metacentric (M) element.

2: (*Below*) the fusion of two basic acrocentric elements (A) producing one super-acrocentric (SA) element and one small metacentric (M) element.

The production of such superacrocentrics measuring up to double the size of the largest acrocentric element of the basic complement apparently is not easy because the original cell nucleus probably is not large enough to handle such supersized acrocentrics during cell-division. An increase of the diameter of the nucleus is a must. Such increase was observed by Post (1965: 77) who found that the diameter of the nucleus of RAC was 11.5μ whereas for n = 24 species he found 6.5–7μ, indicating almost a doubling of the nucleus diameter. For this species with a much reduced chromosome number I found that the longest arm in the metacentrics measured about 1.7 times the length of the largest acrocentric element of the basic complement and this finding seems to be in agreement with Post's finding. Post apparently was not aware of the cause for the increase of the diameter of the nucleus.

During the evolution of these genera (*Roloffia, Aphyosemion* and *Nothobranchius*) in Africa and *Cynopoecilus* (superacrocentrics do not occur in the related "*Austrofundulus*" *dolichopterus*, n = 22) rather short superacrocentrics probably first managed to establish within the changed karyotype. Indeed, at fusion of two of the smallest acrocentrics of the basic complement into a superacrocentric element undoubtedly would not interfere with cell-division, because such "superacrocentric" would not exceed the size of the largest acrocentric of the basic complement. Thus it appears to me, that during the evolution of these genera, there were two interacting series of mutations. One series of chromosome-mutations producing larger and larger superacrocentrics and another series of gene-mutations producing larger and larger cell-nuclei.

If this assumption proves to be correct, informations concerning the evolution of these genera might be had from measurements of the length of the longest superacrocentric element (or longest arm of the corresponding super-metacentric element) of the karyotype, the forms with the longest arms being more advanced than the forms with rather short arms.

GUI, OCC, LOU, LUJ and *Cynopoecilus melanotaenia* have rather short superacrocentrics, the largest measuring 1.4–1.5 times the length of the largest acrocentric element of the basic complement.

Then follows CAL (1.6–1.7 times), BER, GER, ROL, OBS and RAC (1.7–1.8 times), ARN, BUA, CHR-9 and EXI (1.8–1.9 times), RUF, BIV 17 and BIV-19 (1.9–2.0 times), BIV-20 and SJO (2.0–2.1 times), BUR, CHR-15, LAB and GUE (2.1–2.2 times), CIN, GUL and NDI (2.2–2.3 times) and AUS, WAL, GAR and FIL (about 2.1 times). It seems that generally the species of *Fundulopanchax* have very large superacrocentrics, whereas those of *Roloffia* and a certain group of *Aphyosemion s.s.* have rather short superacrocentrics.

Measurements of the individual length of the chromosomes in spermatogonial mitotic divisions strongly indicates (calculations) that some of the superacrocentrics exceed twice the length of the largest acrocentrics of the basic complement. In CIN, FIL, GUL, NDI and GAR, the largest acrocentric element appears to be from 3.3 to 3.5 times the mean length of the basic chromosomes or from 2.5 to 2.6 times the length of the largest basic acrocentrics.

In some species of *Aphyosemion* (BIV, CIN, GAR, GER and WAL) there are two very long superacrocentrics in the haploid complement and these two acrocentrics appear to be very similar or even morphologically identical, although they do not tend to produce quadrivalents during meiosis. In some of these species these two acrocentrics differ markedly from all other chromosomes by size. Although at the present time I am unable to support my theory, I think that one of these superacrocentrics originated from aneuploidy a long time ago. Calculations of the individual and total length of the chromosome complement of these species supports the idea that one of the large chromosomes is an extra one.

I have studied the karyotypes of only a few species of rivulins from South America, mainly to see if among these any development of superacrocentrics might take place. I found no such chromosomes in *Pterolebias peruensis*, *Cynolebias whitei* Myers or in *Rivulus urophthalmus* Günther, whereas in the karyotype of *Cynopoecilus melanotaenia* Regan, there is one superacrocentric element measuring about two times the mean length of the basic chromosomes (for n = 25) and there are two more acrocentrics which may be superacrocentrics. This species has n = 22 in the present aquarium strain; the largest acrocentric element measures about 3.7 times the length of the smallest.

Usually the exact number of superacrocentric elements cannot be found, because this series of chromosomes grades smoothly into the series of basic acrocentrics.

In animals and plants, where superacrocentrics do not occur, the number of major chromosome arms remains constant, independent of the production of large metacentrics and loss of small metacentric elements, providing aneuploidy does not occur. The number of major arms ("NF" or *Nombre Fondamental*, Matthey 1945) is not constant in *Rivulinae* of the Old World. I found these NF figures (some are tentative) so far:

NF = 25: ANN, CHA, DAG and SEX.
24: BIF, DUB, ESE, FAS, GAR-Akure, OCC, SAN and SPI.
23: BIV–20, EXI and GUI.
22: ARN, GAR–Owo, GAR–Eyomojok and GUE.
21: BIV–17, BUA, CIN, ROL and SJO.
20: FIL, GER, LOU, NDI, OBS and RUF.
19: AUS, BIV–19, CHR–15 and WAL.
18: LAB.
15: RAC.
14: CHR–9.

The number of noncentric fusions of acrocentric chromosomes which occurs within a certain species might be estimated from these figures in combination with the haploid chromosome number. GAR–Akure has n = 20 and NF = 24, but the karyotype of this form appears to contain at least two superacrocentrics.

If the geographical distribution of various rivuline species in Atlantic Africa and their haploid chromosome numbers are studied, some ideas concerning the geographic expansion of certain groups of species might develop.

In *Epiplatys*, there seems to be no distinct connection between the chromosome number and the zoogeography, although it might be of some importance that the three species which have n = 25 all live in the western part of West Africa.

The genus *Roloffia* is restricted to the western part of West Africa with a certain concentration of species in southern Sierra Leone. Among these species the highest haploid chromosome number known within *Roloffia*, *Aphyosemion* and *Nothobranchius* occurs in OCC and the very closely related TOD, which both have n = 23. The chromosome number indicates that OCC and TOD probably represent the most ancient species within the three genera mentioned above. The major group of species in *Roloffia* (LIB-ROL-BER group) has n = 21 in the four populations of BER and ROL from western West Africa. All further species of *Roloffia*, *Aphyosemion* and *Nothobranchius* studied so far have 20 chromosome or less. This indicates that both major groups of rivulins from Atlantic Africa (*Epiplatys* and *Roloffia-Aphyosemion*) maintain the highest chromosome numbers in the extreme western part of the West African forest, and that this geographic area may represent the place from where these two groups expanded throughout Tropical Africa.

In *Aphyosemion* eight of the species so far studied had n = 20. Six of these species (BIV, BUR, CIN, GAR, NDI and SJO) are restricted to the forests (and derived savannah) of Nigeria and West Cameroon, the seventh species (BUA) living at about the same latitude in the savannah of East Cameroon. The eighth species (LUJ), however, lives in Central Congo. These eight species also represent almost the whole phenotypic variation of the genus.

This concentration of rather different species all having twenty chromosomes in this rather limited geographic area indicates that an intensive speciation (multiplication of species) once took place in this area without alterations of the chromosome number, but a study of the karyotypes discloses that these are rather different and that much recombination of the chromosome material probably took place during the evolution of this group.

Some of these twenty-chromosomes-species (BIV, BUA and GAR) have given rise to populations (or distinct species) which exhibit a somewhat reduced chromosome number, thus indicating that it is likely that some of these n = 20 species have given rise to independant groups of species.

All known species which have less than twenty chromosomes possibly did not originate from one of the eight species above mentioned. It is very likely that ARN (n = 19) and AHL (n = 18) represent independent forms which gave rise to their own evolutionary lines of descent.

It is a remarkable fact that all but one n = 20 species live north of Equator (north of the 5th latitude) whereas species south of the Equator exhibit much reduced chromosome numbers and appear to represent the major group of *Aphyosemion* species of that area: CHR: n = 9 and 15; LOU: n = 10; LAB: n = 14 and AUS: n = 15.

Although the chromosome conditions of *Nothobranchius* are not as well known as those of *Roloffia* and *Aphyosemion* it seems that n = 19 represents the highest chromosome number of *Nothobranchius* and also that this particular figure occurs in many species of this genus. Post (ob. cit) found n = 19 for the nominal GUE, MAY, MEP and PAL and my own results for nominal GUE, PAL and ORT are rather similar. It might be important that in this genus also a much reduced chromosome number occurs in RAC (n = 8 or 9), a species living at 20° southern latitude. Thus the highest chromosome numbers in *Aphyosemion* and in *Nothobranchius* are rather similar (n = 20 and n = 19) respectively and both genera may have originated from a single ancient form having twenty chromosomes.

The genera *Pachypanchax* and *Aphyoplatys* both have n = 24 with no indications of any reduction of the chromosome number, thus these two genera cannot be incorporated into one of the two major groups of Old World *Rivulinae*. *Pachypanchax* appears to be very closely related to GUI of nominal *Roloffia*, which appears to take in a somewhat intermediate position between *Aplocheilus-Epiplatys* on one hand and *Roloffia-Aphyosemion-Nothobranchius* on the other hand, because in GUI the largest chromosomes are metacentrics, but there is also a single superacrocentric element present in the karyotype of GUI and the longest arm of the two largest metacentrics exceeds the size of the superacrocentric element.

In *Epiplatys* and *Aplocheilus* none of the largest metacentric elements exceeds about 2.6 times the mean length of the basic chromosomes, indicating that these metacentric elements did not originate from fusion of superacrocentric elements (the longest acrocentric of the basic complement measures 4/3 = 1.3, the smallest 2/3 = 0.67 times the mean length of the basic chromosomes), whereas in GUI the largest metacentric element measures about 3.3 times the mean basic length for n = 25.

The tendency for the building up of large metacentric elements from superacrocentric elements, which seems to characterize GUI and which separates this species from other species of *Roloffia*, is also present within the species group in *Aphyosemion* which exhibits much reduced chromosome numbers. LOU has ten large metacentrics, whereas in the n = 9 biospecies of CHR only five of the nine chromosomes are large metacentrics, the largest of these measuring about 4.8 times the mean length of the basic chromosomes.

Hemoglobin patterns for African (A, B, C, D) and Asian (E, F, G, H) rivulin species. The A-pattern appears to represent the basic pattern of African *Rivulinae* as this pattern occurs in all species of *Roloffia* and *Aphyosemion*, in RAC in *Nothobranchius*, in many species of *Epiplatys* and in several species of *Rivulus* in South America also.

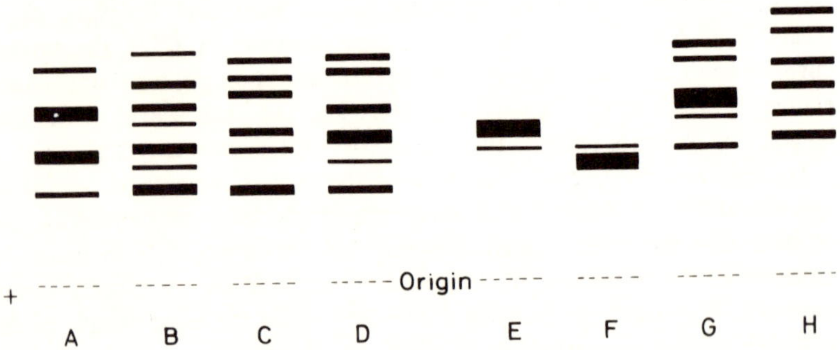

XIII

THE HAEMOGLOBIN PATTERNS

Since 1963, Dr. K. Sick of the Institute of Genetics in Copenhagen has kindly prepared haemoglobin-patterns for most rivuline species kept in my tanks using the techniques previously published by Sick (1961: 894–96 for eel and 1965: 19–48 for cod). Also, a number of hybrids were tested by this technique.

The most simple pattern, composed of two lines, occurs in *Aplocheilus* where LIN and PAN develop rather similar, but different patterns.

By far the most common pattern has four lines and it seems that all species which develop this pattern have identical haemoglobin. All species of *Aphyosemion* and *Roloffia* have this particular pattern but also three species of *Rivulus*, RAC in *Nothobranchius* and rather many species of *Epiplatys* (ANN, DAG, GRA, MAC, BAR, FAS) possess this basic rivuline haemoglobin pattern.

More complicated patterns with six or more lines occur in *Aplocheilus* (BLO and DAY, with different patterns), *Nothobranchius* (GUE) and *Epiplatys*. Within the genus last mentioned there is a large group of species which apparently develop identical patterns. Some of these species undoubtedly are closely related (CHA, ESE, LON and SEX) whereas it is less likely that CHE, which also has this pattern, is a close relative. This pattern also occurs in *Aphyoplatys* (DUB). In *Epiplatys* one of the savannah species (BIF) has a pattern rather different from other species of the genus, but to our surprise we found this pattern to be identical with the GUE pattern in *Nothobranchius*.

Hybrids from crossings between individuals which develop identical haemoglobin patterns develop the pattern of their parents. Hybrids from crossing of species having different patterns (*Epiplatys*) were not viable or almost not viable. The two hybrid individuals from the two reciprocal crossings of SEX and DAG had been preserved before the haemoglobin study had started. (See also under SEX).

It might well be of some importance that ANN and DAG which have 25 haploid chromosomes and which both develop the *Rivulus-Roloffia-Aphyosemion* haeomglobin pattern, by these two biological characters are close to the ancient stem from which the *Roloffia-Aphyosemion* group probably originated, because calculations of chromosome lengths indicate that this group originated from a rivuline species which had 25 and not 24 haploid chromosomes.

Tentative representation of the karyotypes of some rivulin species. The distance between the thin parallel lines represents the calculated length of the largest acrocentric element of the basic complement. Only two of these are figured (the shortest and the longest) but the number of chromosomes not figured is given.

It is assumed that the sum of chromosome lengths of the individual karyotype is constant for all species originating from a "n = 24" species. Corrections for n = 25 and n = 27 have been made. The karyotypes have been based on spermatogonial mitotic divisions and the relative length of the individual chromosome probably is more exact than the position of the centromere.

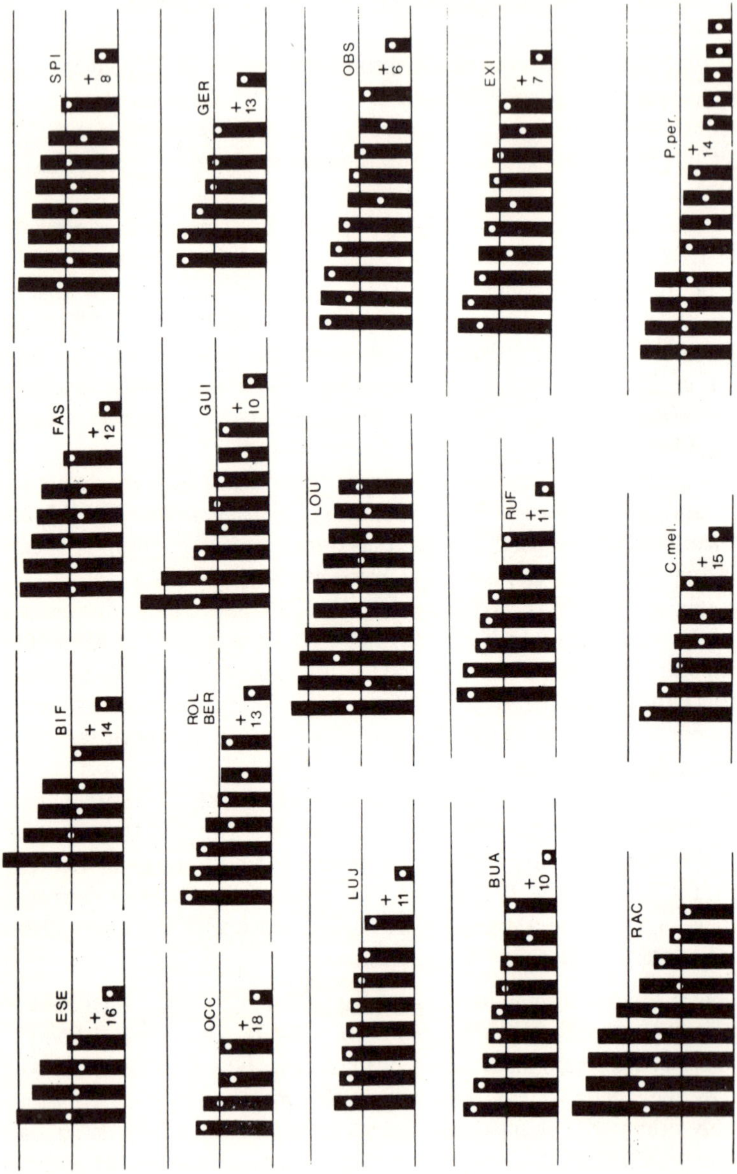

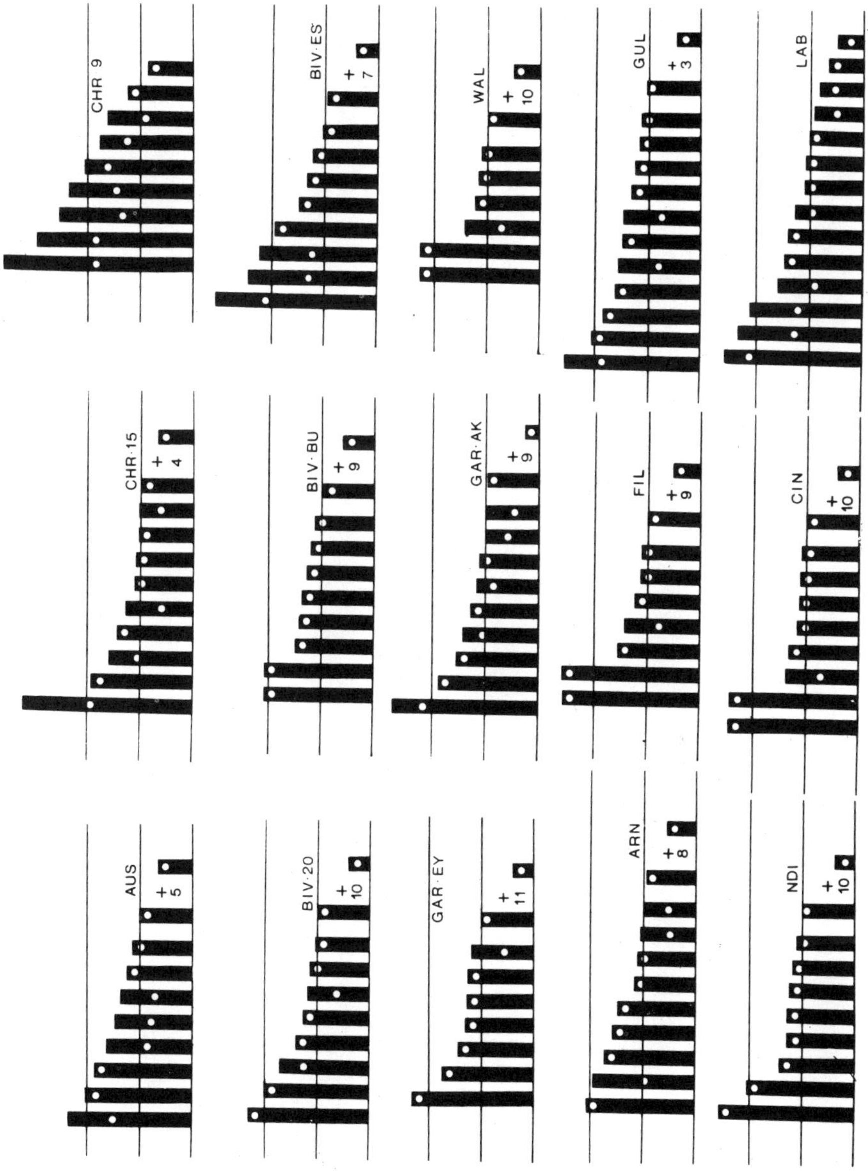

CHR 9
CHR·15
+ 4
AUS
+ 5
BIV·ES
+ 7
BIV·BU
+ 9
BIV·20
+ 10
WAL
+ 10
GAR·AK
+ 9
GAR·EY
+ 11
GUL
+ 3
FIL
+ 9
ARN
+ 8
LAB
CIN
+ 10
NDI
+ 10

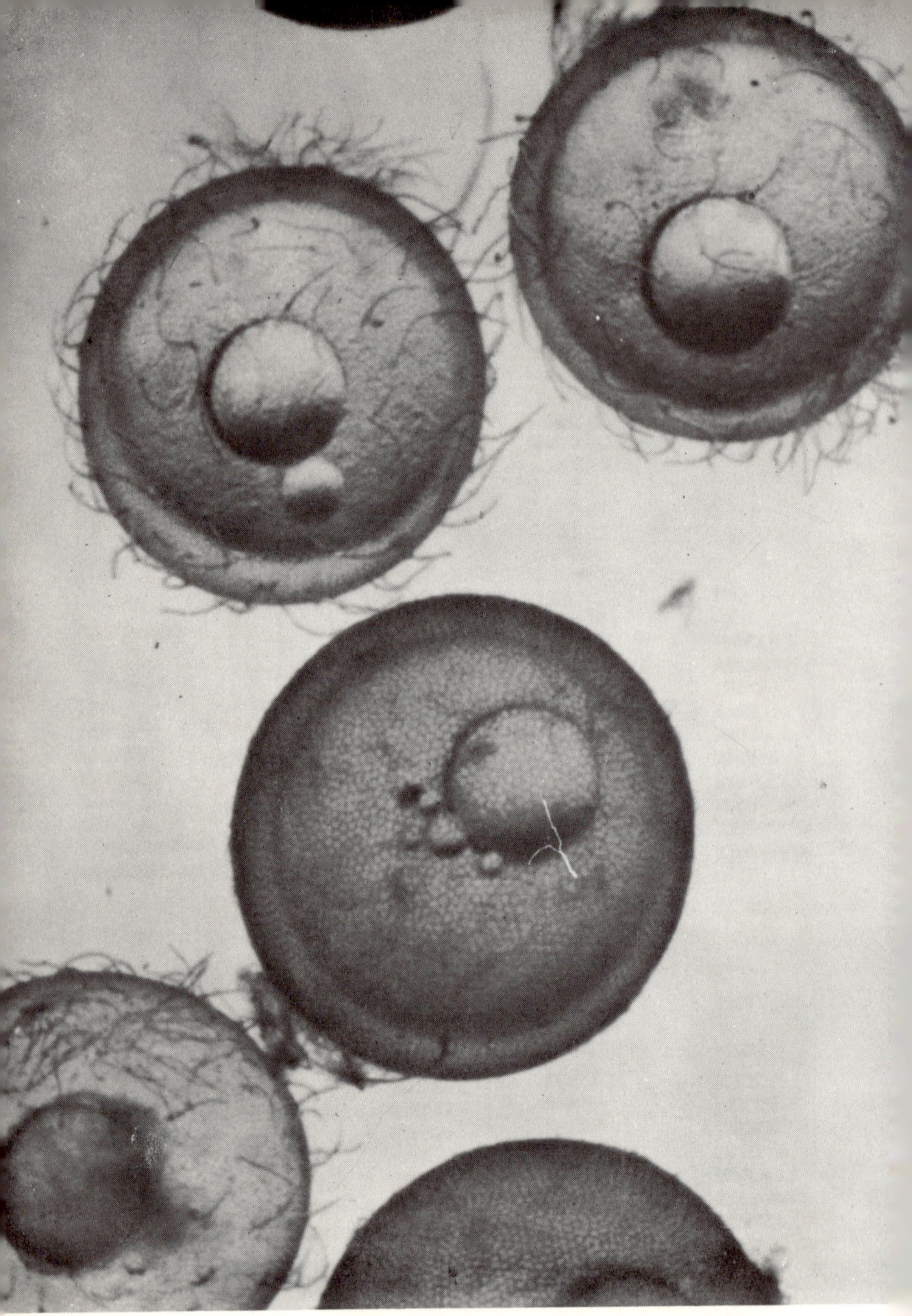

Eggs from FIL (large, without evenly distributed filaments) and from ARN (smaller).

XIV

RESULTS OF CROSSINGS

Rather many crossings have been studied, and the results of these were very difficult to understand before the start of the cytological study. Even at the present time, the results of many crossings are not in accordance with the supposed relationship between the parent species. It should, however, be kept in mind that there seems to be a number of critical phases during the normal development of the rivuline embryo (Peters 1965: 75–87) and apparently some hybrid strains are not able to overcome these developmental difficulties.

The experimental conditions might be kept under control during the development of the embryo within the eggs, but when hatching takes place the life or death of the hybrids is often a matter of chance during the first critical weeks, during which the inspection of the hybrids is very difficult and heavy losses may result. If some of the hybrids live these weeks, they may exhibit much improved viability, and they may even overcome the next critical phase which occurs at sexual maturing.

Individuals which have overcome the sexual maturing usually exhibit much improved viability thereafter and may live for a year or more.

The results of the crossings, schematically, are divided into 16 different groups or phases, and the different crossings are coded using the codes of the parent species. The name of the *male* is given *first*, and thus the AUS/MON crossing has been prepared with an AUS male and a MON female. If different bio-species have been used, these are coded similarly. CHR–9 is the 9 haploid chromosome strain of CHR, whereas CHR is the old (1949) strain with very few red spots on the sides of males (the SCH phenotype).

From breeding experiments, using cytologically sterile hybrid males, it appears that sometimes the unfertilized egg might develop an apparently normal blastodisc, but further development of the zygote of these eggs was highly abnormal and no distinct multidisc developed; instead an irregular mass of cells was seen. No such egg went through gastrulation. Unfertilized eggs from the non-annual species usually die 24 to 36 hours after the spawning, whereas those of annual species may remain alive for up to one week and probably even more. The egg is considered alive when the membranes of the yolk have not decomposed and the yolk ball is visible inside the egg. In eggs from some hybrid females, the yolk ball is abnormally large (or small) and usually it is impossible to determine if such eggs still have unbroken yolk membranes.

The crossing is considered as successful when the egg (eggs) carry through an apparently perfect gastrulation.

GROUP 1: The fertilized egg undergoes an apparently normal gastrulation and does not exhibit further development. No distinct embryo develops, but usually a certain mass of irregularly distributed cells are visible.

GUL/WAL: The development of the eggs was arrested at phase 1b of Peters (1963: 275) and apparently the zygote was hibernating, not being able to overcome the diapause. These two species both exhibit annual characters and they apparently are closely related.

GROUP 2: The fertilized egg develops a small embryo which does not exceed a quarter of the equator of the yolk (90°) and no further development takes place. Black pigments are often visible.

BER/GAR and SPI/PAN: The central part of the embryo did not develop normally, instead a shapeless mass of cells was visible. The head and the end of the tail were quite distinct and appeared to develop normally.

GROUP 3: Same as Group 2, but the space between the two yolk membranes swell and in one pole of the yolk a blister is produced. Black pigment developed on the yolk. The anterior part of the embryo is situated near the top of the blister and a very thin vein usually extends from the head of the embryo to the center of the inner surface of the blister, but no blood elements occur in this vein. Some eggs do not develop the blister.

GROUP 4: The embryo extends over about half the equator of the yolk. The causes for the death of the embryo are unknown, but usually the embryo remains unchanged for many days.

GROUP 5: Same as Group 4, but in most eggs a *very large blister* develops, reaching almost the size of the yolk ball, and these eggs usually burst. In other eggs, a few blood elements occur in veins.

GROUP 6: The length of the embryo exceeds half way round the yolk; many pigments develop particularly near the embryo.

AUS/LIB and CHR/CIN: The development of the embryo resembles that of Group 2. The head and the end of the tail develop normally, but the central part of the body is a shapeless mass of cells. Some blood elements circulate in veins but the circulation soon stops without further development. The embryo is not situated in an equatorial groove on the yolk, instead the body the embryo forms a spiral within one part of the egg.

MON/COG: Apparently the head of the embryo does not develop normally. Blood elements circulate.

GROUP 7: Same as Group 6, but the death of the embryo is caused by abnormal development of the blood system.

Initially the circulation of blood elements is apparently normal. Then blood elements seem to coagulate here and there on the yolk and/or within

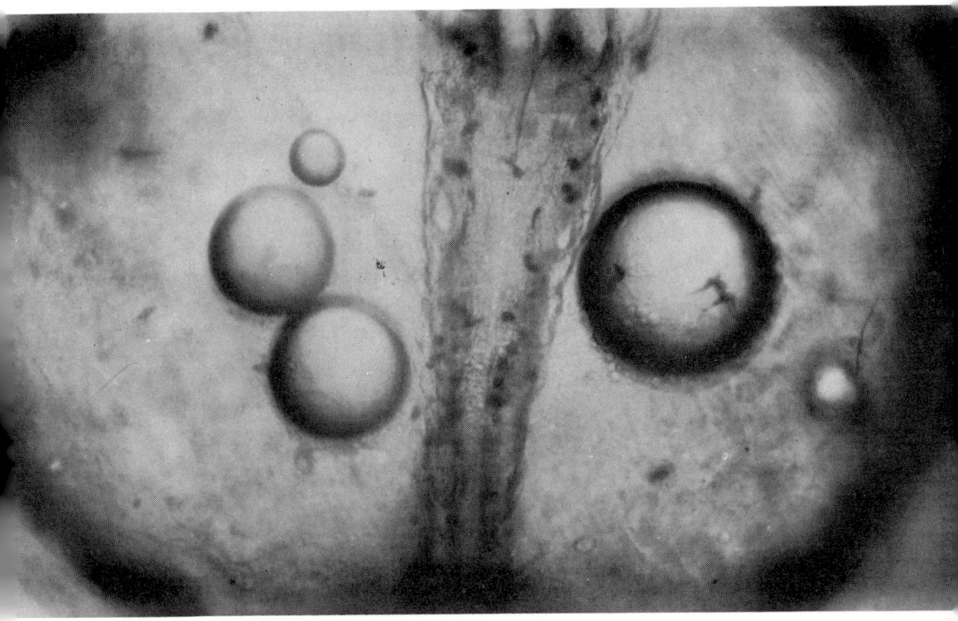

Embryo of BIF from Share a little beyond the "180°" phase. "Superposing photo." Three exposures.

Embryo hybrid FAS/DAG with a large transparent "blister" below head. After the development of the blister, the egg-membrane burst and the embryo hatched out.

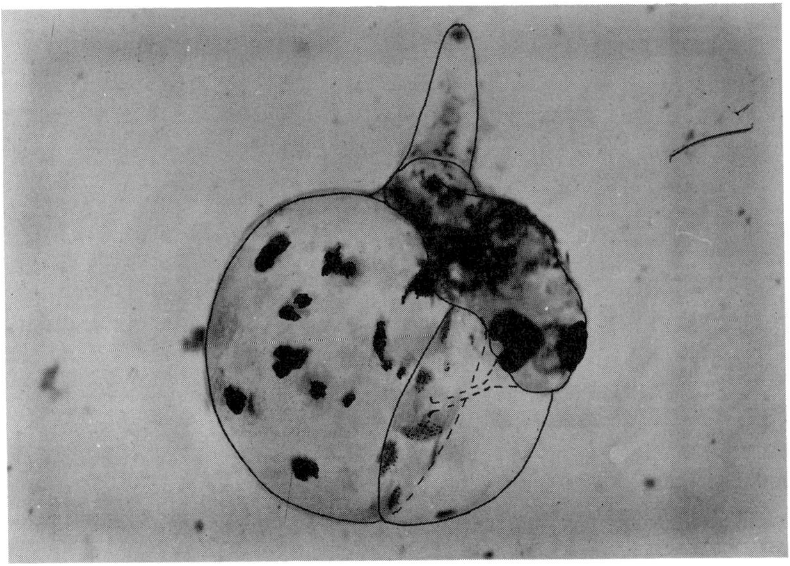

the body of the embryo. Large red accumulations of blood elements occur where arteries branch, but the serum still circulates through these accumulations and more and more blood elements are trapped, until finally no elements circulate. Often, some of the major branches of the system may exhibit normal circulation, whereas in others, the circulation has come to a complete stop. This abnormality is always fatal, but the embryo may remain alive for many days even when the blood has come to a complete stop.

AUS/COG and PET/COG: The development of the embryo was arrested rather early in the development of the hybrid.

LAB/GAR-AK and DAG/LIN: The development of the embryo was arrested when the embryo was almost fully developed in the egg.

GROUP 8: The ripe embryo dies just before or just after hatching from unknown causes. No abnormalities were found. The BIF/SPI hybrids were absolutely motionless, not even the pectorals moved.

GROUP 9: The ripe embryo hatches out and is able to swim, but it dies from unknown causes soon after the hatching.

GROUP 10: The hybrid lives for weeks (or days) after the hatching but it dies before it has reached the phase of sexual maturing. Usually the causes for the death of the hybrid are unknown.

AUS/CAL: These hybrids were extremely slender and they soon lost the air of the swimbladder and were resting at the bottom of the tank.

LUJ/CHR: Rather similar, but the development of the lower jaw was abnormal. The jaw was spoon-like and the mouth could not be closed. Most GAR/COG hybrids were similar.

LUJ/CIN: Hybrids appeared to be normal, but they did not grow although they lived for more than two months after hatching.

MON/CHA: Growth normal, but after some time all hybrids were swimming belly up and were not able to swim normally.

GROUP 11: The hybrid dies during maturing, probably from attack by mycobacteria.

GROUP 12: The hybrid reached sexual maturity and one phenotype (male, female, intersex, non-sex) develops. Often many or even most hybrids have been lost during earlier phases of development. In the case of the MON/SEX and the SEX/MON crossings very many fertilized eggs were obtained but one hybrid only of each combination reached phase 12. Also the viability of these broods of hybrids was very variable. All these hybrids were sterile.

GROUP 13: Similar to Group 12, but two phenotypes (male, female, intersex, etc.) occur. All were sterile.

GROUP 14: Same as Group 13, but one of the phenotypes (usually the female) is not absolutely sterile, but when backcrossed the offspring is not viable. This group is rather variable, as individual differences may occur, one female being partly fertile, the other not.

GROUP 15: Same as Group 14, but a few individuals of the backcrossings do survive.

GROUP 16: F_1 x F_1 spawnings produce a few fully viable and more or less fertile individuals.

VARIATION OF MERISTIC CHARACTERS IN HYBRIDS AND IN PARENT SPECIES

When non-annual species are crossed the hybrids often are intermediate to the parent forms by meristic characters. Crossings between species of *Fundulopanchax* often are rather different and the hybrids develop more fin rays and/or scales than the parent forms.

	D.	*A.*	*Sq-l.*	*D_m.*	*A_m.*	*Sq_m.*	*Indv.*
AUS/AHL	10–12	14–16	30–31	11.2	15.3	30.9	7
ARN/FIL	17–18	17–18	26–29	17.4	17.2	27.2	6
ARN/WAL	17–18	16–18	27–32	17.5	17.2	29.6	9
AUS/GAR	12	15	31				1
BER/ROL	12–14	15–18	27–31	13.0	16.5	30.1	11
BUA-19/EXI	10–11	15–16	28–31	10.9	15.9	29.5	9
CAL/AHL	11–12	15	31				2
CHR/GAR-AK	11	16	30				1
EXI/BUA-20	9–11	14–16	28–30	10.1	15.2	28.9	11
EXI/CHR-9	9–10	15–17	29				2
EXI/CHR-15	9	15	30				2
FAS/SPI	10–12	16–19	27–30	11.0	17.1	28.8	9
GAR-AK/SJO	14–16	16–19	34–38	15.3	17.2	36.2	25
GAR-OW/OBS	12–13	15	30–31	12.7	15.0	30.8	4
GAR-OW/WAL	15–16	17–18	30–31				2
GAR/CIN	13–15	16–17	30–33	14.1	16.7	31.2	10
GUL/GAR-AK	15	19	31–35				1
LAB/COG	9–12	15–17	28–31	10.5	16.0	29.7	15
MON/GRA	10	16–17	28–29	10.0	16.7	28.1	3
MON/SEX	10	17	26–28				1
NDI/CIN	14–16	16–17	31–34	15.0	16.7	32.8	6
NDI/COG	12–14	15–16	30–32				2
NDI/GAR	14–16	16–18	32–33	15.2	17.0	32.3	4
SEX/MON	11	17	28–29				1

	D.	*A.*	*Sq-l.*	D_m.	A_m.	Sq_m.	*Indv.*
AHL	10–12	14–16	30–33	11.1	15.2	31.1	
ARN	15–17	15–18	24–29	16.3	16.4	26.0	
AUS	9–11	14–16	29–32	10.0	15.5	30.8	
BER	11–14	15–18	29–33	13.1	16.4	31.1	
BUA-19	10–12	15–17	28–30	12.2	16.2	28.8	27
CHR	9–11	14–16	27–30	9.7	15.2	28.8	
CIN	13–14	16–17	30–33	13.5	16.3	31.1	
COG	11	15–16	29–30				
EXI	8–9	14	28	8.8	14.0	28.0	
FAS	10–15	15–20	27–31	12.0	17.9	29.4	
FIL	14–17	14–17	24–28	15.7	16.0	26.5	
GAR-AK	13–15	15–17	30–33	13.6	16.1	31.5	30
GAR-OW	12–14	15–17	30–33	13.0	15.8	31.4	18
GAR (PH/AK)	14–15	16–17	30–33	14.1	16.8	31.0	8
GRA	7–11	15–17	26–30	8.4	15.9	27.7	
GUL (SW Nigeria)	15–18	17–19	29–35	16.4	18.0	32.6	
LAB	13–14	16–17	30–31	13.8	16.5	30.7	
MON	8–11	14–17	25–29	9.9	15.4	26.9	30
NDI	14–16	15–18	31–35	15.5	17.1	33.9	
OBS	12–13	15–18	31–34	12.6	16.5	32.2	
ROL	11–14	15–18	29–32	12.8	16.1	30.4	
SEX (Nigeria)	9–13	14–19	27–32	10.8	16.7	29.2	744
SJO	14–18	16–19	31–37	16.4	17.2	33.4	
SPI	7–10	15–19	26–29	9.0	16.7	27.6	
WAL	14–16	16–17	29–32	15.2	16.8	30.5	

RESULTS OF CROSSINGS (morpho-species)

GASTRULA	1. AUS/MON, CAL/ROL, DAY/LIN, FAS/CHA, GUL/WAL, OCC/COG.
EMBRYONAL DEVELOPMENT	2. BER/GAR, ROL/GAR, SPI/PAN, THI/ARN.
	3. CHE/SEX, PAN/LIN, SEX/LIN.
	4. BIF/MON, FAS/DAG.
	5. CAL/SPI, FAS/MON.
	6. AUS/LIB, CHR/CIN, MON/COG, PAN/DAY.
	7. AUS/COG, AUS/PET, CHA/MON, CHR/MON, LAB/GAR, MON/LIN, PAN/FAS, PET/COG, AHL/CHR–9.
	8. BIF/SPI, PET/MON, SEX/GRA.
	9. CHE/BIF, CHR/PET, GAR-AK/LAB, LUJ/COG, SPI/MAC, BIV-19/CHR-9.
PRE-MATURING	10. AUS/CAL, CHR/LAB, LAB/CIN, LUJ/CHR, LUJ/CIN, MON/CHA, MON/MAC.
	11. LAB/CHR.
MATURING	12. ARN/WAL, AUS/GAR-AK, BUA-20/AHL, CAL/AHL, EXI/CHR-9, EXI/CHR-15, GAR-AK/COG, GAR-OW/OBS, MON/GRA, MON/SEX, NDI/COG, NDI/GAR, SEX/MON, WAL/GAR-AK.
	13. AUS/AHL, ARN/FIL, AUS/SJO, CHR/GAR-AK, GAR-EY/CIN, GAR-PH/CIN, GAR-OW/WAL, LAB/COG, NDI/CIN, SEX/SHE, WAL/GAR-OW.
REPRODUCTION	14. BER/ROL, FAS/SPI, GUL/GAR-AK, GAR-AK/SJO.
	15. EXI/BUA-20.
	16. BUA-19/EXI, CHA/SHE, CHR/COG, SEX/LON.

XV

ALPHABETIC REVIEW OF FISHES

Fishes are listed alphabetically by their code name assigned on pages 7 through 11. The scientific name given in the following descriptions is the actual name first given it by the original describer. To find the valid name for the species, refer to pages 7 through 11. This has been done so changes in nomenclature could be readily corrected in future editions without the necessity of changing more than one set of printing plates. After each scientific name you will find synonyms which are applicable.

ACU: ***Haplochilus senegalensis acuticaudatus*** **Pellegrin 1913** **(=SPI)**

Pellegrin (1913: 84) based this subspecies of SEN on four individuals from Guidimoussi, just east of Zinder of the Niger Republic. This locality is within the Chad drainage. Pellegrin thought that ACU differed from SEN by the elongate central caudal fin rays of the former, and the described rounded caudal of the latter. This character, however, is variable in SPI (SEN is a synonym), and ACU for that reason seems to represent a simple synonym for SPI.

Boulenger, however, considered ACU to represent a distinct species (1915: 72, 1918: 426) reporting individuals from the Chari basin within the Chad drainage.

AHL: ***Aphyosemion calliurum ahli*** **Myers 1933** **(Valid sp.)**

AHL is one of the three named species of the CAL superspecies. It seems that AHL is the most ancient species of this group. CAL and AUS which represent two more species of this group probably represent derivatives which independently developed from AHL or a similar ancient form.

The AHL name, unfortunately, was attached to aquarium kept fish of unknown origin and for that reason it is difficult to attach this name to any natural population in Africa. This (?) phenotype first was imported as an aquarium fish in 1910. The 1910-strain was said to originate from Cap

Adult male of AHL from the Mboumboula River of East Cameroon.

Adult male of AHL from the Mboumboula River of East Cameroon.

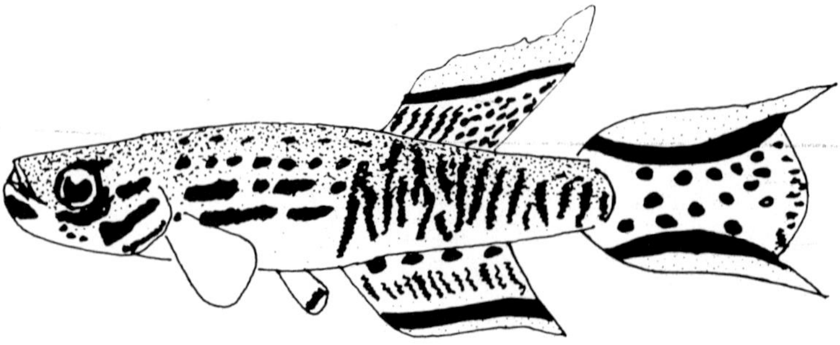

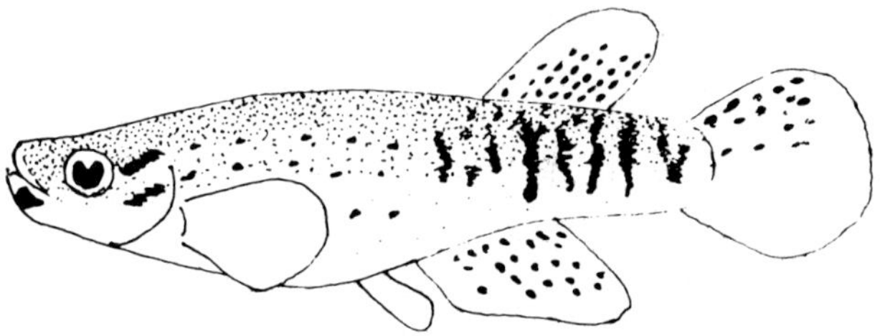

The types for AHL (*Panchax calliurus var. coerulea*) redrawn after Meinken 1929.

The dark crossbars of the female are more marked than in the Mboumboula strain, but in a strain which I took in 1968 just north of the Rio Benito of Rio Muni, this marking corresponded well to Meinken's type.

Lopez of Gaboon and it was described and figured by Arnold (1911: 576) and by Träber (and Thumm) (1911: 773). German aquarists considered the strain to belong to CAL, but Boulenger identified preserved material as CAM. It seems that the 1910-strain differed from Nigerian CAL by minor traits only. Whereas in CAL the red dots on the sides of the body are rather numerous and confluent producing irregular red lines along the sides, at least anteriorly, these spots are less numerous in AHL, and in this species, very thin vertical red lines are produced posteriorly, in some males at least. The dorsal of AHL is larger and higher, and the colors are more pronounced. A similar strain was imported in 1913, and this strain was also said to originate from Cap Lopez. German aquarists placed this strain within CAL, but Boulenger again identified preserved material to belong to CAM. The red dots on the sides of the males of this strain were less numerous than within the 1910-strain and it seems that they did not develop longitudinal or vertical red lines. At the present time the identity of the 1913-strain is still uncertain and this strain may belong to AHL or AUS.

(See AUS.) In 1932 a third strain of similar fishes was imported into Germany and was described and named *Panchax (Aphyosemion) calliurus (Boul.) var coerulea* by Meinken (1932: 66). The name "*coerulea*" was found to be preoccupied by Myers (1933: 183), who renamed Meinken's subspecies as *Aphyosemion calliurum ahli* Myers. The origin in nature of the 1932-strain is unknown.

Populations of *Aphyosemion*, which correspond to the 1910- and the 1932-strains, live in the coastal lowlands of Cameroon. I have seen individuals of this phenotype in between Mbonge, west of the volcano and Kribi in southern Cameroon. These populations seem to be restricted to swamps and to small stagnant and very shallow pools on the forest floor. The natural water measured from 0.0 to 4.0 German degrees of hardness, pH between 5.6 and 6.8 and the electrical conductivity between 14 and 200 reciprocal megohms. The natural water, near the volcano, is rather hard and contains much more dissolved mineral matters than the natural water of East Cameroon. The conductivity of 14 units corresponds to distilled water of medium quality.

The active male is almost black, except for the brilliant golden edges of the anal and the caudal fins. During fighting (in a dark tank) males usually change into a warm reddish-brown color. Pectorals are broadly edged with

Early pro-metaphase meiotic division of AHL male. n = 18.

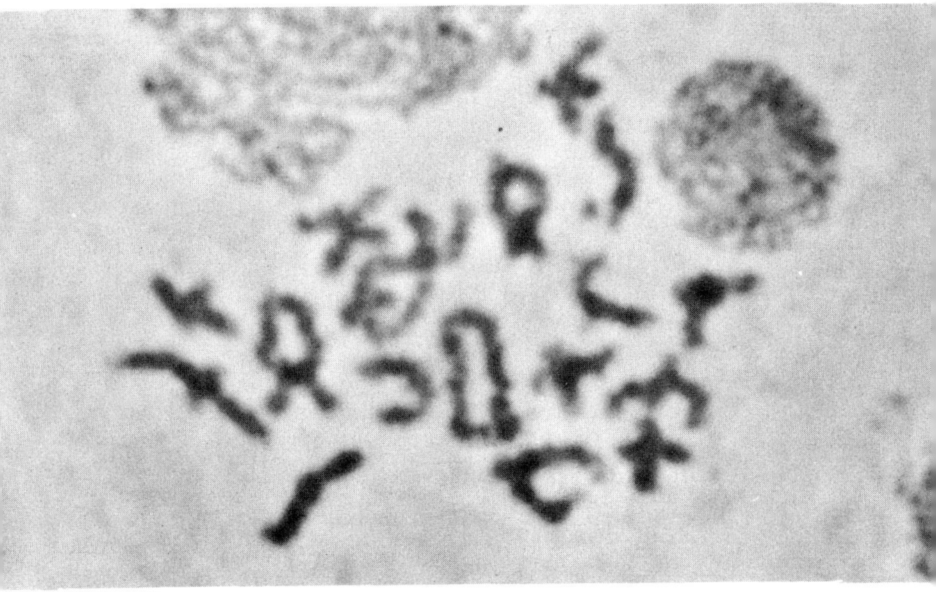

orange-red, the basal part of the fin being milky-white or grey. The thin red vertical lines on the posterior part of the sides are well developed in some males while absent in some from the Mboumboula River (35 km north of Douala, East Cameroon). Phenotypically this strain of AHL is closer to AUS than to Nigerian CAL, and this point of view is supported by the results of crossings. The strain of AHL above mentioned has 18 haploid chromosomes and probably one microchromosome. The details of the karyotype have not yet been worked out in detail. The frontal scalation may contain H-scales in a few individuals.

Although AHL seems to be restricted to swamps and similar biotopes, it reproduces well in hard, alkaline water. Adult individuals are easily kept in hard, alkaline water, whereas juveniles sometimes suffer attacks from mycobacteria. Although AHL, CAL and AUS are rather similar, these three species are reproductively isolated to a very high extent. AHL is probably similarly closely related to the GAR superspecies. PAS may represent a synonym for AHL, and if this point of view proves to be correct, AHL becomes a junior synonym for PAS, because the former was described in 1933, the latter in 1928.

AMS: *Nothobranchius amsingki* Ahl 1928 **(Invalid name)**

This name was created by Ahl (1928: 193) but no description of AMS was published. According to Beyer (1932: 323) Ahl probably intended to attach the AMS name to some aquarium fishes which were collected by Oeser at Cotenou, a locality situated on the coast of eastern Dahomey (where FIL may occur). This locality represents the western limit of the coastal swamps of Nigeria. In 1936 (: 290) Arnold and Ahl used the AMS name, but they placed AMS in *Aphyosemion*, and no description was given.

AND: *Haplochilus andamanicus* Köhler 1906 **(=PAN?)**

Köhler (1906: 388) tentatively named the PAN population of the Andaman Islands, by the name above mentioned. He thought that this population differed from the continental populations of PAN by the higher number of dorsal fin rays (D. 11 versus D. 7–8). Herre (1941) reported PAN from these islands (Bamboo Flot Village, Port Blair and from a brook west of Sipi Ghat (Bhagelsinghpura), South Andaman) but he did not publish any morphological data.

ANN: *Haplochilus annulatus* Boulenger 1915 **(Valid sp.)**

Boulenger (1915b: 203) used individuals collected at Matca of Sierra Leone by Thomas in 1913. This locality is not on my maps. ANN ranges from Boffa of Lower Guinée to Monrovia of Liberia, at least. It seems to be restricted to certain swamp-like biotopes in open land or in ancient forests, and is said to be rare wherever it occurs. The natural water is rather hot, shallow, usually stagnant and probably soft and acid. In all major geo-

Adult male ANN from Monrovia, Liberia.

Adult male ANN from Monrovia, Liberia.

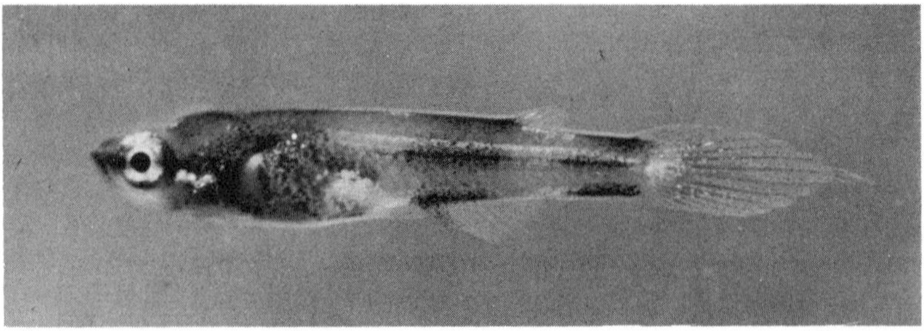

Juvenile ANN. The three thin black lines of the "procatopodine-pattern" have decomposed and are absent in areas which later on will become lightly colored.

graphical areas within its range, it has been found together with individuals of FAS, and in Sierra Leone and in Liberia, individuals of BAR were also present. OCC also occurs in some of the Sierra Leone localities, whereas in Liberia ANN occurs together with LIB.

ANN takes a rather isolated position in *Epiplatys* although it develops most of the major characters of this genus. It is pike-like, with dark cross-bars, with extended central caudal fin rays, this fin measuring more than 35% of the standard length and with a rather deep caudal peduncle. It has

The type for ANN as figured by Boulenger 1916.

Haplochilus annulatus.
Type.

also a rather rudimentary dark longitudinal band in the active female, and in nature it seems to be restricted to the surface of the water. It differs from other nominal species of *Epiplatys* by the patterns of frontal scales and lateral line organs. The frontal squamation is rather irregular, and the scales are small. The supraorbital and the upper postorbital grooves of the lateral line pattern are united, as in *Nothobranchius*, and not separated by a bridge of (F—) scales as in other species. Also the neuromasts along the edge of the preoperculum are housed in open grooves as in *Nothobranchius* and in HOM and not in canals with pores as in other species. Also, the color pattern of the caudal fin is different from other species of *Epiplatys*, as a perfect lyre-pattern develops, as in *Aphyosemion* and *Roloffia*. Also the red flame of this fin is unique and does not occur in other rivulins of the Old World (see also FLA).

The haploid chromosome number is 25, and the karyotype resembles that of DAG, but in ANN, there seems to be more small metacentrics of average basic size. The haemoglobin pattern is of the usual rivuline type (four lines) and corresponds to that of DAG and many more rivulins.

Juveniles of ANN have three very thin black lines along the sides. These lines correspond to the lines, which develop in almost all species of *Procatopodinae*, in FLA, BLO and some strains of PAN. In ANN these lines disappear with age.

The behavior of the juveniles and semi-adults differs from other rivulins in one trait. When disturbed at the surface, the individual make a very quick spiral-movement downwards and suddenly, it is seen standing motionless about an inch below the surface.

Clausen (1967: 30–31) considered ANN to differ sufficiently from other species of *Epiplatys* to be removed from this genus and to be placed in a new genus, which he called *Pseudepiplatys*, and which contains ANN only. Indeed, ANN seems to differ more from other species of *Epiplatys*, than this genus differs from *Aplocheilus*, but this problem might be solved if *Epiplatys* is incorporated in *Aplocheilus* as a sub-genus, and then *Pseudepiplatys* possibly might represent another sub-genus. To the aquarist ANN is a true species of *Epiplatys* (or *Aplocheilus*).

ANN probably is an ancient rivuline species, more closely related to DAG, than to other known species of the genus, but also FLA might be a relative.

The eggs of ANN are the smallest found within *Rivulinae* of the Old World. They measure 0.90 to 0.95 mm and the small size of the eggs is one of the reasons for the rather difficult reproduction of ANN in the aquarium. The newly hatched ANN fry is not able to take brine shrimp nauplii, and during the first weeks other live micro-foods should be offered. Infusoria (*Paramecium*) is generally not taken, whereas various species of *Rotifera* appear to be excellent food. Later, newly hatched larvae of mosquitoes

Adult male of ANN from Monrovia, Liberia.

Primary meiotic division of ANN from Monrovia. n = 25.

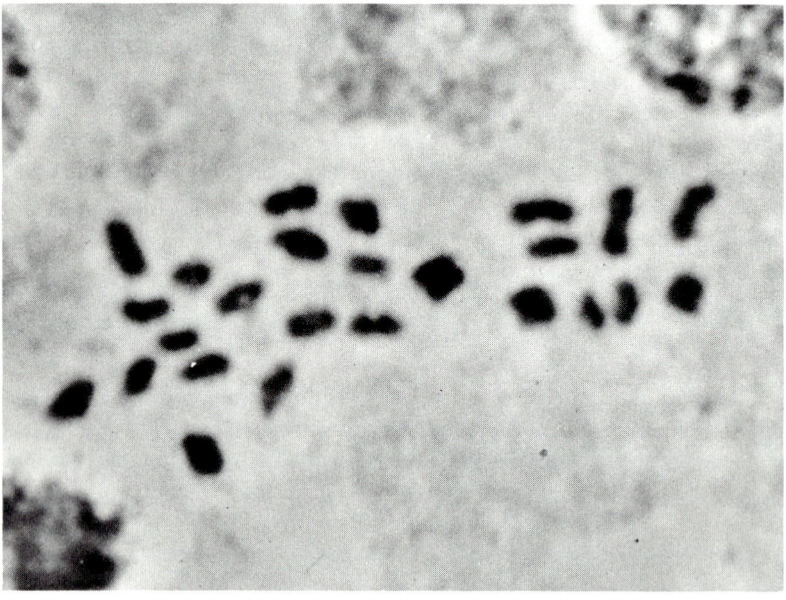

Adult male ANN from Kasewe, Sierra Leone.

(*Culex*) is taken. Using this food, the fry mature within five or six weeks, an unusually quick growth for a species of *Epiplatys*.

Eggs from ANN appear to be rather sensitive, and attacks from various unidentified bacteria often ruin the egg-membrane and the embryo. The embryo hatch out much too early and often die. To prevent such attacks from bacteria, water should contain a minimum of calcium and magnesium, and the pH should be reduced down to 5.0, but eggs might still be ruined. Further improvements are had by adding equal parts of methylene blue and acriflavine (2 ppm of each) to the water, in which the eggs are stored, or to have the eggs spawned into coarse peat (as a bottom layer) or into nylon, which is dried in the usual way for annual rivulins. Some German aquarists do not consider ANN to be difficult in reproduction, but probably the egg-destroying bacteria has not yet spread to their tanks.

Three different strains are kept as aquarium fishes at the present time. One strain originated from localities near Conakry of Lower Guinée (Benl's strain; see Foersch 1966 (163–67) and 1967 (161–163)), another strain comes from Kasewe of Sierra Leone (Roloff's strain, see Roloff 1965a: 5–11) and the third strain comes from Monrovia of Liberia (Clausen's strain; see Scheel 1965: 19–22.) The males from these strains differ constantly by minor traits of the color patterns, but in spite of this, the present author hesitates to create subspecific names for these strains. Hybrids from crossings of individuals from the Kasewe and the Monrovia strains were fully viable and fully fertile. This hybrid strain is more easily kept and bred than the pure strains, and in all hybrid males the red pattern of both strains develops.

ANS: ***Haplochilus ansorgei*** **Boulenger 1911** **(Valid sp.?)**

Boulenger (1911: 262) had two large individuals from the Masoma River, flowing into Lake Agemwe, at Umpokoya, in Gaboon. These names are not on my maps. The Masoma River is part of the Ogowe drainage.

Haplochilus ansorgii.

♂, type.

The type for ANS as figured by Boulenger 1915.

Holly (1929: 425) reported eighteen individuals of ANS from mountain brooks within the Bakoko district of Cameroon. This district is not on my maps, but Holly said that the locality is in SW Cameroon. The Bakoto tribe lives within the southern part of West Cameroon at the present time, and in this area, there are several mountain brooks. If this is the district which Holly had in mind, his individuals of ANS are probably SEX, a common species of this geographic area. Holly also reported SEX from the Bakoko district.

Pellegrin (1930: 210) also reported individuals of ANS from the Sangha-Congo drainage. In 1961 Lambert informed me that he had seen these individuals and that in some of them there were faint dark crossbars on the flanks, whereas in others, traces of a dark longitudinal band was visible. These individuals might represent badly preserved individuals of MUF. Pellegrin (1934: 26) also reported ANS from Man of western Ivory Coast, and he also reported SEX from this locality. These individuals undoubtedly cannot represent ANS, for zoogeographical reasons.

Within the Ogowe drainage there are probably two distinct species of *Epiplatys*, which I have seen (preserved), and which I cannot identify. One of these might represent ANS, but as I have not seen the types for ANS, the identity of ANS remains an open question. (See SAN).

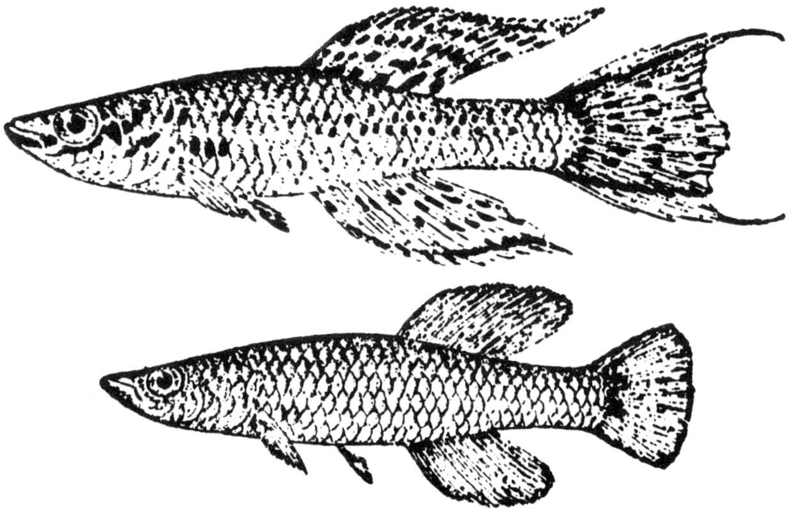

Fundulus arnoldi.

Types, ♂ ♀.

The types for ARN as figured by Boulenger 1915.

ARN: *Fundulus arnoldi* Boulenger 1908 **(Valid sp.)**

According to Arnold (1908: 469 and 1911: 602) the types for ARN originated from Warri of the western part of the Niger Delta. These individuals were caught by a German sailor named Groch, who collected aquarium fishes for Schroot of Hamburg. Eight live individuals were in the first shipment of aquarium fishes from this part of Africa. The shipment also contained live individuals of GRA, SEX and SJO. This historic shipment of aquarium fishes arrived in Hamburg on the 5th of November 1905. One male and two females were bought by Arnold, and the remaining lot was sold to Thumm, who lost them all. The last individual of Arnold's stock died in 1906, and the three preserved individuals were sent to London for identification. Boulenger considered these individuals to represent a new species, which he described (1908: 29) and named. Boulenger said that he had four types, but according to Arnold, there were only three. In August

Adult male ARN from the Wokocha River, near Port Harcourt of east Nigeria.

Adult hybrid male ARN/FIL.

Small hybrid male ARN/WAL.

Adult male ARN from the Wokocha River, near Port Harcourt of east Nigeria.

1906, a second and larger shipment of ARN arrived in Hamburg, and all the twenty individuals were sold to Schneisig of Magdeburg, from where Köhler bought two males which he considered to represent a breeding pair. Using the name *F. gularis* (?) Köhler (1906: 426) published a description and a very fine photo showing his two males. Further importations of ARN took place in 1910 and in 1911, but since then, it seems that no more individuals of ARN were imported until 1961. Various strains, which were imported after 1911, and which were identified as ARN, probably were not this species but various strains of the more variable FIL. Memmler (1910: 358) said that he successfully reproduced ARN, but it seems that he did not raise any individuals to adult size, and all further attempts to reproduce this species were without success, probably because this species is rather feeble in the aquarium and females soon die. Also, the reproduction of ARN is rather annual, and the ripe embryo hibernates in the egg and is difficult to induce to hatch.

In 1961, Ulf Hannerz collected small freshwater fishes suited for the aquarium in several parts of Nigeria. In the Wokocha River, near Port Harcourt (13 miles from Port Harcourt towards Oweri) he caught several individuals of ARN which he brought back to Stockholm. This strain was difficult to keep alive, and Ulf kindly offered me the remaining individuals. On this base an aquarium strain was produced. This strain was also kept by other European aquarists, but they were not able to maintain it. In 1967 my strain became extinct, but at that time, some aquarists in D.D.R. (East Germany) have had live individuals from Nigeria and had established a new strain, which, according to their photos does not differ from Hannerz's strain. See also Scheel 1963 (: 162–166, 277–280).

The two males photographed by Köhler in 1906 correspond to the males of Hannerz's strain. Also preserved individuals from the western part of the Niger delta area correspond to this strain.

ARN seems to be restricted to the Niger Delta area, being replaced by FIL in SW Nigeria, and these two forms probably represent a superspecies very distinctly separated from all further species-groups of *Aphyosemion*. Also, the ARN superspecies may represent a very important group in *Aphyosemion* representing some characters which appear to be intermediate between *Aphyosemion* and *Fundulosoma* (and *Nothobranchius*). The ARN superspecies, *Fundulosoma* and *Nothobranchius* are characterized by properties of the egg membrane, which is not sticky in these species, whereas in all other species of rivulins of the Old World, the membrane and the membrane filaments are sticky. ARN and FIL differ from the species of the two genera above mentioned by the hexagonal pattern of the egg membrane. Furthermore ARN resembles THI by meristic characters, by color pattern and by the evenly distributed short and stiff filaments of the egg membrane. Crossings between individuals from ARN and THI did not indicate close relationships, and all hybrids died very early in the development.

Crossings of individuals of ARN and WAL was more successful, and the hybrids were viable. All ten hybrids were males, which were functionally and histologically sterile.

Hybrids from crossing between individuals of ARN and FIL developed both sexes, and the hybrid females spawned rather regularly, but the eggs did not develop in backcrossings. Hybrid males were sterile also.

Individuals of the ARN superspecies differ from individuals of the GAR superspecies by the juxtaposition of the dorsal and anal fin. In ARN and FIL, the anterior-most dorsal fin ray is always in front of the corresponding anal fin ray; in the GAR superspecies, the dorsal fin ray in question is behind the corresponding anal fin ray. Also individuals of the ARN superspecies have less than 28 scales in a longitudinal series, whereas in the GAR superspecies more than 28 scales are found. This character also separates ARN and FIL from the species of the GUL-SJO group, which usually have 30 or more scales. ARN has 19 haploid chromosomes (not n = 18 as given by Scheel 1966: 147–148) and the karyotype differs from that of FIL by the absence of one of the large double-sized acrocentric elements; instead, a slightly shorter metacentric element occurs.

Adult hybrid male ARN/FIL and female FIL during premating play. Notice the position of the lower fins in male.

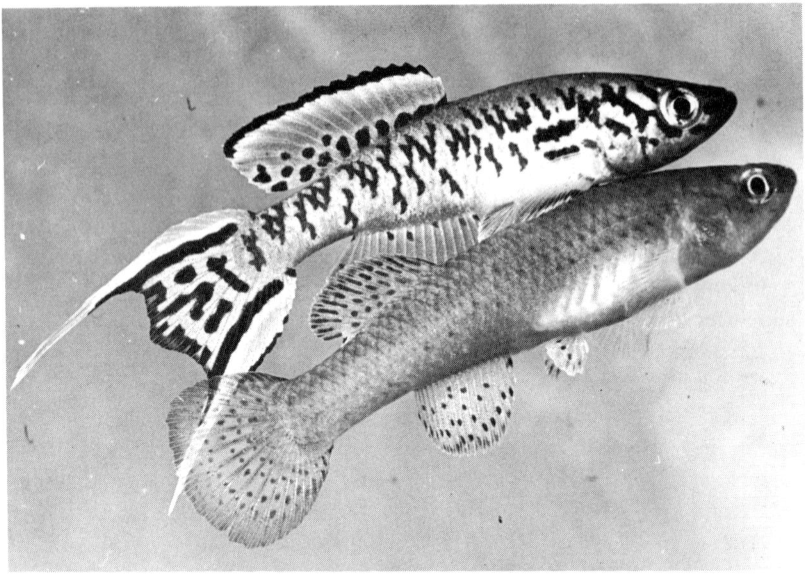

Adult male hybrid ARN/WAL and female hybrid WAL/GAR-Akure during pre-mating play.

Small posing male hybrid ARN/WAL.

Individuals of ARN usually are rather sensitive to attacks from mycobacteria and often are killed by this disease within a short time. It is recommended to keep individuals in clean, hard, alkaline water, unless they are used for reproduction. Annual characters are not well developed in ARN, and under normal conditions the embryo develops almost without delay, but the ripe embryo often hibernates.

AUS: *Haplochilus calliurus australis* Rachow 1921 **(Valid sp.)**

In 1913 (see also AHL) and in 1914 (?) live individuals belonging to the CAL superspecies were imported into Germany. These individuals (1913-strain) were said to originate from Cap Lopez of Gabon. According to Arnold (1913: 701) and Kramp (1913: 665 and 1914: 57) who described and figured these fishes, they were rather similar to CAL and it seems that German aquarists considered this strain to belong to CAL. Boulenger, however, considered preserved individuals to belong to CAM. The 1913-strain also resembled AHL, but in the new strain, the red spots on the sides of males were less numerous, and they did not produce any thin vertical red lines posteriorly on the sides. Most German aquarists did not agree with Boulenger, that this strain belonged to CAM, and Arnold (1913: 701) proposed, that the name "*Haplochilus* aus Kap Lopez" should be used for this strain.

Adult hybrid male AUS/AHL. The pattern of the caudal and anal fins is intermediate to the patterns of the parent species. The black markings on sides occur at times in both parent species.

The present strain, called AUS by aquarists and zoologists differs, however, from the 1913-strain by the pattern of the anal fin. Males of the 1913-strain had a lightly colored border in this fin, whereas in the present strain of AUS, no such lightly colored marginal band develops, although the long streamer posteriorly in this fin is white. The strain, now considered to represent AUS, was, however, kept in Germany in 1915, and a male belonging to this strain was photographed by Milewski (1915: 229), and it is evident, that Milewski's strain and not the 1913-strain (as described and figured by Arnold (1913: 701) and Kramp (1913: 665)) survived the first World War. In 1921, Rachow proposed, that the aquarium strain previously called "*Haplochilus species* aus Kap Lopez" should be called "*Haplochilus calliurus var. australis.*" It seems, that Rachow had Milewski's strain in mind, because this was the strain which survived and needed a separate name. We do not know if Milewski's strain is a simple variety of the 1913-strain or, more likely, belongs to another importation of aquarium fishes prior to World War One. Some years ago Lambert informed me that he had taken AUS near Pointe Noire of the former French Congo.

When individuals of AHL and AUS are crossed, the hybrid males develop a pattern on the anal fin which is intermediate to the broad golden marginal band of AHL and the corresponding dark band of AUS. Hybrid males

Mitosis of male hybrid AUS/AHL. $2n = 15 + 18 = 33$.

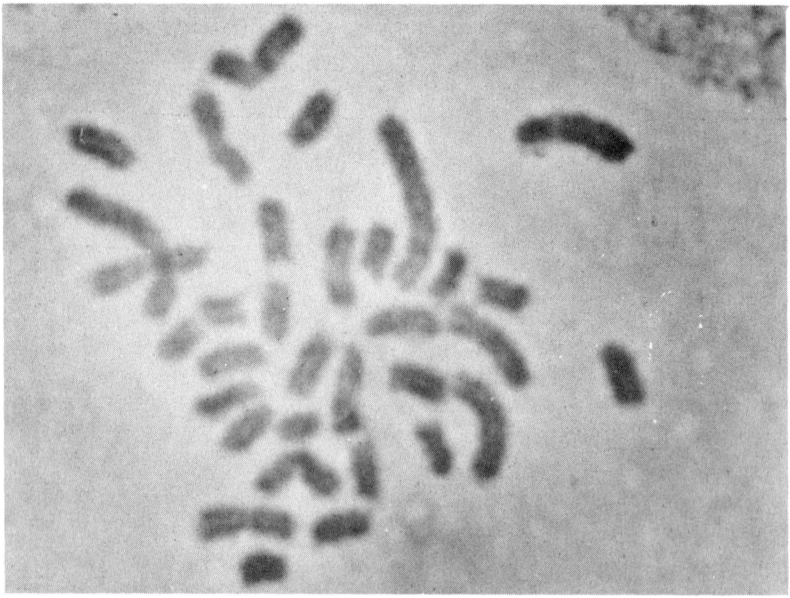

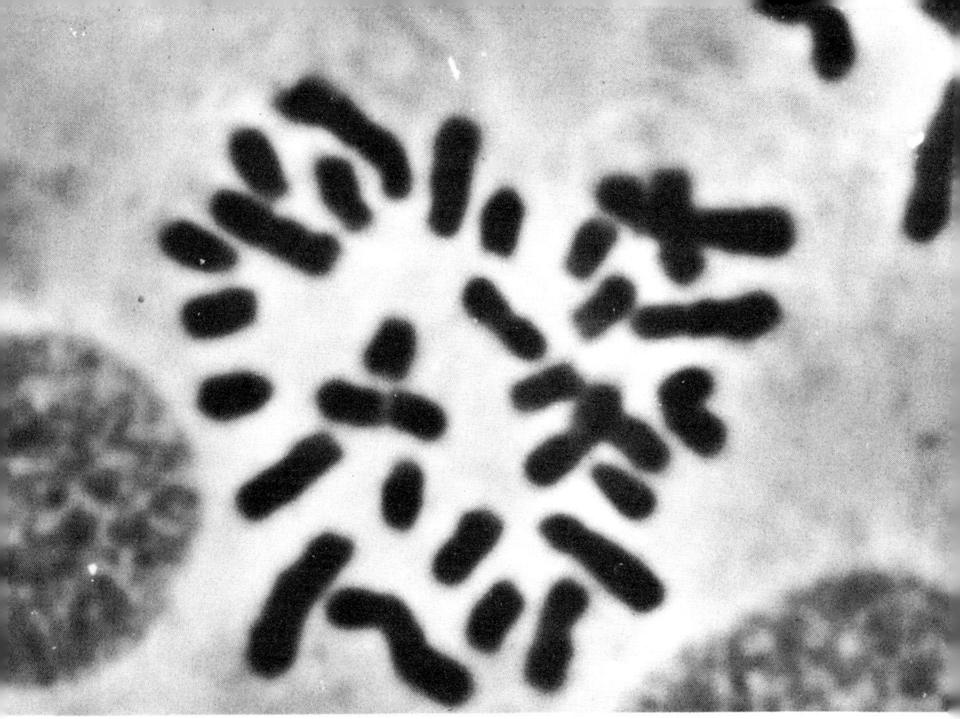

Mitosis of AUS. 2n = 30.

develop a narrow lightly colored border which is pale yellow in young individuals and pure white in older ones.

The present 1915-strain of AUS (Milewski's strain) has 15 haploid chromosomes only. The karyotype, not fully understood at the present time, has a smoothly graded series of chromosomes, the largest being about 3.5 times the length of the smallest, which is acrocentric. The largest chromosome is a submetacentric element (1:2), and there are three more metacentrics longer than the longest acrocentric element of the basic complement. The pairing of the AHL and AUS chromosomes in the male hybrid is so incomplete, that the karyotypes cannot be compared during meiosis. Also the CAL and AUS karyotypes cannot be compared during meiosis because the hybrids die.

Probably AUS is more closely related to AHL than to CAL, in spite of the fact that AHL has 18 (+ 1?), CAL has 16 and AUS has 15 chromosomes.

POL seems to represent a synonym for AUS.

Although AUS probably is restricted to swamps and similar biotopes in nature, it reproduces well in hard, alkaline water. AHL is similar in this respect. The present strain probably has not been mixed with other strains since the first introduction in 1914 (or 1913).

An aquarium-raised mutation, with less black pigments, has been considered a subspecies of AUS, which it is not. (See HJE.)

Adult hybrid male AUS/GAR.

Male of orange aquarium variety AUS first raised by Hjerresen in 1953.

Young male AUS.

Adult male AUS. The streamers of the caudal fin were lost and have regenerated.

Holotype, male BAR from the Barmoi district, western Sierra Leone.

BAR: *Epiplatys barmoiensis* nov. spec.

Individuals of this species were identified by Schultz (1942: 337) as belonging to BIF. Schultz individuals originated from Liberia. I have seen these individuals, which belong to the collections of the USNM (Nos. 118848 and 118850, totalling 15 individuals).

These individuals differ markedly from BIF by the color pattern of the throat and by the red pattern of the sides, which in BIF is an almost perfect red reticulation, whereas in BAR, crescent shaped spots occur. Also, these BAR individuals are larger than individuals of BIF.

Further individuals of Liberian BAR were collected by Clausen near Monrovia in 1965, in a "savannah-swamp" together with individuals of FAS, DAG, ANN and LIB. Unfortunately these BAR individuals did not

Preserved individual of BAR from Liberia. USNM. No. 118848 (11 specimens). Schultz, L. P. (1942: 337).

Allotype, female BAR from the Barmoi district, western Sierra Leone.

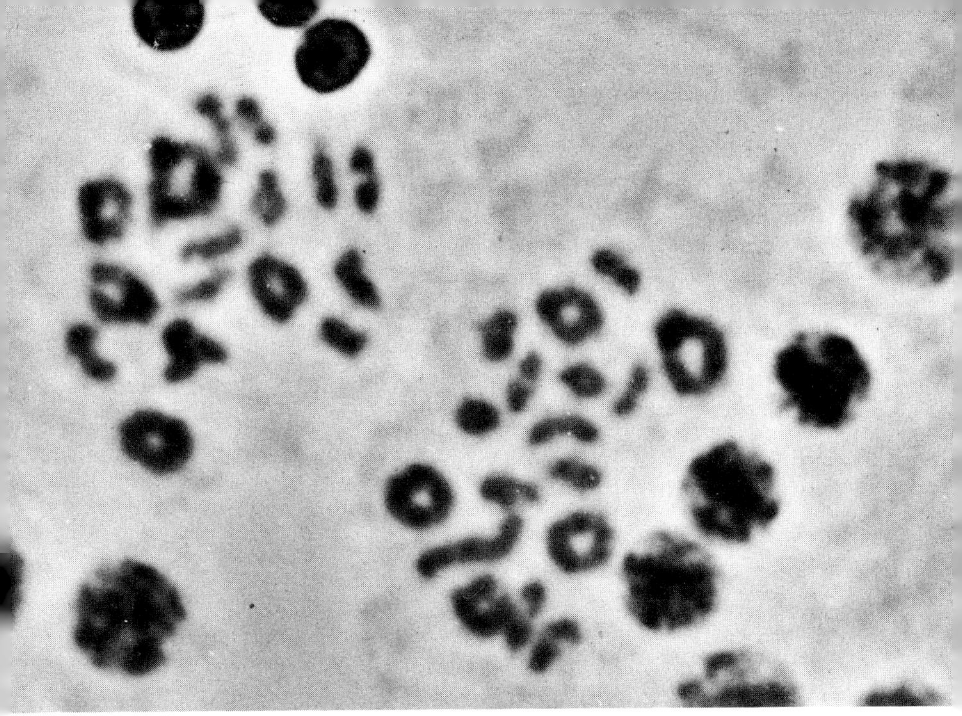

Two primary meiotic divisions of the BAR holotype. n = 17.

survive the air transportation to Copenhagen. In the males of this population, the pectorals are very pointed and long streamers reaching the root of the caudal fin develop.

Other individuals of BAR were collected by Roloff near Barmoi of western Sierra Leone in 1965, and in this area BIF was also found. Roloff kindly sent me an adult pair for study.

The number of fin rays and of longitudinal scales of the 18 BAR individuals correspond to those of BIF (BAR: D.8–9, A.15–18, Sq-l.26–29. BIF: D.7–10, A.14–19, Sq-l.25–29).

BAR has 17 haploid chromosomes only (BIF has 20), but the karyotype could not be worked out because no spermatogonial mitotic divisions, suitable for measurement, were found. The haemoglobin pattern differs markedly from that of BIF and correspond to that of FAS and SPI (four lines). Also the eggs differs from those of BIF, because in BAR, the membrane has a broken hexagonal reticulation, whereas in BIF, the membrane is plain.

BAR and BIF probably are not closely related, and the similarities in external morphology might be a result of convergence. BAR probably should be placed closer to ANN and DAG, although in the external morphology these three forms are rather different.

The data for the holotype are given in the appendix.

Preserved specimen of BAT from Akonolinga, east Cameroon. No. 1913, 10.29, 19–23 of the British Museum. Bates collection.

BAT: *Fundulus batesii* Boulenger 1911 **(Valid sp.)**

Boulenger (1911: 261) based the description of BAT on individuals from Bitye of Eastern Cameroon. This locality belongs to the Dja-Sangha-Congo drainage. In 1915 (: 28–29), he also had individuals from Assobam of the Bumba-Sangha-Congo drainage. I have not seen these, but I have seen the individuals which Boulenger (1915: 28–29) reported under SJO from Akonolinga on the Upper Nyong River of East Cameroon. These individuals actually belong to BAT and not to SJO. Holly (1927: 219) reported individuals of BAT from Cameroon, and Pellegrin (1929: 360) had individuals of BAT from Nyabessan of Cameroon. There are, however, more than one village named Nyabessan in East Cameroon (one in the coastal lowlands and one or more in the Dja drainage), thus Pellegrin's material cannot be fixed geographically.

In 1966 I caught many individuals of BAT between Akonolinga and Abong Mbang of the Upper Nyong drainage. In this geographical area, BAT seems to be restricted to swamps or similar biotopes. During the dry season, we only found adult individuals, whereas the sympatric EXI was breeding during that season. The individuals of BAT were not abundant in these localities and they were rather difficult to catch in most localities because there was too much soft mud and very little water.

Superficially the males of BAT resemble males of AHL, but BAT is larger and more robust. The color pattern is difficult to describe because, as in the closely related SPL, the red pigments are very irregularly distributed, although they are less so than in SPL. The sole regular marking, which is conspicuous, is a broad yellow marginal band in lower part of caudal. There are numerous red markings on the sides, and these markings tend to build thin vertical red lines, often very irregular. Whereas in SPL, all fins of the male develop long streamers, the fins of the adult BAT male do not produce any conspicuous extensions.

BAT has not been introduced as an aquarium fish so far, but probably an introduction will be welcome, because BAT represents a phenotype in *Aphyosemion* which differs markedly from all other phenotypes which we have seen. Males of SPL develop similar phenotypes, but in this species the males are gorgeous, superior to those of BAT.

BAT was found in water of pH 4.2–6.8, with 0.1–1.0 German degrees of hardness, but during the wet season the hardness probably is much lower.

The systematic position of BAT and SPL is obscure. These two closely related forms probably represent a superspecies or a polytypic species, which seems to belong to the *Fundulopanchax* group in *Aphyosemion.* I was not, however, able to discover any prominent ctenoid spines or fin-papillae in the males of BAT and SPL which I have seen.

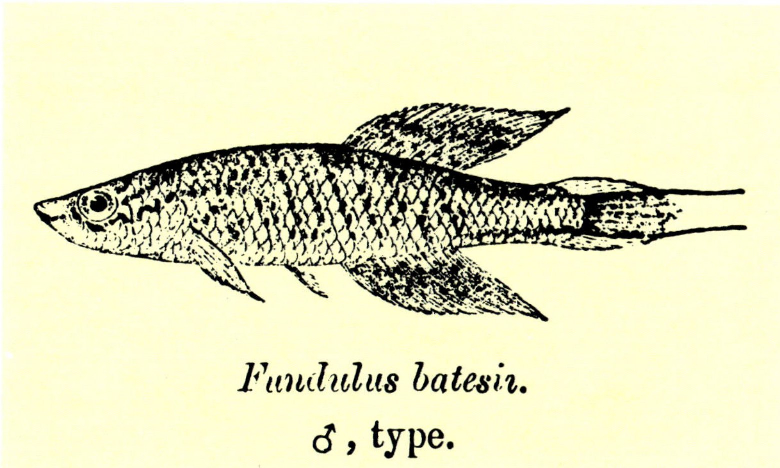

The type for BAT as figured by Boulenger 1915.

Clausen (1967: 13), however, found ctenoid scales and fin ray-papillae in both. The frontal scales of BAT and SPL develop the usual rivuline pattern of the Old World species (G-type, without H-scales), but in both the frontal G- and E-scales are rather small and do not reach the front to the extent seen in *Nothobranchius.*

The BAT-SPL group seem to be restricted to the wooded part of the inland plateau of East Cameroon and northern Gaboon, with BAT in the northern part of the range and SPL in the southern.

BAU: ***Haplochilus baudoni* Pellegrin 1922** **(Valid sp.?)**

Pellegrin (1922: 75) had a single individual from Fort Crabel, within the Gribingui-Shari-Chad drainage.

The meristic data published for BAU correspond to those of BIF and SPI, which both are known to occur in the geographical area in question. The color pattern described for BAU does not seem to correspond to that of BIF, nor to that of SPI. Pellegrin said that there was a lightly colored lateral band in BAU and that the scales of the upper part of the sides had small red dots. BIF and SPI both develop an almost complete red reticulation on the sides of males, and for that reason, re-examination of the type in Paris is needed. See also SAN, which has been taken in the Mbere-Logone-Chad drainage.

BEA: ***Fundulus beauforti* Ahl 1924** **(=BAT?)**

Ahl (1924b: 144) had a single individual, probably a female, from the Upper Lobo River, near Sangmelima of East Cameroon. Holly (1930) inspected the type and found that several of the important characters of Ahl's description did not correspond to the holotype. Holly changed the number of dorsal fin rays from 9 to 15 (!!!) and the number of anal fin rays from 12 to 16, etc. He placed BEA close to BAT and GUL.

Klee and Turner (1962: 514–17) identified individuals of aquarium kept GUL (probably from SW Nigeria) as BEA. Similar individuals of GUL were also identified as BEA by Radda (1963: 70–73), who considered GUS and SRE to represent synonyms for BEA.

The type locality given for BEA, GUS and SRE probably belongs to the Dja-Sangha-Congo drainage, because the Lobo River of this system is easily reached from Sangmelima, whereas the Lobo River of the Ntem drainage, also near Sangmelima, probably is less easily reached from Sangmelima. Individuals of BAT have been reported from localities less than 50 km from Sangmelima. In 1966 we fished intensively in many localities near Sangmelima for one week without finding a single individual corresponding to BAT, BEA, GUS and SRE. The Upper Lobo (Congo) River, very close to our base, was visited several times without results. It might be, that the locality given by Ahl is not the true locality from where the sample on which BEA, GUS and SRE were based originated.

Morphologically BEA, GUS and SRE cannot be separated one from the other and these three forms cannot be separated from BAT, so until more is known, I prefer to consider all three forms as representing synonyms for BAT.

BEL: ***Panchax bellicauda* Ahl 1924** **(=OBS?)**

Ahl (1924: 312) had nine individuals from the Upper Lobo River near Sangmelima of East Cameroon. He placed BEL closest to CHR. Holly

(1930: 212–13) considered BEL to represent a synonym for CAM. In 1966, I collected cyprinodonts at the supposed type locality for BEL (see BEA) and found three different phenotypes of rivulins to be present and common. Two of these could easily be identified as EXI and SAN, whereas the third resembled CAM in most traits, although the red pattern of the males was very poorly developed or almost absent. The Sangmelima form in question also resembled another CAM-like form which I took just west of Yaunde of East Cameroon and which, when it matured fully in the aquarium, proved to be different from CAM and probably representing a different line of evolution. This strain was identified as OBS, and probably the individuals on which Ahl based the BEL name, belong to this species, which also may be identical with STR of northern Gaboon.

BER: *Aphyosemion bertholdi* Roloff 1965 **(Valid sp.)**

The types for BER were caught by Roloff (1965: 362–64) near Kenema of eastern Sierra Leone. BER cannot be separated from ROL by meristic characters, but constant differences appear to occur in color patterns. In the ROL male, the distant edge of the anal fin has a narrow, dark red, marginal band, which may or may not extend into the small pointed lobes produced by the slightly extended fin rays. In BER, the red band of the anal fin is submarginal, separated from the edge by a narrow blue band. Both species develop a lyre-pattern of red bands in the caudal fin, but in BER, the upper and lower edges are brilliant blue, whereas in ROL, these edges usually are golden or, at least the lower edge is golden.

Adult male of BER from the type locality.

Adult male BER from type locality.

Adult male of BER from Totoa of central Liberia. "Müller's Aphyosemion."

Adult male of BER from Sierra Leone.

BER and ROL seems to be sympatric in the Kenema district, although they have not been taken in the same biotopes so far. The ROL phenotype of this geographical area differs, however, somewhat from the ROL phenotype from the Freetown area (type area), and the Kenema population studied might represent another sibling species of the ROL superspecies, to which BER belongs.

The separation of BER and ROL at the species level has been based on biological findings, whereas the separate naming of BER is based on the slight, but constant differences in the phenotypes above mentioned. In 1964 (Scheel 1966a: 135) I crossed individuals of BER from the population from which the types originated to individuals of ROL from the local population, from which the types for ROL originated. The hybrids were rather viable, but the males were functionally sterile and although the females spawned regularly, when young, most of their eggs did not develop in backcrossings, and those which developed produced individuals, which were almost inviable. In 1965 I (Scheel 1966a: 135–36) crossed individuals from the same population of BER to individuals of "ROL" from the Kenema district (Giema). The hybrids were very similar to those of 1964 in phenotype and in fertility. A cytological study of the hybrid males disclosed that the spermatogenesis was abnormal and that a few immobile spermatozoa only,

Adult, sterile hybrid male BER/ROL, Freetown area. Notice pattern of anal fin.

were produced by these males. The pairing of the chromosomes was incomplete. The first division of meiosis contained univalents. On the basis of these experiments we considered BER and ROL to be sufficiently separated to be considered as two distinct species.

The Freetown and the Kenema areas are, at the present time, separated by a wedge of savannah which comes in from the north, and which almost reaches the coast. No population of BER and ROL has been discovered within this wedge, so probably the population of ROL of the Freetown area are not connected to the populations of "ROL" of the Kenema area by an unbroken chain of intermediate populations. It is likely, that small changes of the climates of Sierra Leone may eliminate this gap, during wetter seasons, whereas at other times, the wedge may widen.

The total range of BER is not known at present. In 1965 a certain strain of BER was brought to Germany from Totota of the Central Liberia area. The offspring of this strain was distributed under the name "*Aphyosemion mülleri*", which appears to be an invalid name.

The haemoglobin pattern (four lines), the frontal scalation (G-type, no H-scales) and the chromosome number of BER correspond to those of ROL. It also seems that the karyotypes are similar, as all bivalents of hybrids were symmetrical in primary meiotic divisions with 21 bivalents. Two metacentric elements seem to occur. One of these is larger than the largest acrocentric element of the basic complement, whereas the other is smaller. There are also three "double sized" acrocentrics. Calculations of chromosome lengths in mitotic spermatogonial divisions indicated that BER (and ROL) originated from an ancient form which had n = 25 and not n = 24.

BER is a component of the LIB-group in *Roloffia* (BER, GER, LIB, MAE and ROL) representing the major group of nominal *Roloffia*. The relationships between the described and undescribed species of this group seem to be very complicated, but undoubtedly this group or superspecies represents a group of rather closely related forms.

The species of the LIB-group in *Roloffia* resemble the species of the CAL superspecies in *Aphyosemion*, but crossing experiments of the present author did not support any close relationship between these two groups of species. The species of the LIB-group are semi-annual rivulins and probably some populations reproduce as annuals in nature.

BER is not difficult to keep and to breed. Generally, individuals of BER are less shy than individuals of ROL and they are more colorful. Roloff found BER in rather soft water measuring 0.6 German degrees of hardness; pH 4.2–6.8.

Primary meiotic division of male hybrid BER/ROL, Giema. There seems to be 12 univalents, 13 bivalents and an assemblage of four chromosomes which may represent a quadrivalent or two bivalents.

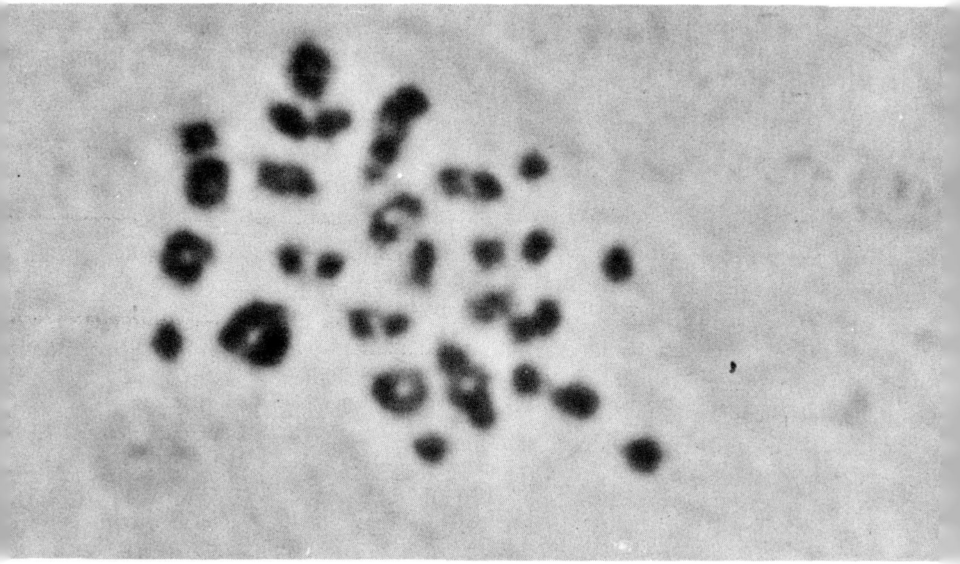

Adult male of BIF from Share, near Jebba of north Nigeria.

BIF: *Haplochilus bifasciatus* Steindachner 1881 **(Valid sp.)**

Steindachner (1881: 199) has described BIF on the basis of specimens from Bahr-el-Seraf and Bahr-el-Gebel of the Nile drainage. BIF has been reported (under different names, see NDE, STE and TAE) from all major and most minor drainages of West Africa and from the Chad drainage. In all these drainages, BIF is sympatric with SPI, except for Sierra Leone where SPI seems to be replaced by certain populations of FAS in savannah biotopes. SPI, however, also occurs in the savannah biotopes of the Congo drainage, where BIF is absent and is replaced by the rather similar CHE, which is not a close relative of BIF. In southern Liberia and in certain parts of southern Sierra Leone BIF is replaced by the very similar BAR, which also is no close relative.

BIF is exclusively a savannah species and has not been reported from the forest. It is characterized by the vertical red lines or bars in the spear-shaped caudal fin. This pattern also occurs in BUA (red) and SPI (almost black), and thus such pattern probably is important to species restricted to savannah biotopes.

NDE from the eastern part of the Chad drainage undoubtedly is a simple synonym for BIF and has been based on juveniles of this species.

TAE described from the Lower Niger area have been considered to represent a subspecies of BIF (Daget, 1954). I have studied many individuals of BIF from various parts of West Africa and from the Nile drainage, and I did not find important differences which could be used to separate the Nile population from the West African populations. Indeed the morphology of BIF remains remarkably constant, compared to other rivulins of the Old World, all over its range.

STE described from the Gambia River was considered to represent a simple synonym for BIF by Daget (1954), and Daget undoubtedly is right.

In the aquarium, individuals of BIF are extremely shy and are rarely seen in open water. The very brilliant colors of the BIF male rarely develop in the aquarium. When many individuals are kept together for many weeks, without the presence of other fishes, some individuals may lose their shyness.

I have kept two strains of BIF for several years. One of these originated from Share, near Jebba (the type locality for TAE) on the Lower Niger River. The other strain originated from SE Ghana and belongs to the Lower Volta population. These two strains were very similar, and the sole difference, which I found, was in the development of a streamer posteriorly in the anal fin. In the Volta strain most males developed a long streamer, whereas in Niger males, this streamer was short or absent. Hybrids from

Adult male of BIF from the Lower Volta drainage of southeastern Ghana.

Young males of BIF from Share, near Jebba, northern Nigeria. Notice the incomplete development of the dark band along the side (NDE-pattern).

Primary meiotic division of BIF. n = 20.

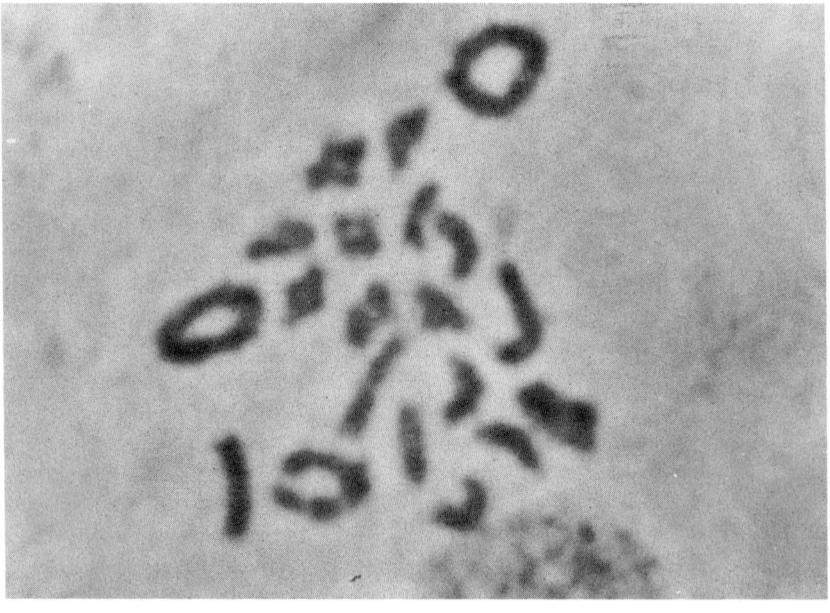

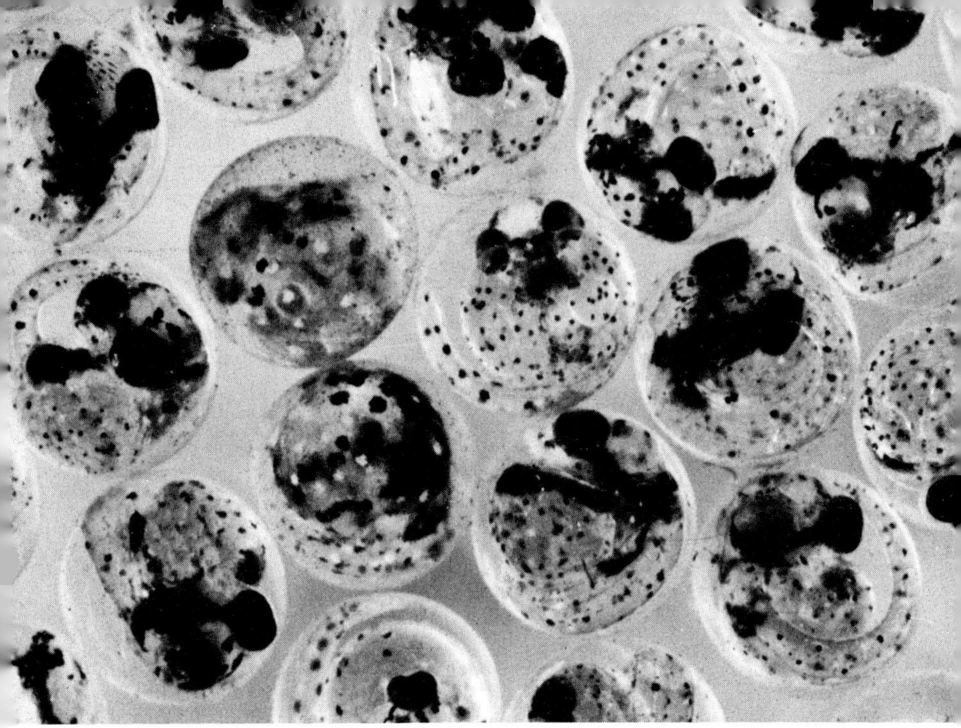

Ripe eggs from BIF from Share.

the crossing of these two strains were fully viable and almost fully fertile, and the phenotype remained constant in the second, third and fourth generations of this hybrid strain.

BIF reproduces well in hard, alkaline water. Individuals are very easily kept and bred, but BIF is not a fish well suited for aquarists, because of its extreme shyness.

Embryo of BIF from Share a little beyond the "180°" phase. "Superposing photo." Three exposures.

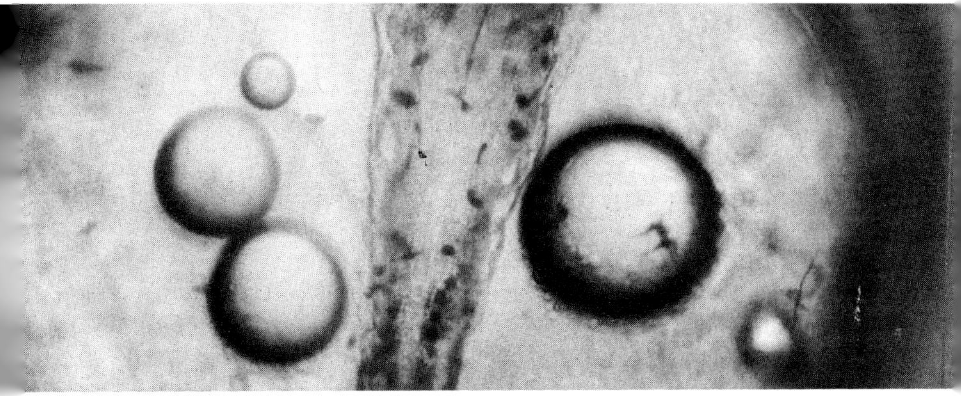

The systematic position of BIF is obscure. The haemoglobin pattern has six lines and differs from other six-line patterns of *Epiplatys* and *Aplocheilus*. A similar, or even identical, pattern develops in GUE, a species of *Nothobranchius*. There are 20 haploid chromosomes and four of these are large metacentric elements, but there seems to be one more metacentric element of size similar to the longest acrocentric of the basic complement, and BIF for that reason may originate from a n = 25 species back in time. I crossed BIF to CHE, DAG and SPI and had fertilized eggs in all combinations, but the hybrids died in the eggs just before or just after the hatching.

It seems that BIF is not closely related to any known species of *Epiplatys*.

BIT: *Fundulus bitaeniatus* Ahl 1924 **(=BIV)**

Ahl (1924b: 141) had four specimens which he said originated from the Niger drainage. He obtained these individuals from aquarists and a professional breeder, Matte of Berlin, who is not known to be a collector of aquarium fishes in nature. For that reason, the type-locality is unknown. Meinken (1930: 221) said that in his opinion the types for BIT were imported into Germany as aquarium fishes, and that it was not noticed that they represented a new species, before Ahl had individuals from Matte. Ahl placed BIT closest to BIV and RUS.

Morphologically BIT cannot be separated from BIV, and thus it should be considered to represent a simple synonym for BIV without exact type-locality.

Old male of BIV from Meko of western Nigeria. Savannah. The streamers in fins rarely develop even in old males of this population.

Adult male BIV from Umudike, near Umuahia of the watershed between the Niger and the Cross drainages.

BIV: *Fundulus bivittatus* Lönnberg 1895 **(Valid sp.)**

Lönnberg (1895: 190) had a single specimen, probably a female, from a small brook near the waterfall of the Ndian River. This locality is within the coastal lowlands according to Sjöstedt's map (1904).

The BIV phenotype extends from Togo in the west, to Spanish Guinea in the southeast, being restricted, generally, to the sedimentary soils. The Togo population is isolated from other populations at the same climates, but from Porto Novo of eastern Dahomey to Spanish Guinea there seems to be an unbroken chain of populations of the BIV phenotype. The meristic data change very little throughout the range of BIV and no differentiation into morphologically distinct subunits (subspecies, sibling species) seems to be possible at present. The color patterns do change within the range, but the differences are not visible in preserved material and in particular not in material preserved in formalin.

The populations of the forests of SE Dahomey and SW Nigeria (west of Ijebu-Ode) are characterized by their red or golden pectoral fins, their red streamers in unpaired fins and a bronze shine in scales below and in front of the dorsal fin. The males from Porto Novo, Lagos and Ijebu-Ode are

Two male BIV from Umudike, near Umuahia, east Nigeria.

Adult hybrid male in BIV. Male from Ijebu-Ode of southwestern Nigeria to female from Lagos, southwestern Nigeria.

slightly different in nature, but not in aquarium raised individuals. The savannah population of SW Nigeria (Meko) develop the characters above mentioned, but it differs markedly in males, which usually do not develop streamers in fins, not even in the caudal fin, and which develop a conspicuous lemon color on the lower part of the body and the head. In this they resemble the males from the western part of the Niger Delta, which, however, develop long streamers in fins, and a grass-green shine on the back.

The populations of the Niger Delta (Benin City in the western part and Umudike, near Umahia, in the eastern part) are rather different, but they have some traits in common, which differs from the western populations. In the males, the pectorals are almost colorless; at least no colors are visible to the naked eye. The scales below and in front of the dorsal fin are grass-green, not bronze. The streamers are red or orange-red. In all these populations, the anterior-most dorsal fin ray is distinctly behind the corresponding anal fin ray.

Cameroon populations (Buea; Mboumboula River, near Douala; Eseka) differ from Nigerian populations by the position of the anteriormost dorsal fin ray, which usually (not in all individuals of the Mboumboula strain) is inserted just above the corresponding anal fin ray. The pectorals are colorless, lemon or orange. The scales on the back are grass-green, tending towards bronze in the Buea males. The streamers generally are bright lemon, but in old Buea males they often are orange and even orange-red.

Further details concerning the nine natural populations (strains), which have been studied by the present author have been published (Scheel 1966: 261–78).

The crossing experiments with these nine strains disclosed that many different populations were not able to exchange genes, at least not directly. It seems that these nine populations may represent six biologically defined species which all develop the BIV phenotype. The crossings between the three Cameroon strains are interesting because in nature these populations are not widely geographically separated. These three strains seem to repre-

Adult male of BIV from Umudike, near Umuahia, Niger Delta area. Notice that the "wound" is almost absent and that the streamers above are much larger than those below.

sent three biologically defined species, and the Eseka strain (n = 17), in particular, was very well reproductively isolated from the two other strains (n = 19), the hybrids being completely sterile.

The different biological species (bio-species) of the BIV superspecies cannot, however, become zoological names because they cannot be separated on the basis of morphological differences in preserved material. For all these species, the BIV name should be used at the present phase of the study.

All populations studied developed similar haemoglobin patterns (four lines of the usual rivuline type). The chromosome number is variable, however. The SW Nigerian and SE Dahomey populations have 20 haploid chromosomes, whereas the Buea and the Mboumboula strains have 19 chromosomes, and the Eseka population has 17 chromosomes. No ctenoid spines or fin ray papillae has been found in BIV. The frontal scalation is normal (G-type, no H-scales).

The BIV name is not the sole zoological name attached to phenotypes similar to Lönnberg's BIV. See BIT, HOL, LOE, MUC, PAP, RUS, SPP and UNS, which are now considered to represent synonyms for BIV, and NII and UNC which also seem to represent synonyms.

BIV seems to stand rather isolated within *Aphyosemion* and no close relatives can be pointed out at present. It probably represents a blind outshoot from the speciation (multiplication of species), which took place at the 20 haploid chromosome level. But within BIV itself, a rather intensive speciation indeed occurs.

Individuals of BIV are best characterized by the two broad black longitudinal bands along the sides. These bands are usually present in juveniles and in females, although females of the eastern and southern populations often do not show these bands, and it seems that the bands are completely

Two fighting males of BIV from Ijebu-Ode of southwestern Nigeria.

Adult hybrid male BIV. Umudike male to Lagos female. Sterile.

Adult male BIV from the Mboumboula River of east Cameroon.

Adult male BIV from a brook about 15 miles north of Eseka of east Cameroon.

Adult male of the old aquarium strain of BIV which usually has been called by the MUC name, although this strain actually belongs to the nominal SPP species. Franz Werner photo.

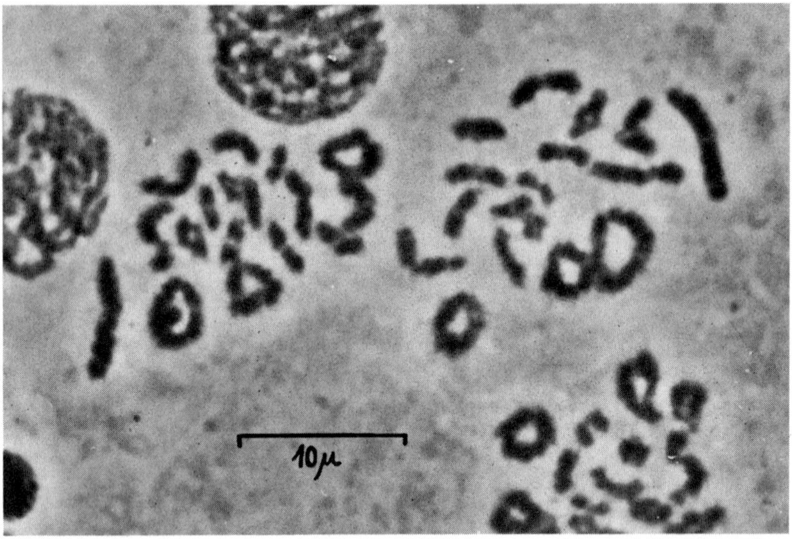

Two meiotic divisions in BIV from Eseka, east Cameroon.

Two meiotic divisions in BIV from Porto Novo, Dahomey.

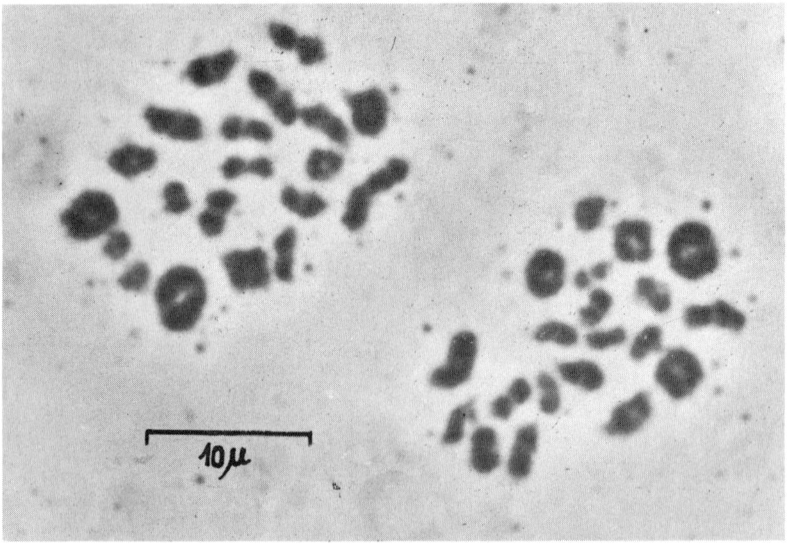

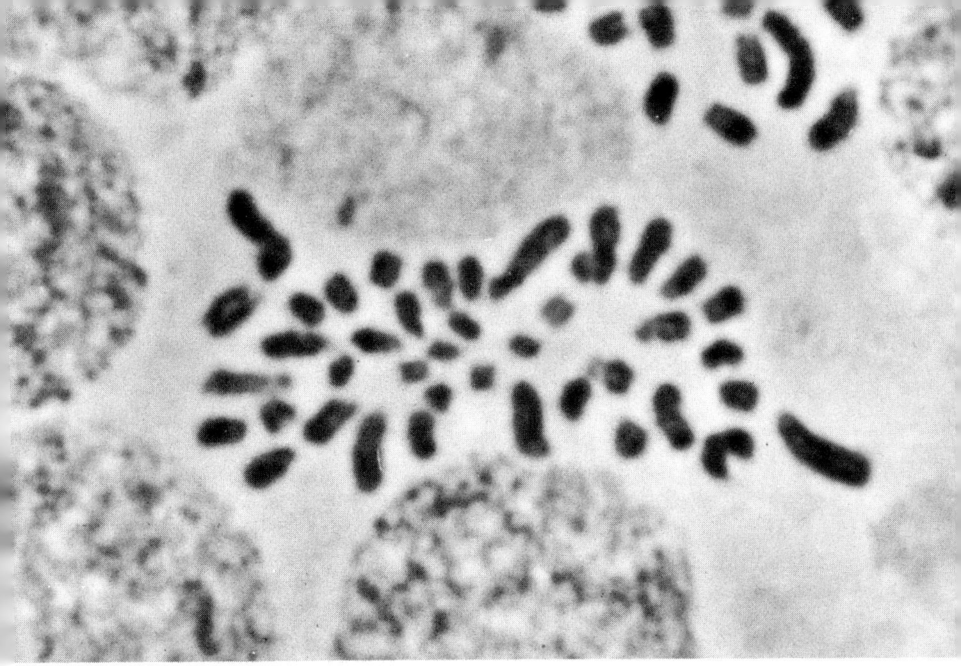

Mitosis of male BIV from Meko, savannah of west Nigeria. 2n = 40.

Mitosis of male BIV from Eseka, east Cameroon. Notice the size differences of chromosomes within this population. 2n = 34.

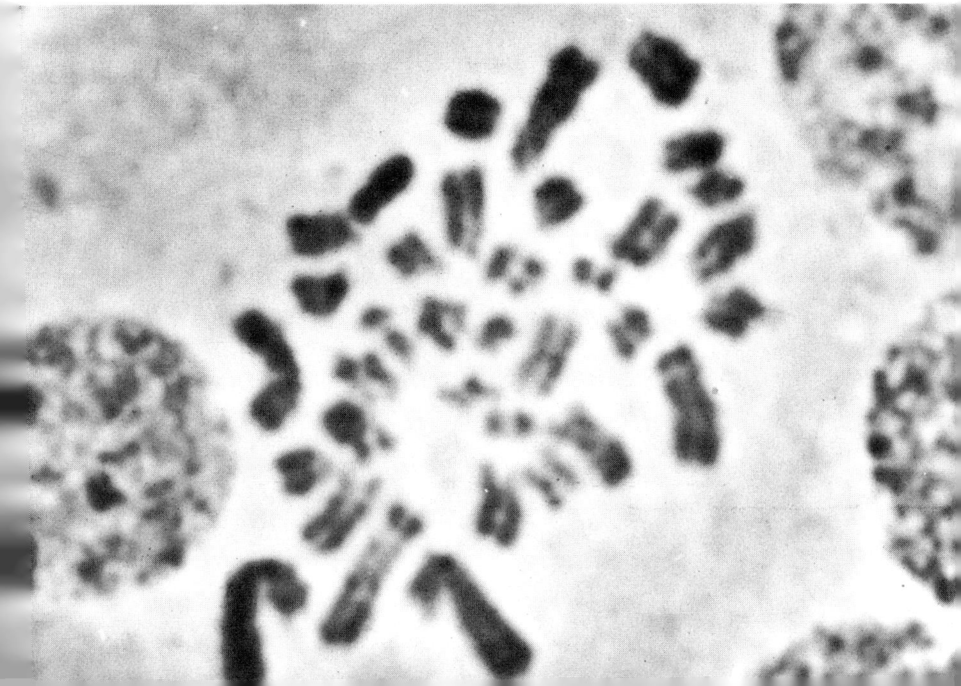

Adult male of BIV from Meko, situated in the savannah of western Nigeria.

Adult male BIV from Benin City, western Niger Delta area.

Adult male BIV from a mountain brook near Buae of west Cameroon.

Adult hybrid male BIV. Benin male to Porto Novo female. Sterile.

Adult hybrid male BIV. Umudike male to Lagos female. Sterile.

Two fighting hybrid males in BIV. Mboumboula male to Ijebu-Ode female. Sterile.

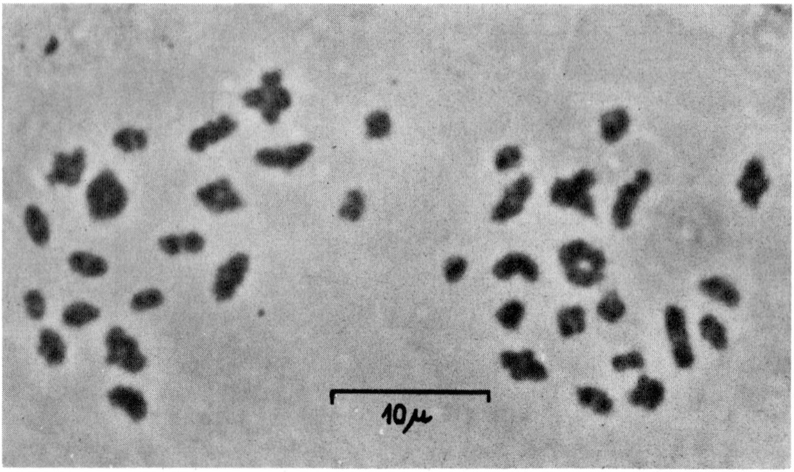

Two meiotic divisions in BIV from Buae, western Cameroon. n = 19.

absent in some populations within the Wouri drainage of East Cameroon. BIV is also characterized by the position of the dorsal and anal fin in connection with a rather low number of fin rays.

BIV has often been placed in *Fundulopanchax*, but undoubtedly it should be placed in a subgenus of its own, as it is not a close relative of SJO and probably not closely related to any known species.

Individuals from the western populations and from the localities close to the Cameroon volcano are rather easily kept and reproduced even in hard, alkaline water. Individuals from the Niger Delta and from the lowland of East Cameroon are not so easily kept and seem to be rather sensitive to attacks from mycobacteria. These strains should be bred in slightly acid and very soft water.

BIV was introduced as an aquarium fish in 1908, according to Arnold (1908: 585). This strain originated from Warri of the western part of the Niger Delta. Up to the war (1914), live individuals of BIV were frequently brought to Germany from this area, but the strains soon became extinct and German aquarists considered these strains to be difficult to keep and breed. It was not until the early thirties that stable aquarium strains were established from SW Nigerian populations (nominal MUC) and less commonly, from a Tiko population from West Cameroon. This population lives in the rather hard, alkaline water of the volcanic area. Both strains are still kept up to the present time (1968), although the Tiko strain seems to be very rare. This strain represents the nominal SPP strain.

For unknown reasons, aquarists have used the MUC name for the Tiko strain, whereas the original MUC strain was called BIV. This exchange of names occurred soon after the importation of these two strains. See Wöhlert (1937: 81) who used the MUC name for the Tiko lot.

Because of the observed almost total sterility of individuals originating from crossings of different populations (strains) of BIV from different geographical areas, it is recommended not to "improve" aquarium strains by introduction of individuals from strains of unknown origin into established strains of BIV. To prevent such crossings I have not released live material of the nine strains above mentioned because the SW Nigerian BIV, now kept as an aquarium fish is superior to all other strains known to me in viability, development of colors and fins of males.

BLO: *Haplochilus panchax blockii* Arnold 1911 **(Valid sp.)**

The first report of BLO was given by Graichen (1909: 517) who said that a new species of *Haplochilus* had been imported from Cochin of India (SW coast of the peninsula) in 1909. He published a fine ink drawing showing the new fish and said that it reached 4 to 5 cm. The scales of the sides were greenish-golden, the lower ones being reddish. The anterior part of the mouth was red, and there were brilliant blue scales on the belly. Anal was green, with red and green spots, a streamer posteriorly and an orange-red marginal band. Caudal with many red spots and streaks and with a narrow black marginal band above and below. Arnold (1911: 672) said, that the Cochin strain of 1909 differed substantially from other strains

Graichen's pair of BLO of 1909.

Adult male BLO. This individual belongs to the strain which develops red spots on the sides in both sexes.

Adult male BLO. No red spots on sides.

Adult male BLO.

of PAN known to him (see PAN) and suggested that this strain should be named *Haplochilus panchax var. blockii,* in spite of the fact that Boulenger had identified preserved material as belonging to PAN. Captain Block, who first caught this fish, had told Arnold, that it had been taken together with RUM (=LIN) in a small pond near Cochin. Arnold also said, that the female BLO differed from the female PAN by the presence of brilliant spots on the sides, such spots being absent in all strains of PAN.

Buchardt (1912: 476) described the dancing movements of the BLO male during pre-mating display. Schreitmüller (1919: 5) reported that the 1909-strain was still kept in Germany, and he described another strain, which was said to originate from Madras (see also PAR), and which in his opinion was not very different from BLO and probably represented a color variety. This strain probably represents the nominal PAR, thought to represent a synonym for BLO. He noted that German aquarists considered both strains to be difficult to breed because the fry were very small. He also said that Arnold and Gerlack had been informed by Boulenger that BLO was not valid because it was not different from PAN, and that the differences in general appearance and in colors were unimportant. Schreitmüller did not agree with Boulenger in this, however.

Myers (1933: 181) considered BLO to represent a valid species, and this point of view was generally accepted. He suggested, that PAR represented a synonym for BLO. Job (1940) also considered PAR to be a synonym for BLO and said that the Indian name for BLO was *Pachai munda kanni*, and that this species represented the smallest rivuline species in India. It lives in stagnant waters always under the cover of trees and in the shade. He also said, that BLO probably lives near Madras only, on the eastern coast of the peninsula, and that it had been reported from Ceylon also. He said, that BLO resembled the more northernly PAN, but also, that BLO should be considered to represent a distinct species. Eggs sometimes remain attached to the vent of the female after spawnings. Siegal (1963: 80–82) deals with the keeping and breeding of BLO.

Two different strains of BLO were imported to Denmark directly from Bombay of India in 1965. One strain corresponded well to the old strain (1909), whereas in the other, all adult individuals had rows of red spots on sides. Both strains were breeding true to these characters, but their hybrids were fully fertile and viable, and the observed differences in color patterns thus do not reflect important genetic differences. In the opinion of the present author, such minor differences in color patterns should not be used to create subspecific names and in particular not when the exact origin in nature of the different color varieties is unknown. See, however, PAN, where quite similar differences have been used for the creation of various nominal subspecies.

The pre-mating display ("dancing") of the BLO male is rather unique within *Rivulinae* and resembles the behavior of DUB and many species of *Procatopodinae*, but BLO also exhibits further procatopodine traits. The shape of the scales on the sides are more procatopodine (exceptions occur in *Procatopodinae*) than rivuline. The thin black median line and the corresponding line above the root of the anal, which characterize most species of *Procatopodinae*, is present even in (some) adult individuals of BLO, as in FLA, whereas in ANN and in some strains of PAN, these lines occur in juveniles only.

BLO has 24 haploid chromosomes. No large metacentric element is seen, and the karyotype probably correspond well to the basic karyotype for the family *Cyprinodontidae*. The haemoglobin-pattern is very complicated and very different from that of PAN (and LIN) and seems to be closest to that of BIF and GUE. The pattern of lateral line organs of the head appears to be rather variable compared to other rivuline species, because the fleshy lobes often are vestigial, and because more than one pair of neuromasts occur on the dorsal part of the snout. It seems that the pattern is "closed" or intermediate to "closed" and "open," in some individuals. Also the frontal scalation is rather variable, ranging from the E-pattern, which is the usual pattern, to the G-pattern with or without prominent H-scales. In spite of the rather limited material from the two strains above mentioned, it seems that the lateral line pattern and the frontal scalation is more variable in BLO than in any further species of *Rivulinae* known to me (see however HOM).

An intense lemon color develops on the surfaces of the anterior part of the head in individuals (males) which have just been preserved in alcohol, but after a few days the color disappears completely. Similar temporary lemon coloration occurs in males (very rarely in females) of some rivuline species (see DAG).

Ventrals are short in both sexes. The dark marking in the basal part of the dorsal is always present in females, in juveniles, and often in males. No throat pattern has been observed in BLO, but in the male there are more black pigments on the throat than in females.

BOU: *Haplochilus boulengeri* Pellegrin 1926 **(=MUF)**

Pellegrin (1926: 206–07) had one specimen from Kidada, three from Mongende and three from Tshimbu, all from the Central Congo drainage. Pellegrin placed BOU closest to GRA, CHI and SUP.

Poll (1952: 162) placed BOU as a synonym for MUF, and undoubtedly he is right.

BRI: *Nothobranchius brieni* Poll 1938 **(Valid sp.?)**

Poll (1938: 170–71) had twenty-four specimens from Bukama on the Lualaba River of Katanga, Congo. These individuals originated from pools connected to the river by channels. Nieuwenhuizen (1960b: 154–60) published some information concerning BRI which he had received from Graindorge of Jadotville, Katanga, Congo. Graindorge found BRI in the huge swamp area between Jadotville and Elisabethville, the swamps being connected to the river by channels. In this area many temporary, often isolated, pools occur during the wet season, which begins in November and ends in May. BRI was found in these shallow and stagnant pools. He measured about 5 German degrees of hardness, pH 6.3–6.5 and 19–28°C. A photo showing the BRI male was also published. Nieuwenhuizen (1963a: 302–309) received eggs from BRI from Graindorge and raised an aquarium strain. He pictured the male in an excellent color photo (1963b: 157). Magis (1963: 174–85) studied some populations of BRI in nature for 1½ years. These populations were found in temporary pools and in ditches near the road between Jadotville and Mwadingusha (east of Jadotville). He found strong indications that BRI reproduced as an annual fish in these biotopes. The individuals studied were smaller than those described by Poll from Bukama. Magis' individuals did not exceed 32 mm, the females being smaller than the males.

BRI develops a rather dark red reticulation on the sides. The red color seems to contain much dark blue color. The males photographed by Graindorge and Nieuwenhuizen have a prominent, regular red reticulation, whereas those described by Poll appear to possess crescent shaped reddish-brown spots. The large dorsal has a very dense pattern of dark red spots and streaks and a narrow blue edge. The anal has a low number of dark red bands parallel to the root, the end being bright orange in Nieuwenhuizen's male. The pattern of the caudal fin corresponds to that of TAP and RAC and seems to contain no orange pigments.

BRI is probably more closely related to TAP and RAC than to any group of species of *Nothobranchius*, and if Nieuwenhuizen's strain represents Poll's species, it seems to differ sufficiently from TAP and RAC to be considered as distinct.

BRU: *Haplochilus brucei* Boulenger 1911 **(=GAR?)**

Boulenger (1911: 263) had eight specimens from a pool at Udi of the Cross drainage. In 1915 (:26) he considered BRU to represent a synonym for GAR. In the original description of BRU, Boulenger said that there were 30 to 34 scales in a longitudinal series, whereas for GAR (1915: 26) he published the figures 26–32 scales, so probably (?) the BRU types had no more than 32 scales. I have studied many individuals of GAR and found that 34 scales frequently occurs in this species. Boulenger's point of view probably can be accepted.

Adult hybrid male RUF-Ndop/EXI.

Adult male BUA from Bamkin (Kimi) of east Cameroon.

Adult male BUA from swamp near Ndop of west Cameroon.

BUA: ***Panchax bualanum*** **Ahl 1924** **(Valid sp.)**

Ahl (1924: 54) had a single individual from "O-Sanga Uham, 1200 m hoch, Buala, Grasland, Kamerun." The present town Buala (Bouala) is situated near the watershed between the Upper Sangha and the Uham drainage, which also is the border between the Chad and the Congo drainages. The holotype is probably a male, because it had a large number of narrow lightly colored crossbars on the sides. Red pigments often are lost after preservation (they are always lost after preservation in formalin) and red crossbars thus become lightly colored bars. Ahl also said that the fins were dark. Ahl placed BUA closest to LUJ and CAM. Holly (1930: 211–13) considered BUA to represent a synonym for CAM. This point of view cannot be accepted for zoogeographical and ecological reasons. (See also CAR.) Holly probably based this point of view on meristic characters, which rather often fail to separate species which by other criteria are considered to be very distinct.

Adult female BUA from crossing of the Ndop male to the Bamkin female.

In 1966 I collected rivulins about 100 km from Bouala, which is now situated in the Central African Republic, within the headwaters of the Mbere-Chad and the Mambere-Sangha-Congo drainages and in both major river systems a species of *Rivulinae* very similar to Ahl's BUA was frequently found. This species is common in all savannah biotopes of the huge Sanaga drainage at an altitude of 600–1200 meters all over Eastern Cameroon, whereas in the Congo and the Chad drainage it probably is restricted to the headwaters of the affluents being replaced at lower altitudes by BIF and SPI.

It seems that generally BUA is restricted to savannah biotopes, being replaced within the wooded part of the inland plateau by the rather similar and closely related EXI. These two forms probably are parts of a superspecies.

Populations of BUA are found under rather different biological conditions. Some populations are living in cool and quick running mountain brooks probably up to the 1200 meter level, whereas other populations live in stagnant, hot and turbid water in the marshes of the plains. We found the hardness of the natural water to vary between 0.5 and 20 German degrees, being 2–3 degrees in most localities, the the pH figure varying between 5.5 and 7.8, being almost neutral (7.0) in most localities.

The phenotype of the BUA populations studied in nature appeared to be rather variable, and probably some local populations are sufficiently different from other populations to be considered to represent subspecies, or, if reproductively isolated from all other populations, to be isolated as distinct species.

Individuals from two populations were brought alive to Copenhagen for further study of biological differences. These two strains are morphologically so similar that preserved and live material cannot be differentiated. The two populations are, however, reproductively isolated to a rather high extent, and they also differ by the haploid chromosome number.

The strain from Ndop has 19 haploid chromosomes. This population lives in an isolated valley at the 1200 meter level on the eastern slopes of the Cameroon Mountain chain. From this very limited population the types for RUF were taken (see RUF). No other species of *Cyprinodontidae* occurs in this biotope. The biotope is a swamp in highland savannah, pH 6.4 and 3.0 German degrees of hardness during the last part of the dry season. This population is physically isolated from other populations of BUA of the lowlands by steep slopes.

The strain from Bamkin (or Kimi) was found in almost stagnant, very turbid, fish culture ponds within the town of Bamkin (see also CAR). Adult males were very weakly colored, but in the aquarium they are very similar to those of the Ndop strain, although generally, they are somewhat darker and the streamers in fins are shorter. This strain has 20 haploid chromosomes.

The karyotypes of the Ndop and the Bamkin strains are rather similar, and both are characterized by a rather smoothly graded series of acrocentric elements. Both karyotypes have a single metacentric element corresponding in size to the supposed largest acrocentric of the basic complement. Cytological studies of the hybrids between Ndop and Bamkin individuals indicated, however, that the karyotypes are more different than indicated by the differences in haploid numbers. Generally, there are less than the expected 19 "bodies" in primary meiotic divisions, indicating trivalents occur. These meiotic divisions are, however, not sufficiently distinct ("woolly

Adult male produced by crossing of two hybrid strains: BUA-Ndop/BUA-Bamkin to BUA-Ndop/EXI. The male was almost fully fertile.

chromosomes") to permit a study of the combinations of the individual chromosomes. The hybrids above mentioned are not completely sterile, but most F_1 **x** $F_1 = F_2$ individuals are too feeble to be raised to sexual maturity. A certain exchange of genes thus is possible under aquarium conditions, but probably not, or at least very rarely, in nature. On the basis of the crossing experiments I consider the Ndop and the Bamkin populations to represent two distinct bio-species, but I have not been able to detect any constant diagnostic differences in the phenotypes and both strains are called BUA in the present study. RUF is considered to represent a synonym for BUA. (See also BIV.)

The EXI superspecies (EXI+BUA) inhabits the 600–1200 m. inland plateau of East Cameroon, extending into northern Gaboon and the western parts of the Chad and the Central African Republics. Within this huge geographical area, BUA lives in the savannah in the northern part of the area and EXI in the wooded southern part of the area.

When EXI (Sangmelima strain, n=18) is crossed to the two BUA strains, the differences between the two strains last mentioned are easily seen. With the Ndop (n=19) strain, EXI produces hybrids which are very viable and almost fertile in both sexes, whereas with the Bamkin strain, the hybrids are viable, but almost completely sterile, at least the F_2-individuals (F_1 **x** F_1 spawnings) are almost inviable. This finding is not too surprising, because the Bamkin BUA probably represent the BUA genotype, which all the way along the border between savannah and forest is in contact with the EXI populations. For this reason, it is likely, that the EXI and the BUA populations of the 600 meter plains have developed reproductive isolating mechanisms, which prevent an exchange of genes between these two closely related species. The Ndop population is not, however, in physical contact with any other population of cyprinodonts, and selection thus has not favored an accumulation of isolating mechanisms towards other populations.

The observed isolation from the Bamkin population thus probably is accidental.

Adult hybrid male BUA/EXI. This strain has 19 + 18 = 37 chromosomes in ordinary cells.

The results of the three crossings above mentioned are important. The genetic differences seems to be minimal between the Ndop and the EXI strains and at a maximum between the Bamkin and the EXI strains, the differences between the Ndop and the Bamkin strains being intermediate. The karyotype differences seems to be of the same magnitude and probably cannot be judged by the difference of the haploid chromosome numbers. The phenotype differences between the Ndop and the Bamkin strains appear to be nil, whereas between these two strains and EXI, several important differences occur, involving meristic characters as well as differences in

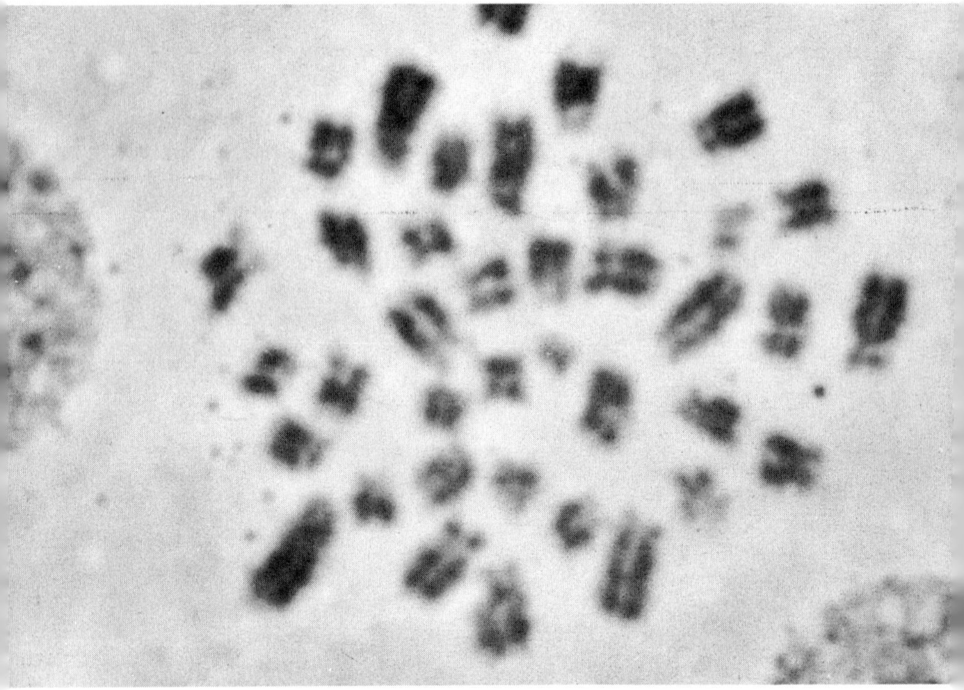

Mitosis of BUA from Ndop. 2n = 38.

colors, size and development of fins. It might be that these differences are the result of the differences in the choice of biotopes only and do not reflect the true genetic differences, because two of the populations studied live under rather similar biological conditions.

The fact that at least some of the BUA populations have 20 haploid chromosomes probably is highly important in the study of the phylogeny of *Aphyosemion*. All other species of *Aphyosemion* which have 20 haploid chromosomes (none has more than this figure) are true *Fundulopanchax* (SJO) or closely related to *Fundulopanchax s.s.* (GAR, CIN, NDI, BUR) or to BIV or LUJ. The only true species of *Aphyosemion s.s.* which has 20 haploid chromosomes is BUA (Bamkin). This important species lives within the upper reaches of a number of major river systems, and thus it is likely that it has given rise to some of the groups of species which exhibit a reduced number of chromosomes. The chromosome numbers of BUA and EXI indicate that BUA is the most ancient species and that EXI originated from some BUA-like ancient form, which, from the savannah, extended into forest biotopes. Further evidence supporting this point of view is given under EXI. Although only a few crossing experiments have been studied so far, it is likely that the major group of *Aphyosemion s.s.* of the Central Congo drainage are derivates of the EXI superspecies. These Congo species are

CAS, CHR, COG, CON, DEC, ELE, MAR, SCH and maybe LUJ and OGO (Gaboon) as well. They are components of the ELE group or superspecies.

The populations of the EXI superspecies are characterized by the vertical red bars or lines in the caudal fin of males, by the brown vertical lines posteriorly on the sides of females, by the absence of submarginal red bands in the dorsal and anal and by the almost marginal red bands in the caudal.

The two BUA strains reproduce well in hard, alkaline water, and individuals are very easily kept and bred. Juveniles are slow-growing compared to other species of the genus.

The names ELB, TES and RUF are synonyms for BUA.

BUR: "*Aphyosemion burundi*" **(Invalid name?)**

In 1964 (?) a strain of *Aphyosemion*, very similar to the "yellow" strain of GAR, was distributed as an aquarium fish in Germany under the name "*Aphyosemion burundi*." At the same time, another strain of *Aphyosemion* was distributed under the name "*Aphyosemion ruwenzori*." The strain last mentioned belongs to FIL. Although the BUR strain seems to be rather close to GAR, it cannot be incorporated into this species because of the rather low number of dorsal fin rays. In the many individuals of GAR, which I have studied from very many localities, I never found less than twelve dorsal fin rays, whereas in BUR (limited material only) I found ten such rays. Also, the pattern of the anal fin is somewhat different, and the submarginal band is oblique in BUR. No zoological description of the BUR phenotype apparently has been published, and certainly this form probably does not originate from Burundi of East Africa, but probably from some part of southern Nigeria, (Niger Delta area ?). This strain should not be called by the name above mentioned, whereas the "Burundi-Aphyosemion" might be used.

In 1966 (Scheel 1966: 134, 147) I thought that the males of BUR developed ctenoidy in scales. Further study of this species disclosed that the structures seen did not represent true ctenoidy, but series of small organs situated along the exposed edge of the scales. These bud-like structures might be neuromasts. I have seen such structures in BUR males only. BUR has 20 haploid chromosomes and the karyotype is characterized by a large acrocentric element measuring more than three times the mean length of the basic chromosomes. There is also a somewhat smaller metacentric element.

CAB: *Aphyosemion (Aphyosemion) calabaricus* Ahl 1935 **(=LIB)**

For the description of CAB, Ahl (1935: 402) used aquarium kept individuals of unknown origin. These individuals originated from F. Mayer (1936: 257) of Hamburg, who bought them from O. Winkelmann of

Adult male of the "*Burundi-Aphyosemion.*"

Altona. Winkelmann was told by the professional collector that these fishes originated from Calabar of the Niger Delta area. This information undoubtedly is incorrect because CAB, according to the pattern of lateral line organs of the dorsal part of the snout, belongs to *Roloffia* which does not seem to extend beyond Ghana eastward (also see DOR).

In 1965 Clausen caught one male and three females of a form which in all characters corresponds to CAB. These individuals were taken about 20 miles from Monrovia, Liberia. Ahl's types for CAB thus undoubtedly originated from western Liberia and probably from the Monrovia area. See Clausen and Scheel (1966: 82).

The morphological data published for CAB correspond sufficiently to those given for LIB, which was based on females only. The type locality for LIB is Monrovia. Thus CAB probably is a simple synonym for LIB. Further information concerning the aquarium strain called CAB are given under LIB.

CAL: *Haplochilus calliurus* Boulenger 1911 **(Valid sp.)**

In January 1908 a shipment of aquarium fishes arrived in Hamburg from Africa. This shipment contained two very small specimens which were bought by Arnold (1908: 314), who raised the individuals into a breeding pair, which he bred. A young male from the offspring was preserved

Adult male CAL from locality near Ijebu-Ode of southwest Nigeria. Grey males also occurred in this population.

Adult male CAL from the aquarium trade 1964/65.

Haplochilus calliurus.

♂ ♀, types.

The types for CAL as figured by Boulenger 1915.

and sent to Boulenger for identification. Boulenger identified this male as ELE, a Congo species. After having received further preserved material and a color sketch prepared by Arnold, Boulenger considered Arnold's strain to represent a new species, which he (1911: 265) described and named.

Arnold was told that the individuals originated from "fresh-water pools near Sierra Leone," but in the description, Boulenger gave Liberia as the type locality. The labels for two subsequent shipments from Arnold to Boulenger were probably exchanged. (See DAG.) Arnold, however, probably was misinformed by the collector (or the dealer) because the types for CAL belong to *Aphyosemion* and not to *Roloffia*, and the genus first mentioned apparently does not range beyond Ivory Coast (see WAL). See Clausen (1967: 24–25). Clausen (1967: 24–25) also suggested that the types for CAL did not originate from the January 1908 shipment, but from another shipment of CAL from Nigeria, and that the individuals which Arnold bought in January 1908 were not CAL, but one of the species of *Roloffia* which occur in Sierra Leone and Liberia, and which superficially resembles CAL. This point of view probably is not correct, because it is known from Arnold (1908b: 654) that the second shipment containing live individuals corresponding to the CAL phenotype arrived in Hamburg in November 1908, and that before that time, Arnold had already sent the first preserved specimen and the color sketch to Boulenger showing the breeding male. On the 25th of May 1908 Boulenger wrote Arnold that after having seen the color painting he considered Arnold's strain to belong to ELE. Arnold's sketch definitely pictures a male of *Aphyosemion*, not a male of *Roloffia*, because in the anal fin the male has a broad yellow marginal band (not present in any known species of *Roloffia*) and there were also long fin streamers, which are unknown in the *Roloffia*.

The strains of CAL which were kept by aquarists from 1908 to about 1914 originated from the second and later shipments and not from Arnold's first breeding pair, and the offspring from this pair was lost by the dealer who bought it (Milewski, 1913: 566).

In 1913–14 the aquarium strains of CAL gradually were replaced by the more colorful strains of AUS (and AHL), but in 1924, CAL again was imported, according to Rachow (1924: 753), but this strain apparently soon became extinct.

Since 1963 live individuals of CAL have frequently been exported from Lagos, Nigeria and several aquarium strains have been established. These strains appear to be rather different, at least in color pattern. In 1962 Clausen caught some live individuals of CAL south of Ijebu-Ode of SW Nigeria, and these individuals were placed in my tanks and bred. The two adult males of this shipment were rather different in color patterns and in basic colors, but both undoubtedly belonged to the SW Nigerian phenotype of CAL. The Ijebu-Ode strain correspond to the most common strains of CAL now kept by aquarists all over the world, but I have seen individuals from other strains which seems to differ so much from the common strains that the basic material probably originated from other populations of CAL or of CAL-like species in Nigeria.

The common strains of CAL are often called VEX by aquarists. Morphologically VEX is a simple synonym for CAL. (See VEX.)

CAL is a component of the CAL superspecies, which comprises CAL, AHL and AUS so far. There are probably more than these three biologically defined species which belong to the CAL superspecies and which have the haploid chromosome numbers 16, 18 and 15 respectively, but it is also likely that future studies of more populations of this superspecies will eliminate the apparently sufficient morphological differences, which permit an identification of preserved material of all three species, and thus perhaps in the future, a single morphologically definable species (CAL) will remain.

Adult hybrid male CAL/AHL. Sterile.

The CAL name frequently has been used for Nigerian strains of GAR. These two species are easily separated by the number of dorsal fin rays. The highest number found in CAL, so far, is ten, whereas in GAR the lowest number appears to be twelve.

Although no population of the CAL superspecies having twenty haploid chromosomes has been discovered, I do not think that these three morphospecies are derivates of any of the known 20 chromosome species in *Aphyosemion*. The color pattern of the males of the CAL superspecies has certain, probably ancient, traits (the dark basic color, the throat pattern of both sexes), which are not present in any of the known species of *Aphyosemion*, except, in part, for CIN. The most ancient known component of the CAL superspecies, which probably is AHL, might be a derivate of an ancient species also closely related to GAR and CIN.

It is a remarkable fact, that crossings between individuals of the CAL superspecies and individuals from various species, which are rather similar in morphology and general appearance (CHR, COG, LAB of *Aphyosemion* and LIB, ROL and PET of *Roloffia*) did not produce viable hybrids, whereas with BUA fully viable hybrids of both sexes developed. On the other hand, crossings between individuals of the CAL superspecies and individuals of the *Fundulopanchax* group in *Aphyosemion* (GAR and SJO) produced highly viable, but sterile hybrids. These results of crossings

Arnold's breeding male of CAL as figured by Arnold (in color). This male is the holotype for CAL or the father of the holotype.

indicate that the CAL superspecies may represent an independent line of evolution in *Aphyosemion* originating from some extinct species of *Fundulopanchax* (or vice versa).

An exact reconstruction of the karyotype for CAL (Nigerian strain) is not possible at the present time, but it seems that the karyotype is characterized by two large and very similar sub-metacentric elements, measuring about three times the mean length of the basic chromosomes. There is also a large metacentric element measuring about 2/3 the length of the long submetacentrics.

The haemoglobin and the frontal scalation patterns are normal (4-lines, G-type, no H-scales). Individuals which are afraid, develop a rounded black marking anteriorly on the throat. I have not been able to discover ctenoid spines on scales of males, but Clausen (1967: 15) found extremely minute ctenoid spines on scales of large males.

CAM: *Haplochilus cameronensis* Boulenger 1903 **(Valid sp.)**

Boulenger (1903: 435) had eight individuals from the Kribi River of southern Cameroon and one specimen from the Dja River also of southern Cameroon. These individuals were caught by Bates. The Kribi, now the Kienke River, is within the coastal plains for almost its whole course, whereas the Dja River is within the wooded part of the inland plateau. These two geographical areas have different cyprinodont faunas, no species occurs in both areas, probably. Individuals corresponding well to Boulenger's CAM occur, however, in both faunas, but most of these individuals do not belong to CAM.

CAM.

The types for CAM as figured by Boulenger 1915.

CAM is best characterized by the color pattern of the male. Boulenger said: "... *with carmine spots, which are more profusely distributed in males, in which they often form two longitudinal bands, one close to the lower outline, from below the pectorals to the root of the caudal, the other on the upper border of the caudal peduncle, fins grey, more or less spotted with carmine, anal edged with black and white and lower border of caudal white.*"

Males from most populations of *Aphyosemion* from the two fauna zones of the wooded part of East Cameroon do not develop these two dark red lines and the fin pattern described by Boulenger. The population of AHL from the coastal zone corresponds well to CAM by meristic characters (see AHL) and easily could be taken for CAM. From a locality west of Yaunde on the inland plateau I have a strain of *Aphyosemion* which also corresponds to CAM by meristic characters, but which differs by color patterns. This strain probably represents OBS, which may be a synonym for STR. I have seen nine preserved individuals identified as CAM by Boulenger and caught by Bates within the Nyong River drainage in 1904. In the males of this sample, the two red lines described for the CAM male are very conspicuous, and they are listed under CAM in the Catalogue (Boulenger 1915: 47–48).

At the present time, it seems to be impossible to determine if CAM is a species of the lowland fauna or of the inland plateau or if it occurs in both zones.

The meristic characters of CAM are very similar to a number of nominal species of *Aphyosemion* and *Roloffia*, and for that reason CAM has been reported from many widely separated localities of Atlantic Africa. Under AHL, it has already been mentioned, that Boulenger identified individuals of AHL and (?) AUS as belonging to CAM. He also (1915: 47–48) reported CAM from Mayanga of the Lower Congo area. These individuals might be LAB, or LOU or other similar southern forms. Pellegrin (1935: 466) reported CAM from Kissidougou of High Guinée. These individuals are GUI. (See GUI.) Norman (1932: 185) reported CAM from Sierra Leone. These individuals also are GUI. Poll (1952: 164) also reported CAM from the Lower Congo area. Among rivuline species from Atlantic Africa CAM cannot be separated from BEL, BRU, CAR, CAS, CHR, CIN, CON, ELB, ELE, ESC, FER, FLV, GER, JAU, LAB, LIB, LOU, LUJ, MAE, MIC, NIG, NOR, OBS, OGO, PAS, POL, PRE, TES and VEX by meristic characters.

Preserved male of CAM from the Nyong River, east Cameroon. No. 1904, 7.1, 214–22 of the British Museum. Bates collection.

The best thing to do at the present time is to restrict the range of CAM to southern Cameroon and to consider the male CAM to develop the two red lines on the sides, which Boulenger described and figured (1915). The NDI species from East Nigeria develop a very similar color pattern, but has more fin rays and scales. NDI may represent the form from which CAM derived.

The name CAM has frequently been used in the aquarium literature, although live individuals of this species have never been kept in the aquarium. Boulenger's identification of individuals of AHL as CAM was not accepted by German aquarists. Together with the 1913 importation of AHL (?) from Cap Lopez of Gaboon, individuals of another species of *Aphyosemion* was imported. These individuals were caught in Gaboon and were sold as *Rivulus urophthalmus* (or *R. rubra*, which is not a zoological name), but Arnold identified this strain to belong to CAM. Arnold's point of view seems to have been generally accepted, although the males of this strain definitely do not develop the color pattern described for CAM. These males are very similar to OBS and most probably they represent STR, described from northern Gaboon. Drawings and photos of CAM in aquarium literature figure this and similar strains.

Holly (1930: 221) considered BUA, BEL, CAR, MIC, NOR, OBS, PAS and PRE to represent synonyms for CAM. I do not agree with Holly in this. See these nominal species, except for MIC which together with ESC probably represent synonyms for CAM.

CAP: *Fundulus capensis* Garman 1895 (New World ??)

Garman (1895: 113) apparently had a single, not well preserved individual, labelled "False Bay, Cape of Good Hope, S. Africa. D.13, A.13, Sq–1, 36. Anal originating below middle of dorsal. Caudal convex. A lightly colored band in front of eye. A silvery band from gill cover to root of caudal. Five or six broad brownish blotches across the flanks with alternating lightly colored bands. A band crosses the caudal near its base. Teeth slender, pointed, outer series larger, hooked . . . resembling those of *Heterandria formosa*."

Boulenger (1915: 39) refers to Garman's description without comments, but in spite of this I do not think that CAP could possibly represent a rivuline species of the Old World, or any rivuline species. Probably CAP represents some funduline form.

CAR: *Panchax carnapi* Ahl 1924 (=OBS)

Ahl (1924: 303–304) had a single specimen from Yaunde of East Cameroon. Pellegrin (1929) reported CAR from Banyo, Foumban, Bamkin and Yoko of East Cameroon. These localities are situated in the savannah and probably all these individuals are BUA. It has already been said under BUA, that one of my strains of this species originated from Bamkin. There were no signs of further species of *Rivulinae* in this geographical area. I have also collected individuals of BUA at Banyo and near Foumban, and in these localities BUA seems to be the sole rivuline species.

Holly (1930: 211–13) considered CAR to represent a synonym for CAM, but probably this is not correct. In the Yaunde area, there seems to be two species of *Aphyosemion* only: EXI and OBS, and probably CAR is a synonym for the latter.

CAS: *Aphyosemion castaneum* Myers 1924 (=CHR?)

Myers (1929: 2–4) had nine specimens from Stanleyville and Medje of the northern part of the Central Congo drainage. This species is the type-species for *Aphyosemion.*

Myers considered CAS to differ from related species by the lack of red spots on the sides and red markings in fins, except for the anal fin, which had an orange line close to the outside edge. He reported, however, lightly colored spots on the sides and in fins. Probably, the red pigments of these specimens have been almost lost, perhaps as a result of preservation in formaline or of bad preservation in alcohol, and that the decomposition of the red pigments had been arrested by a change of the preservation fluid.

Pellegrin (1930: 202) reported CAS from Liboumbi of the Upper Ogowe area. Poll (1952: 166) considered CAS to represent a synonym for SCH. Lambert (1961: 33–36) considered SCH to represent a synonym for CHR, and indicated, that probably CHR might represent a synonym for LUJ.

After having seen the types for DEC, I do not doubt that CHR is a synonym for DEC, but also ELE, representing an even older zoological name, seems to be so close to DEC and CHR, that these forms cannot be sufficiently separated by meristic characters.

In 1966, however, I received two aquarium strains of the CHR phenotype from friends in Germany. Morphologically, I was not able to separate the individuals of these two strains, though the color patterns were similar. One strain, however, had 15 haploid chromosomes; in the other strain 9 such chromosomes were found. I crossed individuals from these two strains and found the hybrids to be highly inviable, as all hybrids died just before or just after hatching. A study of the morphology of the two karyotypes disclosed that apparently only three chromosomes were common to both strains. This means, that the DEC (or ELE) morpho-species is composed of a number of bio-species. The CHR phenotype, however, is rather variable, and different strains breed true to their characteristic color pattern. This means the present CHR morpho-species might be split up into several morpho-species if a biological study has disclosed that strains which differ by color patterns are distinct bio-species. But first an extensive biological study of different natural populations is needed to limit the range and morphological variability of different bio-species. Such study probably will not be carried through in the near future, if it is ever carried through, but until it has been prepared I think that it is desirable to fix the nomenclature now in use for the strains of this superspecies. These strains are usually referred to as CHR.

CHA: *Haplochilus chaperi* Sauvage 1882 (Valid sp.)

Sauvage (1882: 323–325) had eight individuals from Couacrou, situated at the bottom of the Assinie Lagoon in eastern Ivory Coast. This geographic-

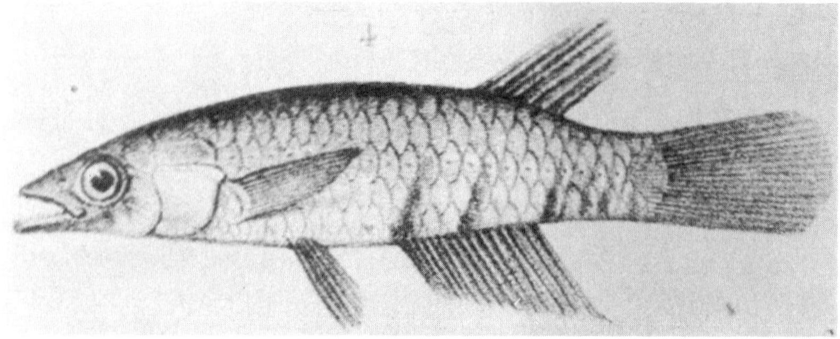

The holotype for CHA as figured by Sauvage 1882.

al area was within the Gold Coast when Sauvage received the material.

The CHA name has frequently been used in literature, but it seems that before 1964 this name has been attached to individuals which actually are DAG. The reason for all these misidentifications is due to the fact that Boulenger in 1908 identified some aquarium fishes from Monrovia, Liberia as CHA, and because in 1915 (:56–57) he figured and described (in part) one of these aquarium kept individuals under CHA.

Individuals of CHA were not, however, unknown to zoologists and the British Museum has a very large collection of preserved individuals from various localities of Ghana. All these individuals were, however, identified as SEX by Boulenger and others.

It was not until 1964 that the identity of Sauvage's CHA was pointed out by Daget and Arnoult (1964: 469–471) who examined the type material in Paris.

CHA seems to range over a large geographical area of West Africa. Generally, it seems to be restricted to forest biotopes, but populations also have been reported from the savannah (see SPM). The eastern limit of its range is found on the western slopes of the Togo Hills of Ghana and the western limit is in Ivory Coast or in Liberia. I have studied many preserved individuals of a very similar form from N'Zérégoré of High Guinée, which probably represents a western subspecies of CHA. (See OLB.)

CHA appears to be composed of a number of distinct subspecies, which are well separated by their color patterns, *not by their meristic characters.* The geographical range of the nominal CHA subspecies seems to be rather limited. This form occurs in the extreme SE Ivory Coast, though near the coast another subspecies occurs. It also occurs in the corresponding part of SW Ghana and seems to reach the coast near Cap Three Points. This phenotype is characterized by having four dark crossbars on the sides of almost all known individuals, and it breeds true to its characteristics. The four crossbars are all behind the anteriormost anal fin ray. This phenotype is also characterized by its basic color of sides and of fins. In the young male and in some adult males the sides are grass-green, changing into a brilliant yellow. The red spots on the sides are arranged in regular rows along the sides, but they are inconspicuous and small. The fins (except for pectorals) of these males are bright lemon, changing into golden. There is no row of large red spots below the root of the anal fin. In some of the older males, the yellow color of sides and fins change into orange or even orange-red.

The females of this subspecies develop a warm reddish-brown color on the back.

The SHE subspecies (see SHE) lives south of the range of the CHA subspecies, and in Ghana it is restricted to the extreme SW of the country. In Ivory Coast it expands northwards, north of Abidjan and seems to reach the northern limits of the forest and even the savannah (see SPM). This phenotype also has been taken north of the forest in western Ghana. The SHE phenotype is closer to the OLB phenotype than any other major population of CHA. It is characterized by the cold bluish-violet, brilliant metallic shine on sides and in unpaired fins. The red pigments always occur concentrated into rows of red spots on the sides and spots or streaks in fins. These red markings often are edged with black, particularly on the sides of the male. The spots on the sides are larger and much more conspicuous than in the CHA subspecies. Just below the anal fin, there is a row of large, prominent red spots, which are edged with black. Yellow pigments are not visible. This phenotype has five dark crossbars on the sides of almost all individuals. The four bars are situated as in the CHA subspecies, whereas the fifth bar is situated just behind the root of the pectorals.

An unnamed subspecies lives north of the range of these two subspecies in western Ghana. I have studied live specimens from Kumasi. The male of this type is very dark all over, and almost no metallic shine develops on sides and in fins. The red pigments are not concentrated into distinct markings generally, but rather evenly distributed on sides and in fins, as yellow, orange or red pigments of the CHA subspecies. The upper edge of the dorsal and, particularly, the caudal, is brilliant white to bluish-white. Only in a dark aquarium are these white markings seen. There are the usual four normal black bars behind and above the anal fin, and usually a similar bar above the root of ventrals. Very rarely a similar bar may develop behind the root of pectorals at least on one side of the body. Generally, the Kumasi phenotype is more variable than the SHE and the CHA phenotypes, and in central and eastern Ghana there are populations of CHA which cannot easily be incorporated into any of the three major phenotypes above described. These populations are intermediate to the Kumasi and the CHA phenotypes, more or less. Additional black crossbars may develop in females (very rarely in males) of all populations studied, but in the Kumasi females, these bars develop much more frequently than in other strains.

Adult male CHA from Angona, southwest Ghana.

Adult male CHA from Abra of southwest Ghana.

Adult male CHA from Kumasi of Ghana.

Crossings between the three major phenotypes produced hybrids which were fully viable and fully fertile. No abnormalities occurred in the further reproduction of these hybrid strains. The observed geographical distribution of the different phenotypes (subspecies) and the distinctness of these (all are breeding true to their characteristics) cannot be explained at the present time, but probably the different phenotypes do not cross to any extent in nature, and thus the different subspecies might be classified as distinct species (see also LON and SEX).

The haploid chromosome number seems to be 25 for all aquarium kept strains. There are no indications of the presence of large metacentric elements, and CHA probably possess the ancient karyotype for *Rivulinae* of West Africa. The haemoglobin pattern correspond to that of SEX, LON, ESE, DUB and CHE (six lines).

The pattern of lateral line organs on the dorsal part of the snout differs markedly from that of SEX, ESE and LON. CHA develop the closed pattern of snout neuromasts, which is the common pattern for *Epiplatys* and *Roloffia.*

Hybrids from crossings of CHA and SEX were rather viable, although some individuals could not be raised to adult size. Both sexes developed, but these were sterile. The pattern of lateral line organs of the snout of the hybrids were of the CHA type or intermediate to SEX and CHA.

Three of the paratypes for CHA. M.A.4424 of the Paris Museum. Notice the absence of dark crossbars behind pectorals.

CHA and SEX probably are rather closely related and are extremely similar in morphology, biology and ecology. These two forms replace each other in their respective geographical areas.

Half grown males of all the strains of CHA which have been kept in the aquarium are very handsome and interesting, but adult individuals are not ideal aquarium fishes. They are brutal, aggressive and do not move much. They stay at the surface of the water for hours.

The male holotype for CHA. M.A.4423 of the Paris Museum.

Young male CHE from Stanley Pool, Congo.

CHE: ***Haplochilus chevalieri*** **Pellegrin 1904** **(Valid sp.)**

Pellegrin (1904: 222) had three individuals from Brazzaville, (French) Congo. He placed CHE closest to SIN and CAM. Poll (1939:36) rediscovered CHE in a collection of fish from Stanley Pool of (Belgian) Congo, and in 1952 (:296) and 1959 (:140) he considered CHE to be endemic to the Stanley Pool area. Lambert (1961: 29–30) considered CHE and NIC to be so similar that he suggested that NIC represented a subspecies of CHE. He separated these two forms by the relative depth of the body (NIC being more slender than CHE). Poll, however, placed CHE closest to MAC. The present author thinks that CHE resembles BIF more than it resembles other species of *Epiplatys* (except for BAR). (See BIF.)

The throat patterns of CHE and BIF are rather similar and quite different from that of BAR. The two dark crossbars of the throat are of the usual *Epiplatys* type, and separate CHE and BIF distinctly from MAC, which also develops a different hemoglobin pattern. CHE and BIF develop traces of a third crossbar or cross line on the lower outline of the head below the center of the eye. This marking is very rare in the genus, and seems to occur in BIF, CHE and MUF only, but in the species last mentioned this marking is more conspicuous and larger. Also the dark pattern on the belly, which Poll described for CHE, is present in BIF and in MUF. In CHE, the dark pattern of the throat of females is more conspicuous than that of the male. In BIF no such sexual dimorphism develops whereas in MUF the dark throat pattern of the male is more conspicuous than that of the female.

The red spots on the sides of the male CHE are perfectly round, in BAR they are crescent shaped and in BIF an almost regular red reticulation develops. MUF is rather similar to CHE in this respect, but in the former, the spots often are oblong.

Young male CHE from the aquarium trade.

Throat pattern of preserved CHE male from Stanley Pool, Congo. Notice the dark marking just below the center of eye, resembling the pattern seen in BIF.

BIF and CHE appear to be restricted to savannah biotopes, whereas MUF lives in the forest.

CHE and NIC are rather similar by meristic characters, but so are many distinct species of *Epiplatys*. Further information is given under NIC.

Individuals of CHE and BIF were very difficult to cross, but after much effort a few fertilized eggs were obtained. One egg only developed an embryo which hatched out and died. In crossing to SEX, which develop a similar hemoglobin pattern (6 lines) also, the individuals were not willing to spawn, but after a long time some eggs were produced. These hybrids died rather early in the development within the egg. The systematics of CHE remain obscure.

CHI: *Panchax chinchoxcanus* Ahl 1924 **(=ORN or MAC)**

Ahl (1924: 305–306) had a single specimen from "Chinchoxo, Müdung des Quillu," and he placed CHI closest to GRA. The locality given is not on my maps, but just south of the Gabon River there is a place named Chinchoua. There are also two rivers named Kwilu in Congo, but both names given by Ahl have not been found within a single geographical area.

If CHI originated from the affluent of the Lower Congo named Kwilu it might represent MAC. If it originated from Gaboon, it might represent ORN or a similar form of the GRA superspecies.

Adult male CHR from the Kinshasa area. This male actually represents the SCH phenotype in CHR.

The caudal fin of CHI was pointed and much longer than the head. It was yellowish brown, each scale with a red spot, which was edged with black. The spots were arranged in longitudinal and vertical rows, the transverse rows being oblique. Fins colorless, dorsal and anal with rows of small dark spots. Similar small spots in upper part of caudal.

CHI cannot be separated from many species of *Epiplatys* including GRA, ORN and MAC by meristic characters.

CHR: *Haplochilus christyi* Boulenger 1915 **(=DEC?)**

Boulenger (1915: 46) had ten specimens from the Lindi River, near Stanleyville, Congo. As I have already pointed out under CAS, it seems that within the Central Congo drainage there are a number of populations of the CHR phenotype which are very similar, but which differ considerably by their karyotypes and chromosome numbers, and that some of these populations, at least, represent distinct bio-species. The geographical distribution, allopatric or sympatric occurence and characteristics of color patterns and/or meristic characters are not known at the present time, and it is recommended that we fix the nomenclature used at the present time for aquarium strains, which generally are called CHR.

The CHR phenotype has been kept as an aquarium fish since 1949, the first importations to Belgium originating from Leopoldville. The males of this particular strain had few red spots on the sides. The strain first was called SIN, then SCH. In 1951, a strain having very many red spots on the sides and similar markings in fins was imported also from the Leopoldville area and was described and named COG by Meinken. (See COG.) I crossed individuals of these two strains and found the hybrids (eleven females) to be fully viable, but sterile in backcrossings to the COG males and semi-sterile with the CHR males. The few F_2 individuals, which could be raised to adult size, all developed the CHR phenotype. (See also COG.)

The type for CHR as figured by Boulenger 1920.

Haplochilus Christyi

Adult male CHR. The U.S. strain of 1960/61 which was known under the COG name.

Up to 1962 (?) these two strains were the sole representatives of the CHR group in Europe, whereas in the USA, some strains, which develop intermediate numbers of red spots on the sides and fins of males, were kept. In 1962 I received eggs from one such strain and raised an aquarium strain.

In 1966 I received adult individuals of two very similar strains, with rather many red spots on the sides, from Germany and found that these two morphologically similar strains were reproductively isolated to an extreme extent (F_1 individuals all were inviable). The two strains had n=9, 2n=18 and n=15, 2n=30 respectively. These five strains all bred true to color patterns. (See CAS.)

The mitotic spermatogonial divisions were very distinct in both 1966 strains, and the reconstruction of the karyotypes could be prepared rather exactly. It seems that three of the haploid chromosomes are common to both species. The n=15 strain is characterized by a very large metacentric element, measuring about 4.3 times the supposed mean length of the basic chromosomes (for n=25), the largest acrocentric element being much smaller and measuring about 2.5 times the mean basic length. The karyotype also has three smaller metacentrics. The n=9 strain is characterized

Young male CHR from the U.S. aquarium trade. 1961/1962. This strain usually was called COG.

Adult male of CHR. This male was cytologically studied and had nine haploid chromosomes.

Adult hybrid male CHR/GAR-AK. The male used for this crossing represented the SCH phenotype in CHR. Hybrids were sterile; W. Foersch photo.

Adult hybrid male CHR/LAB and an adult male (below) of CHR from the strain used for the crossing. Hybrids and photo produced by Franz Werner.

Adult male CHR. n = 9 strain.

Very early primary meiotic division of CHR. n = 9.

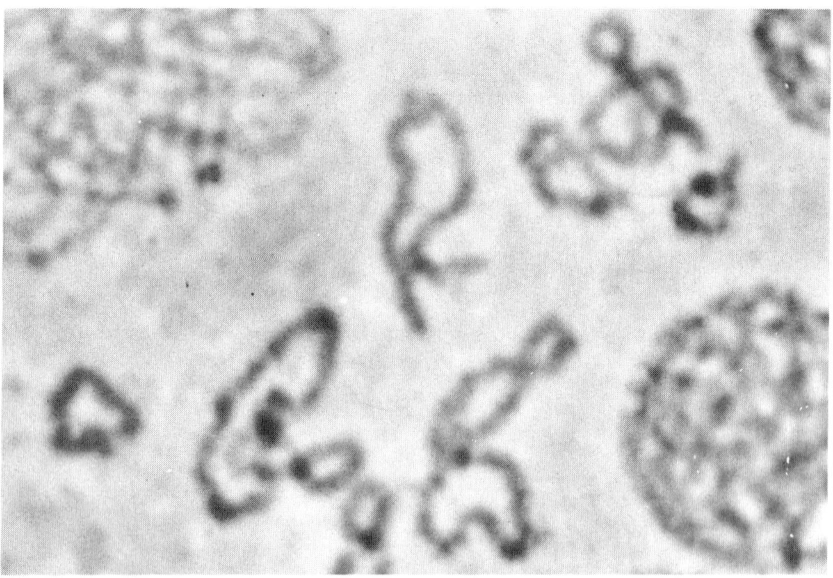

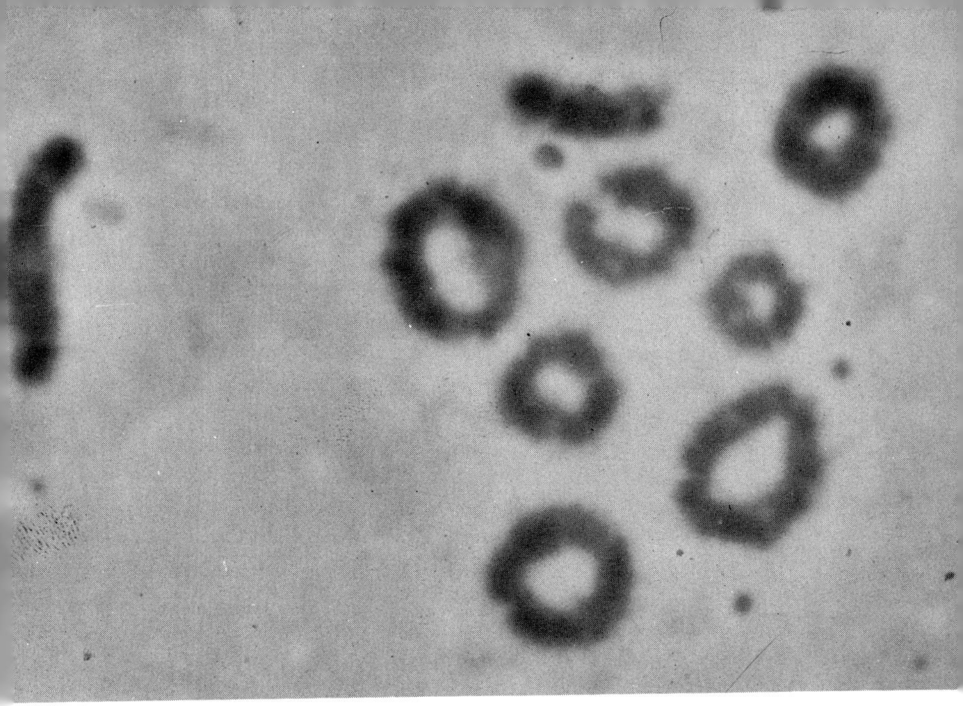

Primary meiotic division of CHR. n = 9.

Secondary meiotic division of CHR. n = 9. The short arms of all chromosomes take up the strain and are genetically active.

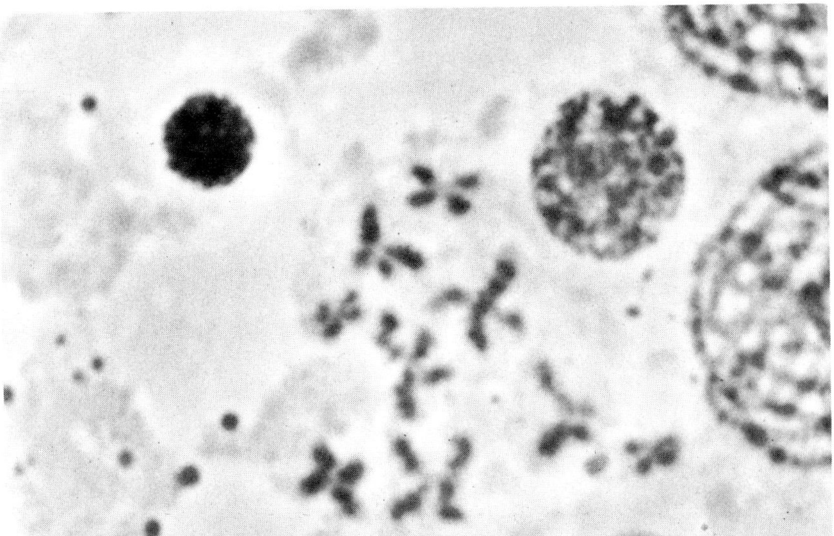

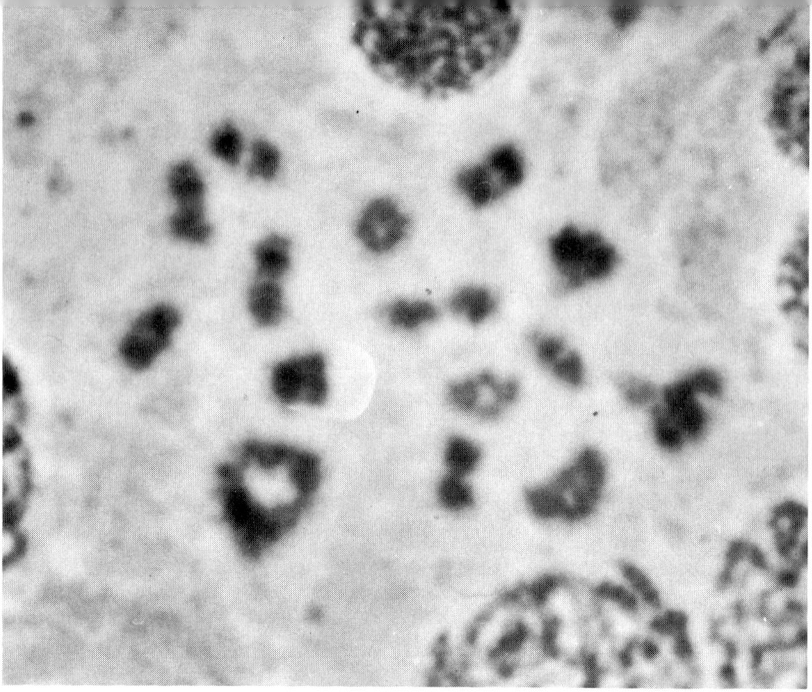

Meiosis of CHR male. n = 15.

Mitosis of CHR male. 2n = 30.

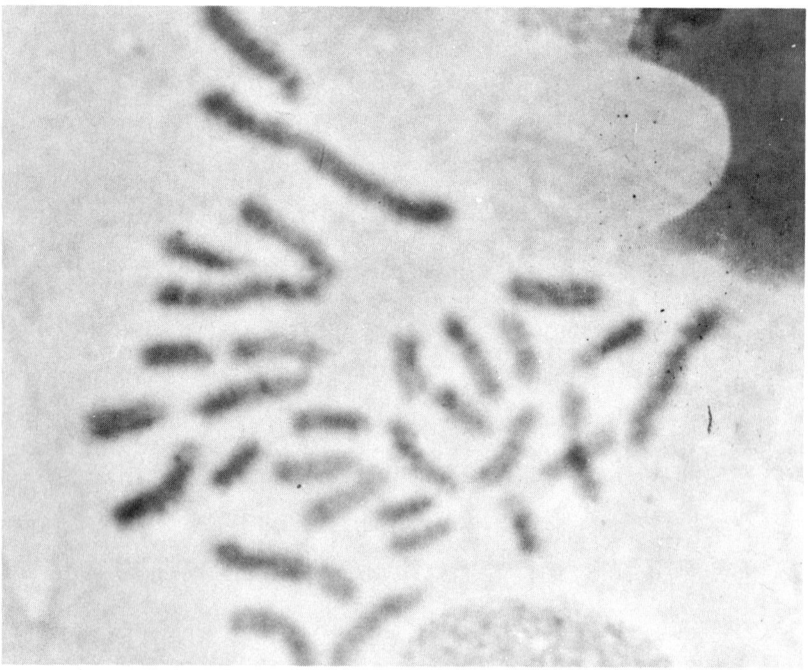

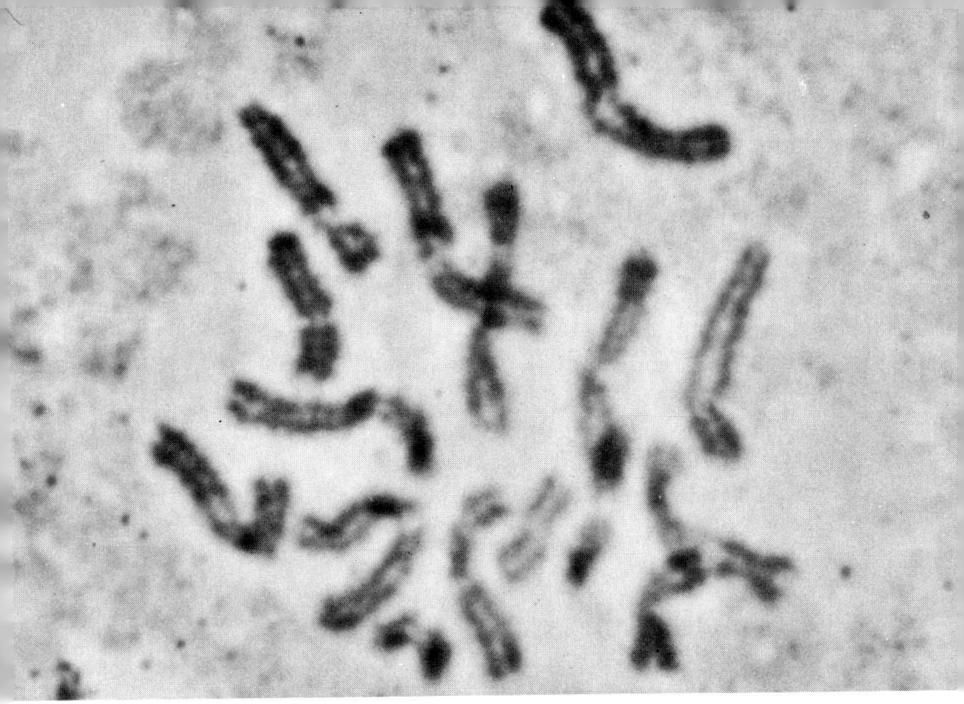

Mitosis of CHR. n = 9, 2n = 18.

Converging primary and (center) secondary meiotic divisions of male CHR, n = 9, 2n = 18, showing the very many ring-bivalents which occur in this species.

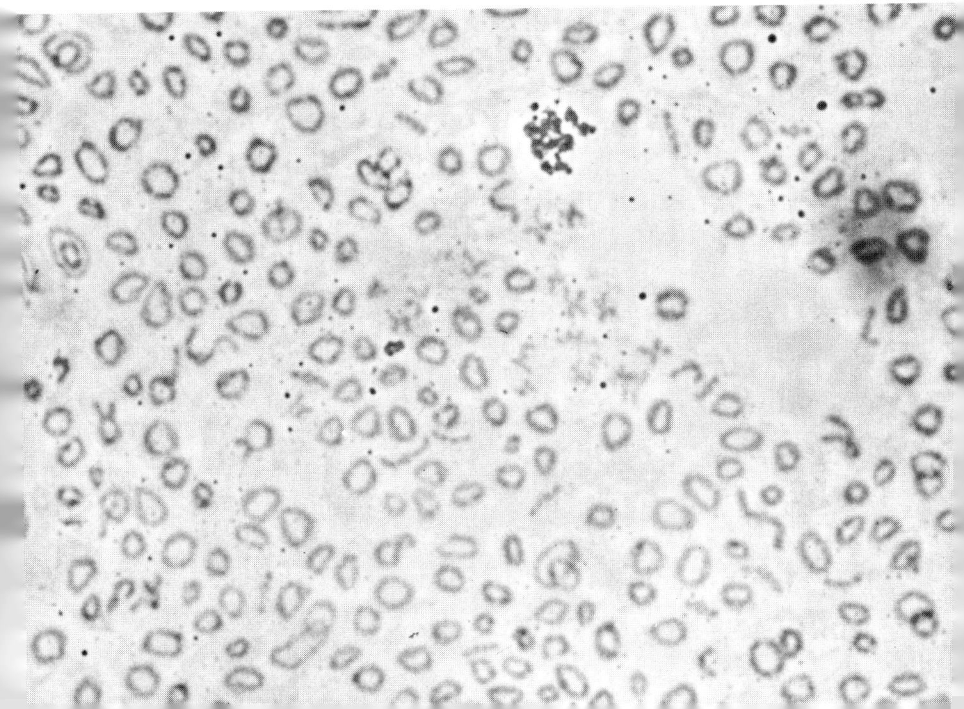

by having only one (acrocentric) chromosome of the basic complement left unchanged. There is a smoothly graduated series of chromosomes, the four largest being distinctly metacentrics and the complement has four acrocentrics only.

A further study of populations (strains) of the CHR phenotype including cytology and crossings is sorely needed to discover if the observed differences in color patterns (number of red spots on the sides of males) reflect important differences or not.

According to Lambert (1961: 33–36) who studied the types for CHR, this species is characterized by the very high number of red spots on the sides and fins of males. On the other hand, the Leopoldville strain of 1949, which is characterized by the very low number of such spots, corresponds well to the SCH phenotype, so we might continue to use the SCH name for this particular strain. See also CAS, COG, CON, DEC, ELE, LUJ, OGO, MAG and SCH.

CIN: *Aphyosemion (Fundulopanchax) cinnamomeum* Clausen 1963 (Valid)

Clausen (1963: 201) had six males and two females from a brook 44 miles north of Kumba of Western Cameroon. I visited the type locality in 1966 and found the geographical position of CIN to be highly interesting. CIN is found on the top of a low range of mountains between the main block of the Cameroon Mountains and Mount Roumpi, west of the chain. The top of the small range is rather flat. Towards the south there is a very steep slope which rivuline species are unlikely to be able to overcome. Towards the north, the slope is less steep, but the brooks here are inhabited by a different species of *Aphyosemion*, still undescribed. Towards the east and the west there probably are no biotopes suited for CIN-like fishes, which are not good swimmers. Just below the southern slope, in the Kumba plains, there was a CIN-like species of *Aphyosemion* which differed from CIN by the red spots on the sides of males.

Holotype, male CIN from west Cameroon. No. 351699 of the Copenhagen Museum.

Pair of CIN from the type locality in west Cameroon.

I had my first live individuals of CIN from Clausen in 1960 and distributed eggs to friends in various countries just after the description, in 1964. It seems, however, that CIN is not a common fish, not even among killifish-fanciers. The reason for this probably is not the reproduction, which is not too difficult, but the fact that males rarely show their fine colors. Too often they are rather similar to the females, except for the pale yellow marginal band in the anal and caudal fins. Also the males mature rather late in life, and the broods often contain a large surplus of one sex. I raised 20 individuals in 1967 and all were males.

CIN do not develop any red pigments on the sides or in the fins, not even in the adult male. The throat has no color pattern, but in the active male, the throat is jet black. This throat signal is very rare in *Aphyosemion* (except for AHL and AUS no other species develop this pattern) but very common in *Roloffia*. The frontal scalation is normal (G-type, no H-scales). The karyotype has 20 haploid chromosomes, and CIN is characterized by two very long and apparently identical acrocentric elements measuring 3.3 times the mean length of the basic chromosomes. Other chromosomes are about half this length or smaller, and one of the largest of these is a metacentric element.

Adult male CIN from type locality.

CIN probably is an ancient rivuline species which at the present time maintain a stronghold on the top of the 20 × 20 km plateau in West Cameroon. The water of the type locality measured 1.5 German degrees of hardness, pH 6.8. It is found together with *Procatopus glaucicaudis* Clausen belonging to the coastal group of *Procatopus*, the subgenus *Procatopus*. No species of *Epiplatys* were found on the plateau or north of it, whereas just below the southern slope, SEX was common.

Crossing experiments indicate that CIN is rather closely related to the GAR superspecies. Hybrids are of both sexes, and very viable, but they are sterile. These hybrids are very colorful and handsome, but they are aggressive and brutal.

When kept in hard, alkaline water, CIN is rarely difficult to keep, whereas in soft and acid water, mycobacteria and *Oodinium* attack. Individuals of CIN usually are rather shy and remain near the bottom of the tank, where eggs are deposited.

Further information is given by Scheel (1964: 324–27, 399–400, 411–12) and Franz (1966: 11–14).

COE: *Fundulus gularis var. coerulea* Boulenger 1915 **(=SJO)**

The first aquarium-kept individual of COE was taken by Groch near Warri in the western part of the Niger Delta. This specimen, a male, was in the first shipment of aquarium fishes from Atlantic Africa, according to Arnold (1906: 319). The shipment arrived in Hamburg on the 4th of November 1905. This male was kept by Arnold for one year and reached a length of four inches. A fine photo of this particular male was made by Oelze, who published it in 1908. In August 1906 several individuals of COE were imported into Germany, but all died soon after. In December 1906 a female was imported and taken over by Arnold (1908b: 517–20). On the basis of the 1905 male and the 1906 female Arnold recognized that these individuals corresponded to GUL in morphology, but not in colors. In September 1907 twenty-eight live individuals were imported, and Arnold received a pair of each of two color varieties (COE and GUL). He prepared a color sketch showing the male and the female of each sort (and a pair of ARN). This sketch, published in 1908 (b: 470–71), was sent to Boulenger together with the 1906 female. Boulenger identified COE and GUL as GUL only. In March 1908, reproduction of COE was carried out by Arnold, who noticed that the central rays of the caudal fin of juveniles produced as in *Epiplatys*. Up to 1911 GUL and COE frequently were imported into Germany and were called "var. B (yellow)" and "var. A (blue)" respectively, as recommended by Arnold (1908b). Both forms were bred by several aquarists. German aquarists did not agree with Boulenger that these two forms belonged to GUL and they considered the two "varieties" to represent distinct species. Several preserved individuals were sent to London, but Boulenger did not change his mind. In April 1911, Boulenger himself visited Arnold in Hamburg and adult live individuals of both species were placed before him. He was informed that they were sympatric, that they did not cross in the aquarium and that both forms were breeding true to the color patterns observed. Boulenger did not change his mind. In 1915 (: 623) he gave some credit to Arnold's opinion and under GUL he published this note: "In a large *var. coerulea* (*var. A. Arnold*) the body of the male is bluish with red vertical bars and the caudal is yellow in the middle and blue above and below."

Ahl (1924: 52–53) considered COE to represent a distinct species on the basis of the differences in the number of scales and in colors.

Myers (1924: 4) made COE the type species for the subgenus *Fundulopanchax* in *Aphyosemion.*

Clausen (1966: 331–41) considered COE to represent a synonym for SJO. I have seen the types for SJO and they correspond in morphology and in color pattern to the form which we have called COE since 1915 and GUL-var-A-(blue) before that time. (See SJO).

COG: ***Aphyosemion cognatum*** **Meinken 1951** **(Valid sp.)**

Meinken (1951: 255–57) based COG on aquarium kept specimens, which were said to originate from Kinshasa, Congo. Poll (1952: 168) reported COG from Stanley Pool and the Leopoldville area and in 1959 (: 140) he had further individuals from the Ndjili and the Binza Rivers, both near Stanley Pool, and he suggested that COG was endemic to this area.

Adult females. CHR (SCH phenotype) above, COG below. Notice the development of the dark reticulation on the sides of these two species.

In his description of COG, Menken did not say by which characters he separated COG from CHR and similar species from the Central Congo area, and it seems impossible to separate preserved specimens of COG from the phenotypes of the CHR group.

My first strain of COG was bought at Aquarium Hamburg in 1954 and probably this strain was identical with the strain on which Meinken based COG. At that time it was easy to distinguish between individuals of the COG strain and the strain of the CHR (or SCH) strain imported in 1949, by the number of red spots on the sides of males (very many in COG, very few in SCH), but after 1962 (in Europe) when the strains of CHR with many red spots on sides and in fins were introduced, it became very difficult to separate COG from CHR. Males from the original (?) strain of COG differed from the CHR strains with very many red spots on sides by some minor traits. In COG, the upper and the lower borders of the caudal, when fully spread during fight or sexual display, are almost parallel, whereas in CHR, the fin edges are much more separated posteriorly than anteriorly.

There might be further small differences in the color patterns of dorsal fins, and in COG the caudal fin streamers usually are considerably shorter than in CHR. As has already been mentioned under CHR, the strain of this species, which was introduced in 1949, was almost fully reproductively isolated from the old strain of COG. COG belongs to the very difficult group of CHR-like species inhabiting the Central Congo drainage. See CAS, CHR, CON, DEC, ELE, LUJ, OGO, MAG and SCH. The systematics of this group are obscure.

The "true" COG seems to have become extinct as an aquarium fish. It seems that all strains now called COG by aquarists are CHR. Females of COG differ from females of CHR by conspicuous dark brown reticulation on the sides. This reticulation is very faint in CHR or completely absent.

CON: *Panchax congicum* Ahl 1924 **(=CHR?)**

Ahl (1924: 307) had thirteen individuals from "Congo." He placed CON closest to SPI (SEN). The published meristic data for CON place this form within the CHR-group, but the color patterns place CON very close to LAB, which however, has more dorsal fin rays. It is a well-known fact that Ahl too often published data for types of "new species" which were not in agreement with the data of the individuals used as types, so the data for CON may also be wrong. Until the type material has been studied by a qualified person, I prefer to incorporate CON in the CHR group, although indeed, CON may be a senior synonym for LAB.

DAG: *Epiplatys dageti* Poll 1953 **(Valid sp.)**

This species was described by Peters in 1863 (: 396) under the name *Poecilia sexfasciata*. Peters had a single specimen from Liberia. In 1868 Peters created the genus *Lycocyprinus* with this species as the type species. Peters apparently was not aware that the name "*sexfasciatus*" was preoccupied by Gill's *Epiplatys sexfasciatus*. Günther (1866: 313) said that perhaps Peters' species represented the young state of INF (= SEX) and also Günther probably was not aware of Gill's description of SEX. Garman (1895: 131) placed Peters' species (see SEF) as a synonym for *Haplochilus spilauchen*, now *Aplocheilichthys* in *Procatopodinae*.

Boulenger (1915: 56) and Myers (1933: 182) placed Peters' SEF as a synonym for CHA, and Myers also considered *Lycocyprinus* to represent a synonym for *Epiplatys*.

Adult male COG from "Aquarium Hamburg's" strain of 1953–54. Notice the shape of the caudal fin.

The first live individuals of DAG were introduced as aquarium fishes in Germany in 1908. The shipment originated from Monrovia, Liberia. Arnold (1908: 296) bought a pair from this shipment (together with the types for LIB), and the remaining lot was bought by a dealer in Hamburg, who lost them all. Arnold bred his pair, and as he was not able to identify it, he preserved the breeding pair and sent it to London for identification. Boulenger identified Arnold's individuals as belonging to CHA. Arnold consulted the original description of CHA and found that his strain was not in agreement with Sauvage's color description. He sent further preserved material to Boulenger, who did not change his identification. Arnold then distributed the offspring of his strain under the CHA name. Boulenger (1915: 56) figured Arnold's breeding male under CHA and said that this individual originated from Sierra Leone. Arnold, however, said that the strain originated from Monrovia, Liberia. (See CAL and LIB.)

The offspring from Arnold's breeding pair became a popular aquarium fish, very easily kept and bred, and this strain has been maintained as an aquarium fish up to 1965 without further importations.

Daget and Arnoult (1964: 596–98) recognized the identity of this strain and placed it within DAG as a subspecies named MON.

Sheljuzhko (1952: 312) collected live fishes for Werner of Munich in SE Ivory Coast. In a small pool near the airport of Abidjan (Port Bouet) situated on a narrow island off the coast, he found some small individuals of a species of *Epiplatys* which resembled Arnold's strain of 1908. Live individuals from this population were distributed in Germany, but this strain did not become popular, and it soon became extinct. Some individuals were bought by Roloff who realized that this strain differed from Arnold's strain of 1908. He sent individuals to Poll, who found himself in agreement with Roloff and described Sheljuzhko's strain as DAG (1953: 262).

In 1962 Clausen found a quite similar strain at Awiebo, SW Ghana and individuals of this population were placed in my tanks and an aquarium strain was raised.

At the present time DAG has two subspecies. Further information about the Liberian subspecies is given under MON.

The DAG subspecies lives in SW Ghana and SE Ivory Coast. The western limit of its range is unknown at present. It lives in swamps and similar biotopes, always near the coast. It differs from the MON subspecies

Two male DAG from Awiebo, southwest Ghana. These males belong to the DAG subspecies.

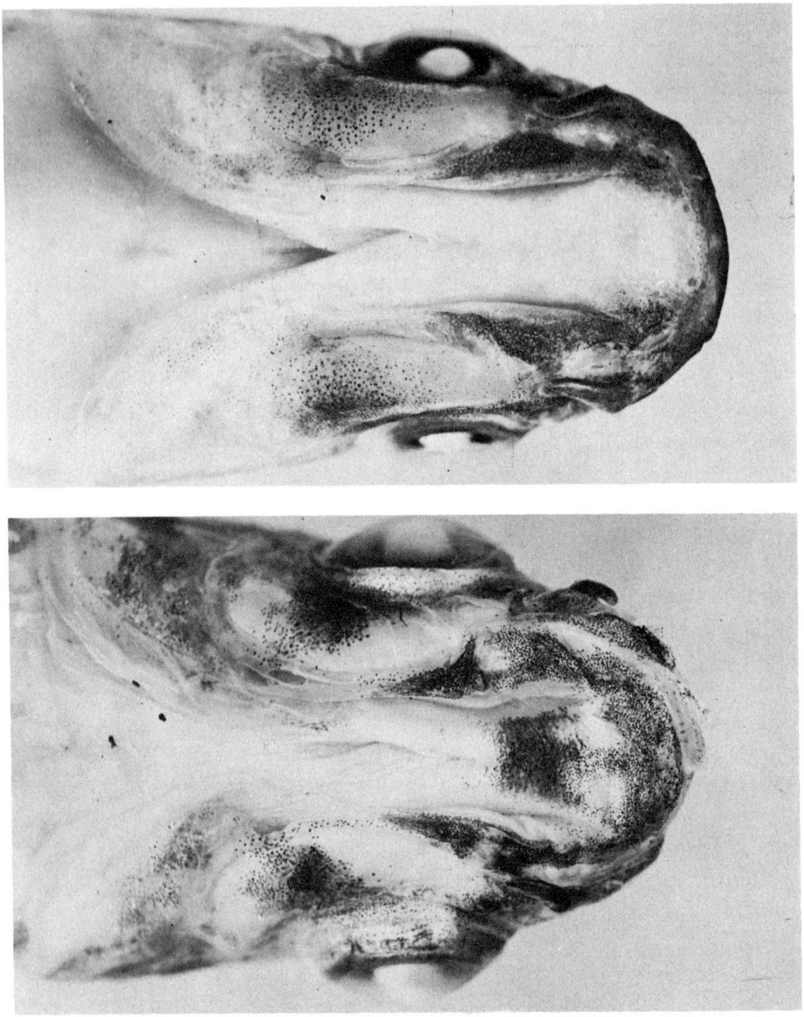

Pattern of dark pigments of the throat of preserved individuals of DAG DAG from Port Bouet, Ivory Coast. Male (above) and female (below).

by the color pattern only. In MON, the anterior part of the throat of males is orange-red, whereas in the DAG subspecies, the anterior part of the throat is white, grey or pinkish. The MON form has five dark crossbars, four above and behind the anal and one behind the root of pectorals. The DAG form has six bars usually, the sixth bar being situated above the root of the ventrals. The sides of the male of the DAG subspecies are bronze, whereas

Adult male DAG DAG and adult female (below) DAG DAG from Awiebo, south-west Ghana. In background adult male of DAG MON of the old aquarium strain.

in MON males they arc greenish-blue. Just after preservation in alcohol, a conspicuous lemon color develops anteriorly on the throat of males of the DAG subspecies covering the area, which is orange in the live MON male.

Both subspecies have 25 haploid chromosomes, the largest of these being about double the size of the smallest. There seems to be two metacentric elements of medium size. The haemoglobin pattern is of the usual rivuline type (4 lines). The frontal scalation is normal (G-type, no H-scales). The unreduced chromosome number (n=25, which also occurs in ANN and in CHA) indicates that DAG probably is an ancient species, although it differs markedly from all other species by the throat pattern of both sexes. Only one close relative can be pointed out at present. This species is GRA, which replaces DAG ecologically in Nigeria. With GRA, rather viable but sterile hybrids (female-like intersexes) are produced, whereas with other rivuline species the hybrids were inviable or almost inviable. With SEX most hybrids are inviable, but a few may be raised to adult size. One hybrid individual of both possible combinations was raised. They were sterile males or male-like intersexes, very feeble and slow-growing.

With CHA, inviable hybrids were had in spite of the fact that DAG appears to be closer to CHA than to SEX in external morphology and in chromosome number. Similar results were had with FAS, LIN and MAC, and all hybrids died in the egg rather late in the development. Also, the results of the crossings to CHR, COG and PET were similar, whereas in those with BIF and AUS, the hybrids died early in the development in the egg.

Further information on DAG and MON are given by Clausen and Scheel (1966: 77–88).

DAY: *Haplochilus (Panchax) dayi* Steindachner 1892 **(Valid sp.)**

Steindachner's material (1892: 376) originated from Ceylon and DAY has been reported only from this country.

DAY was introduced as an aquarium fish into USA during 1937 and live individuals were sent to German aquarists. This strain apparently soon became extinct. Further importations took place in 1952 and since that time, DAY has become a common and rather popular aquarium fish.

Superficially, DAY resembles LIN, but undoubtedly these two forms are not closely related. DAY has 24 haploid chromosomes and there seems to be no large metacentric or double-sized acrocentric elements. Thus DAY probably has the ancient cyprinodont karyotype. The hemoglobin pattern is complicated and quite different from that of LIN. There are five lines (two in LIN), and the pattern differs from all other patterns of rivuline-species so far studied. The frontal scalation is stable and normal (G-type, no H-scales). The pattern of lateral line organs on the dorsal part of the snout is of the usual *Epiplatys* type ("closed") often with large and flat fleshy lobes, resembling the condition found in PLA. Ventrals are pointed in both sexes, but longest in the male.

Adult male of DAY. Usual aquarium strain.

The short dark crossbars on the sides are often absent in the adult male. The temporary color pattern of fear resembles the permanent pattern of dark lines of BIF, but in DAY, the posterior parts of the two dark bands along the sides are very faint or even absent. The lower band, which often is interrupted, may occur in the fighting female, in combination with the dark crossbars, which then are rather faint. The throat of the male does not seem to become darker during fighting or pre-mating display, but in the female, the anterior part of the throat may develop a rather faint, but quite distinct dark color, similar to that of the CHA female, and within this dark area two short longitudinal bands of the DAG type sometimes are seen.

The largest of the types for DEC in the Paris Museum. No. 04.61.

DEC: *Haplochilus decorseii* Pellegrin 1904 **(Valid sp.?)**

Pellegrin (1904: 223) had five specimens from Bessou (Oubanghi) situated in the northern part of the Central Congo drainage. He placed DEC closest to *Aplocheilichthys spilauchen* Dumeril.

Myers (1924: 2–4) placed DEC in *Aphyosemion* with some reservations, whereas Poll (1952: 160–61) placed DEC in *Epiplatys*.

I have seen the five types and found them to correspond well to the CHR phenotype, but the type locality for DEC seems to be separated from the Central Congo drainage by some rapids, and thus DEC might be isolated from the CHR group within the central drainage. (See CHR.) Pellegrin's data for the types is not quite correct as far as scales in a longitudinal series are concerned. There are 28–32 scales and not 25–28; the mean number being about 30 scales.

DEF: *Panchax grahami decemfasciata* Pellegrin 1934 **(=SPI)**

Pellegrin (1934: 2–4) had one specimen (23 + 7 = 30 mm) from Bobo-Dioulasso of High Volta and two individuals (23 + 7 = 30 and 27 + 8 = 35 mm) from a brook 100 km from Bougouni towards Odiennes of Ivory

Coast. I have seen one of the individuals last mentioned and it corresponds to SPI in all visible details. SPI is known to occur in the geographical areas in question. Also the description of DEF corresponds well to that of SPI (or SEN). DEF was considered a synonym for SEN by Daget (1957: 159).

DOM: *Aplocheilus panchax dorsomarginatus* Klausewitz 1957

Klausewitz (1957: 196–99) based DOM on three individuals which were caught by Roloff near Patalung of southern Thailand.

The basic color of the live male is light brown to red, with large diffuse bluish-green markings on the gillcovers and above the root of the pectorals. The posterior part of each scale is bluish-green and the anterior half is greyish-brown to brilliant red. The lower part of the head is yellowish, the belly being white. The outer part of the dorsal and caudal fins are yellow. The anal has four or five broad red cross-lines near base and on a faint yellowish basic color. This fin also has many small red spots and a broad orange-red marginal band. A black marking near base of dorsal. Ventrals faint orange to transparent. Female has less brilliant colors. The basic color is grey with a faint blue shine. Anal with a narrow orange marginal band, caudal similarly colored above and below. Dorsal with a dark marking near base, the marking is edged with white in front and behind. Other fins colorless.

Klausewitz placed DOM close to the Thailand PAN described by Smith (1945: 422–23) differing from these by the brilliant bluish-green color of the gillcovers, the broad dark marginal band in the dorsal, which is absent in the populations studied by Smith. In DOM the orange marginal band in the anal is more narrow, he said.

Roloff (1957: 302–03) described and figured DOM. He said that DOM was considerably smaller than other Thailand PAN, and also more colorful.

The strain of DOM, which Roloff brought to Germany, apparently has become extinct.

DOR: *Epiplatys dorsalis* Mayer 1936 **(=FAS)**

Mayer (1936: 257) bought two live specimens from Grelk of Altona, and he was told that these individuals originated from Old Calabar, Nigeria. Both were males, and they resembled FAS in size and in general appearance, but as he was not sure of this identification of his individuals, he sent one male to Ahl, who (1938: 55) described and named the strain. Ahl used the same name for this strain as Mayer used two years before. I consider Mayer and not Ahl to be the author of DOR, because Mayer first used the name, and his description and ink-drawing characterize DOR just as well (or bad) as Ahl's description. Ahl placed DOR closest to SEX and FAS and said that it differed from both by the higher dorsal fin ray count and by the position of this fin. This is true for SEX, but apparently not for FAS

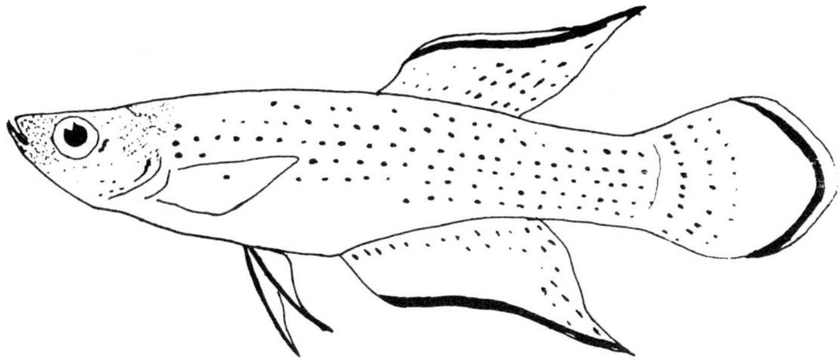

The holotype for DOR redrawn after Mayer 1936.

(FAS: D.10–15 and DOR: D.14). Also other characters of DOR agree with those of FAS. The "type-locality" given by a dealer, who did not collect the fish himself, is of no importance and should not be taken into serious consideration. I consider DOR to represent a synonym for FAS. (See CAB. The type-material of both forms was said to originate from Old Calabar.)

DUB: *Epiplatys duboisi* Poll 1953 **(Valid sp.)**

Poll (1953: 298) had eleven individuals from the Stanley Pool area of Congo. DUB is a dwarf species, and does not seem to exceed 30 mm total length. It has not been reported from other localities in Africa.

DUB resembles other species of *Epiplatys* by several characteristics, but it differs from all other known species by certain features. The caudal fin is pointed, but the central rays do not produce a distinct lobe. Certain strains of SPI are, however, similar. The central rays of the caudal are rather short and intermediate between *Aphyosemion* and *Epiplatys*, 30–32% of standard length.

The chromosome number is n = 24, 2n = 48, placing DUB closer to *Epiplatys* and *Aplocheilus* than to *Aphyosemion*. The karyotype seems to correspond to the basic cyprinodont karyotype. No large metacentric or double-sized acrocentrics are seen. There is, however, one metacentric element somewhat smaller than the largest acrocentrics, and this element might have originated from a fusion of two of the smallest acrocentrics of the basic complement. The basic chromosome number thus cannot be pointed out.

The hemoglobin pattern also is of the *Epiplatys* type, similar to that of SEX, LON, ESE, CHA and CHE and different from all known species of *Aphyosemion* and *Roloffia*.

The lateral line organs of the dorsal part of the snout are rather vestigial,

Adult male DUB from the Kinshasa (Leopoldville) area of Congo.

and also the fleshy lobes are highly reduced or absent. The pattern seems to correspond to that of SEX, LON and ESE and to be similar to *Aphyosemion* and *Nothobranchius*. The frontal scalation is normal (G-type, no H-scales).

The color pattern of the throat is very simple and correspond to that of ANN and BLO and the procatopodine species. In the male there are slightly more melanophores than in the female, but no patterns develop. Just after preservation in alcohol, the whole anterior part of the head of the male becomes brilliant lemon (see DAG, BLO and PAN) but this remarkable color soon disappears. The active female may develop a rather faint, irregular, dark, median band along the posterior part of the side, indicating a relationship with *Epiplatys* and not to *Aphyosemion*. The fin pattern of the male is not of the *Epiplatys* type and there are thin submarginal red lines in all unpaired fins, resembling the color pattern of *Aphyosemion*, *Roloffia*, FAS and ANN (more or less). It seems that the hindmost ray of the dorsal is inserted in front of the corresponding anal fin ray in males, whereas in females the juxtaposition of these fins is normal.

The male "dances" in front of the female during the pre-mating display. This behavior resembles that of the BLO male and the males of many procatopodine species. The display of fighting is, however, of the usual rivuline type (folding and unfolding of fins). Eggs are deposited within plants.

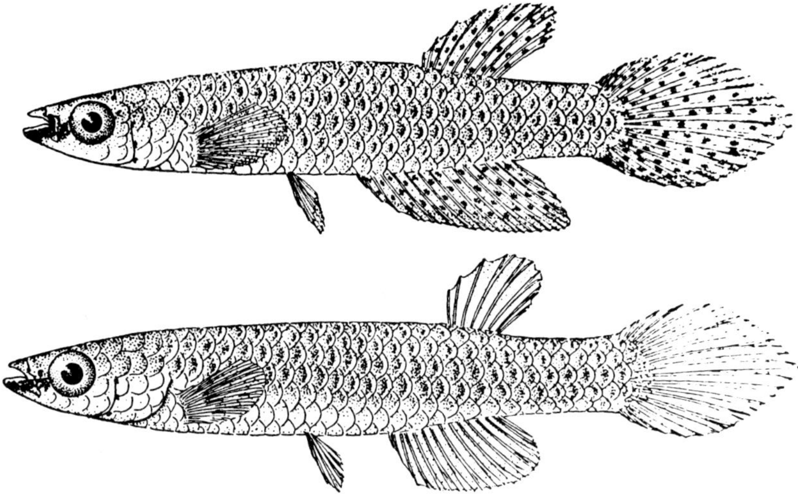

Fig. 4. — *Epiplatys duboisi* sp. n., type ♂ (au-dessus) et paratype ♀. Léopoldville (× 3).

The holotype and allotype for DUB as figured by Poll 1953.

Primary meiotic division of DUB. n = 24.

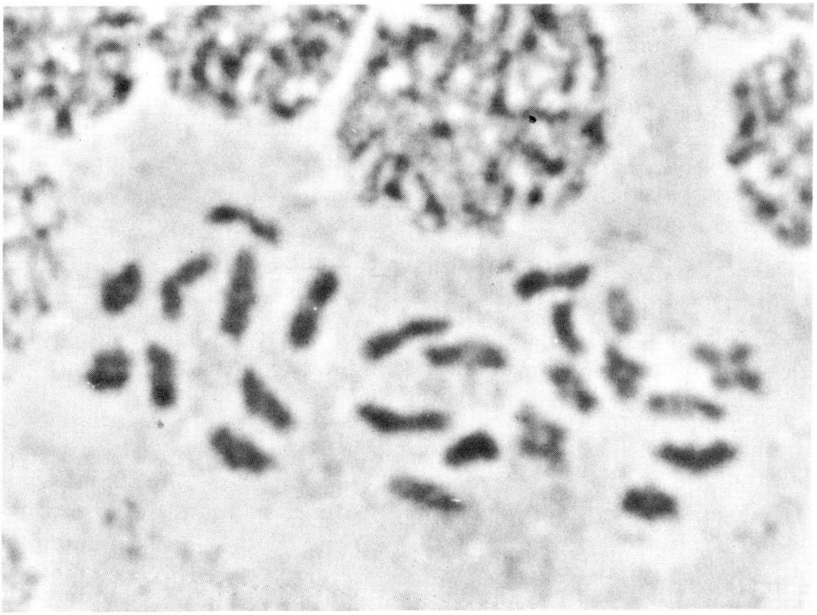

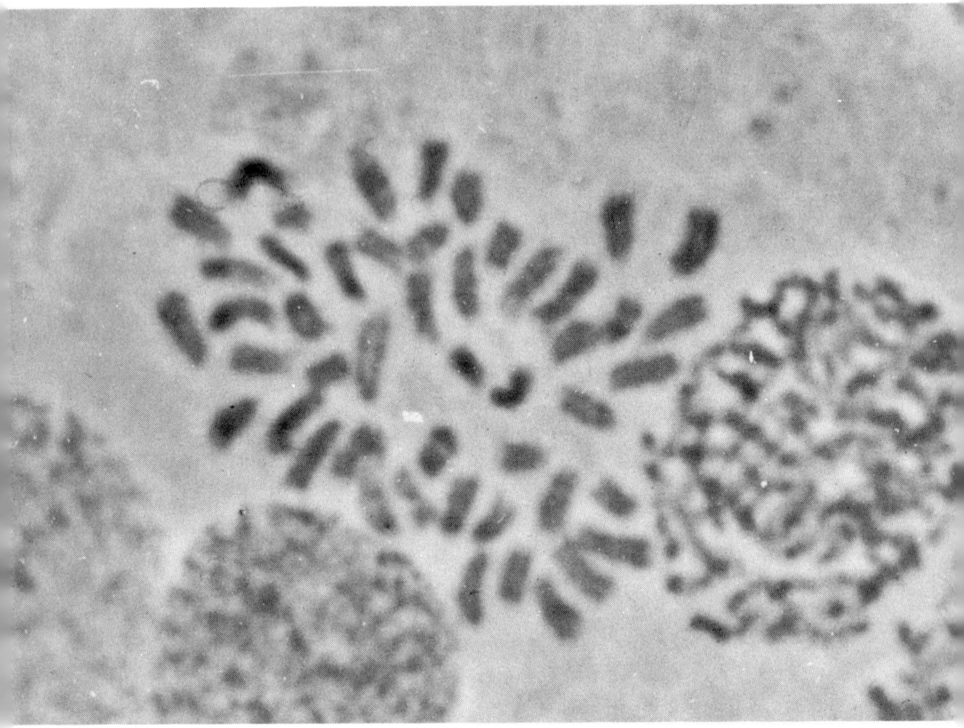

Mitosis of DUB from the Stanley Pool area of Congo. 2n = 48.

The pattern of the membrane is simple, being almost plain with numerous small "tubercles."

Clausen (1967: 32) separated DUB from *Epiplatys* and created a new monotypic genus for this species. He named this genus *Aphyoplatys*.

EIM: *Nothobranchius eiminki* Ahl 1935

Ahl (1935: 125–26) had a single individual, probably a female from Kongoran-Botto of Tanganyika. This locality is not on my maps. The hindmost part of the body and the unpaired fins were dotted with dark spots.

ELB: *Panchax elberti* Ahl 1924 **(=BUA)**

Ahl (1924: 309–10) had three individuals from the "Jade Plateau, Lebo Fluss, Ost-Kamerun." These individuals were caught by Dr. Elbert. The Jade Plateau is a large geographical area ranging between the western part of the Logone-Mbéré drainage in the west and the Bahr-Sara-Uam in the east. In the south the plateau reaches the upper reaches of the Sanaga (Mambere and Nona Rivers) River and the Lobai River. This means that the plateau is the water-shed separating the Chad, the Congo and the Sanaga drainages. The Lebo River is not on my maps.

In 1966 I found populations referrable to BUA in the upper reaches of all these drainages within the Jade Plateau.

Ahl's original description of ELB does not correspond to that of BUA, whereas Holly's redescription (1930: 210–11) of the type material placed ELB within the known variation of BUA. Ahl said D. 6–7, A. 11–12, but Holly said D. 11–12, A. 16–17 for the types of ELB. Ahl's description of the colors of ELB correspond well to those of BUA. He said that there were rather many red spots on the sides, and that these were merging on the hindmost two-thirds of the side, producing red crossbars.

Ahl placed ELB closest to STR, but Holly considered ELB to represent a synonym for ESC. I do not agree with Holly in this. ESC was described from a locality near Libreville, Gaboon and is a species of the tropical forest, whereas ELB is a species of the highland savannah. ELB undoubtedly is a synonym for BUA.

ELE: *Haplochilus elegans* Boulenger 1899 **(Valid sp.)**

Boulenger (1899: 113) had three individuals from Bikoro, Lake Tumba of the Central Congo drainage. These individuals were caught by Delhez. In 1908 (see CAL) he identified some aquarium fishes from West Africa as belonging to ELE, but in 1911 he removed these individuals from ELE and placed them in CAL. In 1911–12 (223) he referred some individuals caught by Luja at Kondue, Kasai, Congo to ELE, but these individuals probably are the ones which he used as types for LUJ. Nichols and Griscom (1917: 725) reported several individuals of ELE from Medje, Faradje and Stanleyville. Some of these individuals were used by Myers 1924 as the types for CAS. Poll (1952: 164) had numerous individuals of ELE from affluents to the left of the Congo River (Mondimbi, Flandria, Eala, Kunumgu and Bikoro) and also from Genema of the Ubangui drainage. He considers ELE to be very slender, the body being about 6 times as long as deep, and with long streamers in fins.

Matthes (1964: 127) had 82 individuals of ELE from Lake Tumba. He considered CHR (including CAS and SCH) to represent a subspecies of ELE, the latter having a central distribution with Coquilhatville as the center of distribution, and with the former living in the affluents to the right of the Congo River and in Kasai, thus with a more peripheral distribution pattern. He published some meristic data for 30 individuals of ELE and 54 of CHR. The body was 4.75 to 5.75 times as long as deep in ELE and 4.0 to 5.25 in CHR. ELE: D. 7–11, A. 13–16, Sq-l. 29–31, Sq-t. 19–24 and CHR: D. 8–11, A. 13–16, Sq-l. 28–32, Sq-t. 21–25. His description of the colors of the male ELE correspond to those of the strains of CHR with many red spots on the sides. The information published did not, in my opinion,

Haplochilus elegans.
♂, type (A. M. C.).

The type for ELE as figured by Boulenger 1915.

support any differentiation of ELE into two subspecies. (See CHR and other species of the CHR group.)

He also studied the stomachs of many individuals and found that the major food was insects (ants) and insect larvae living in the water. He said that ELE prefers shallow water, with slow currents in forests.

ESC: *Panchax escherichi* Ahl 1924 **(=CAM?)**

Ahl (1924: 369–70) had thirteen individuals from Attogondema, Nga-Zuflüsse, Kamerun. These individuals were caught by Escherich. This locality is not within the present Cameroun Republique, but is situated within the narrow northern strip of Gaboon, just south of Rio Muni (Spanish Guinea) which was taken over by the Germans in 1911 as a result of the treaty between the French and the German governments. The present town Atogondama is situated at the upper reaches of the Nduya River, an affluent of the Muni River and on the slope, which separates the coastal lowlands from the inland plateau.

Ahl placed ESC closest to CAM, but he did not indicate by which characters he thought ESC to differs from CAM. Holly (1930: 210–11) revised the types and changed some of the data given by Ahl. He found considerably more anal fin rays (A. 15–17) than did Ahl (A. 12–13). He considered ESC to represent a distinct species, different from CAM and he placed ELB (= BUA) as a synonym for ESC. (See ELB.)

Some details of the color pattern of the ESC male were described by both authors. Carmine spots on the anterior part of the side, these spots merging posteriorly, and red lines were produced. Rows of red spots in fins, and a red band posteriorly in pectorals.

Small male caught by the author in 1968 in the hills east of Bata, Rio Muni. This form resembles CAM, but it differs somewhat in color-pattern and it might represent ESC. The strains from two different rivers are different in respect to color patterns. In one of these the upper part of the dorsal is rust-red in males and females.

The holotype for ESE. Adult male from a brook 15 miles north of Eseka of East Cameroon.

It seems that ESC and CAM cannot be separated by meristic characters or by color patterns. A study of the Attogondema rivulins is needed because this locality is rather close to the type locality for the CAM-like STR.

ESE: *Epiplatys esekanus* sp. nov. **(New Species)**

The data for the holotype are given in the appendix.

This species was discovered in a small brook about 15 miles north of Eseka, near the junction of the Eseka, Edea and Yaunde roads. The type locality is situated just at the foot of the steep slope, which separates the coastal lowlands of East Cameroon from the inland plateau. This brook also contained the 17-chromosome population of BIV and a CAM-like species of *Aphyosemion*. ESE was also found in a larger brook just south of the junction above mentioned, but in this brook BIV was absent. The geographical range of ESE appears to be very limited, because some miles north and south of the two localities mentioned, ESE was absent and was replaced by SEX, the usual *Epiplatys* of the brooks of coastal Cameroon. On the inland plateau ESE also appears to be absent, being replaced by SAN.

Primary meiotic division of ESE (holotype). n = 21.

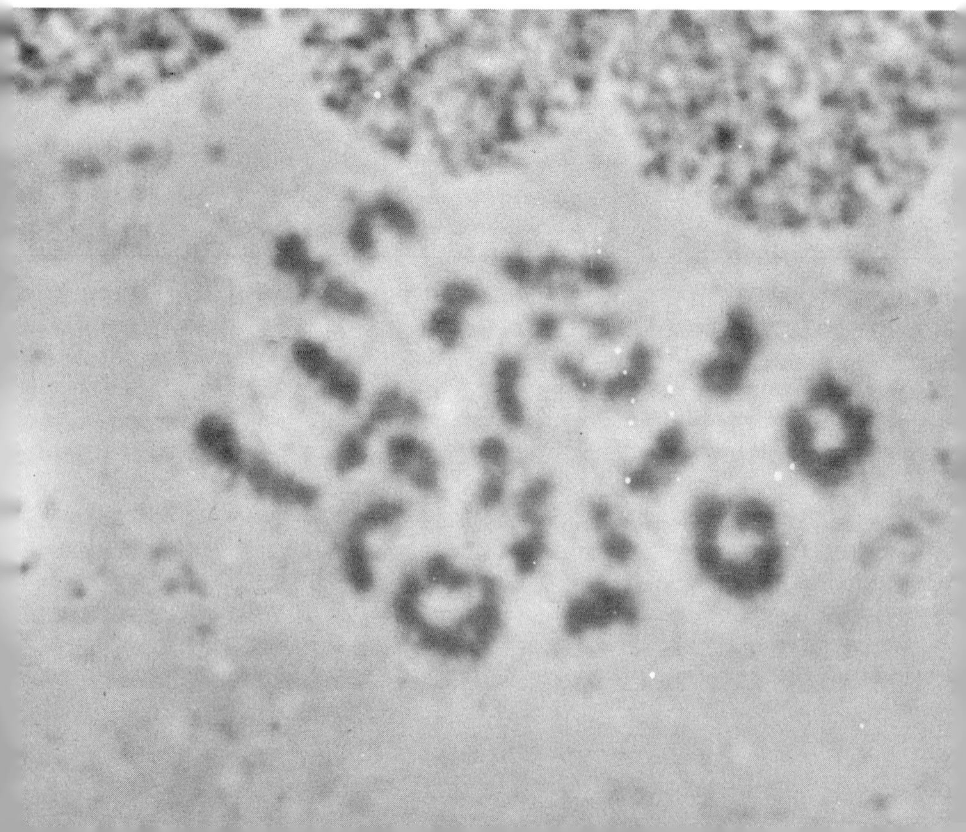

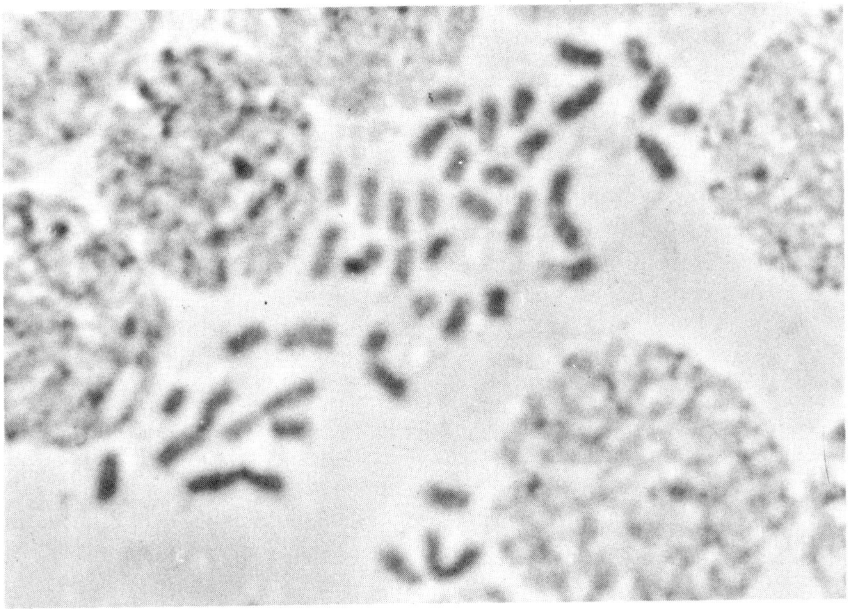

Mitosis of holotype ESE. 2n = 42.

ESE resembles SEX in most morphological characters including the pattern of lateral line organs on the dorsal part of the snout. The hemoglobin patterns and egg-types are also similar. It differs markedly by the haploid chromosome number (ESE: n = 21, SEX: n = 24) and by details of the color pattern. I was not able to cross wild individuals of ESE and Cameroon SEX, as no sexual response was observed.

Mitotic spermatogonial divisions have six large metacentric elements measuring between 2.1 and 2.8 times the mean length of the basic chromosomes, indicating that these metacentrics are composed of basic acrocentrics of above-average length, and that the ancient species, from which ESE originated, probably had n = 21 + 3 = 24.

ESE probably is an ancient species, now restricted to some particular biotope-type probably in connection with the slope above mentioned. It is difficult to understand why SEX do not occur in the localities occupied by ESE, because, apparently, there are no physical obstacles to separate these two populations.

Adult individuals of ESE seem to be somewhat smaller than similar individuals of SEX in nature and in the aquarium.

ESE was mentioned by Clausen (1967: 27–28) who placed it in the subgenus *Epiplatys* together with SEX, LON and one more doubtful species from the Mamfe lowlands. (See SEX.)

EXI

The type for EXI as figured by Boulenger 1915.

EXI: *Haplochilus exiguus* Boulenger 1911 **(Valid sp.)**

Boulenger (1911 : 265) had a single specimen, a male, from the Nyong River of southern Cameroon. No further report on this species has been published under the EXI name.

In 1966, I found this species to be very common in all small freshwaters of the wooded part of the inland plateau of East Cameroon. It lives in brooks, swamps and other freshwaters within the Sanaga, Nyong, Ntem, Ogowe and the Congo drainages in Cameroon. Different populations did not seem to differ and there were no indications of differences in different major drainages. EXI seems to be restricted to the forest being replaced by BUA in the adjacent savannah biotopes. In its biotopes, EXI prefers the clearings, where the vegetation of the banks (*Tradescantia* and other plants) dip into the water, offering a hiding place. In the Raffia swamps I found it only where there was sufficient light to support a growth of green algae. The hardness of the natural water varied between 0.1 and 1.5 German degrees, pH being 4.2–6.8. In these localities EXI is sympatric with SAN and OBS the latter species being replaced by the larger BAT in the swamps of the Upper Nyong drainage.

Young males of EXI do not develop the handsome overall yellow-orange color of old males. They resemble young males of BUA in most details, including the brilliant blue shine on the sides, but they differ by the absence of red spots and streaks in the distant third of the anal fin. Females of most strains have thin dark brown cross-bars posteriorly on the sides, but these bars are absent in some females and in all females of some populations. (See JAU.)

Adult male of EXI from brook 30 km north of Sangmelima of east Cameroon.

Adult male of EXI from brook 30 km north of Sangmelima of east Cameroon.

Primary meiotic division of EXI. n = 18.

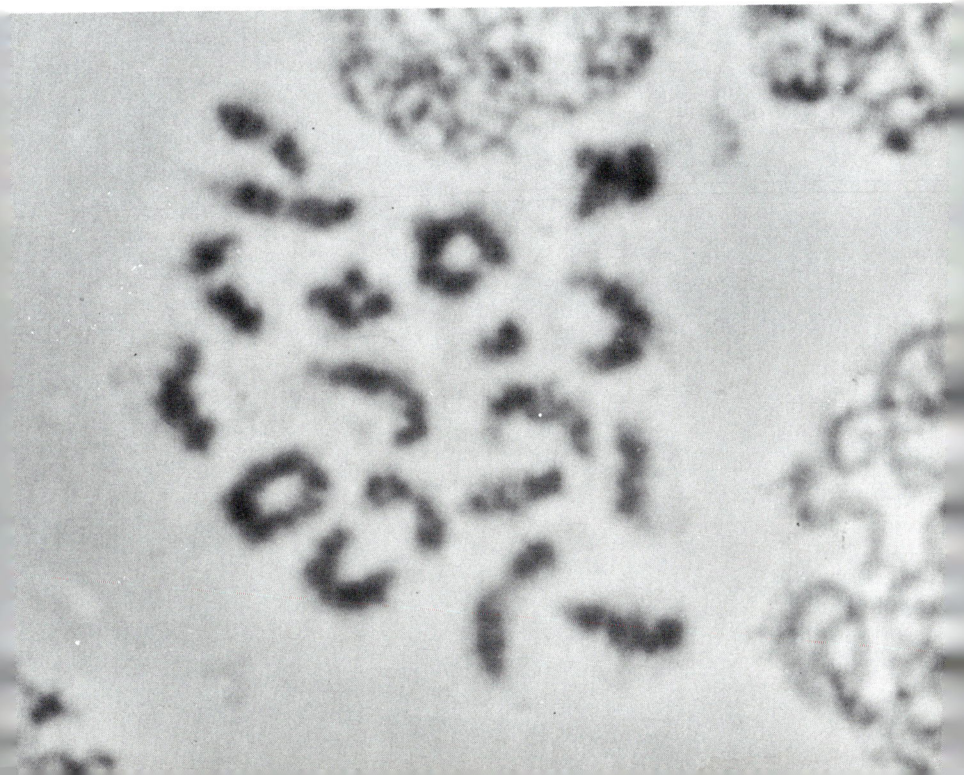

Adult male EXI from brook about 30 km north of Sangmelima towards Mbalmayo, east Cameroon.

Mitosis of EXI. $2n = 36$.

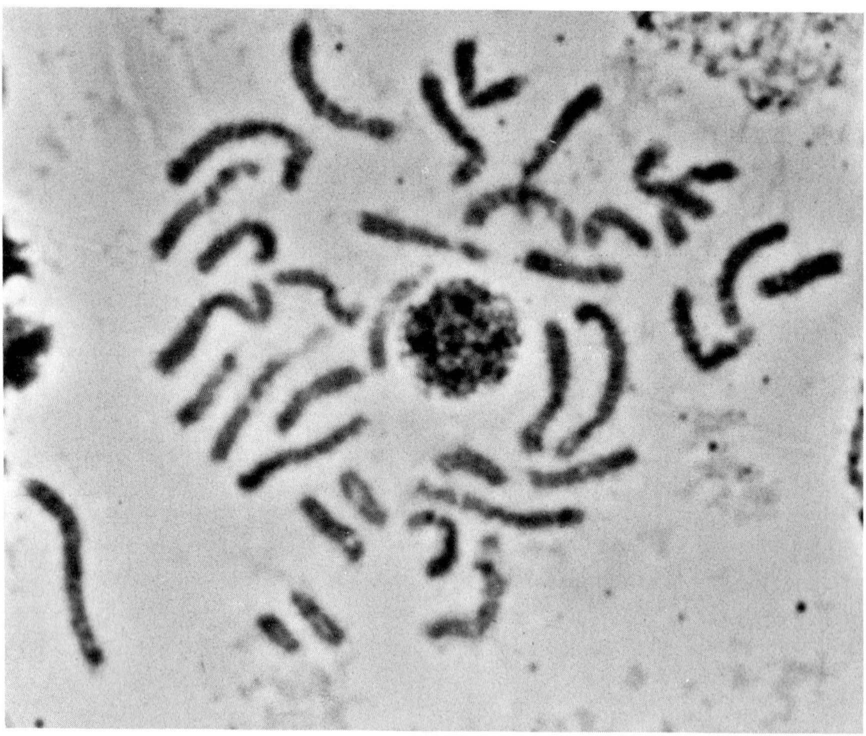

EXI from a locality 30 km north of Sangmelima of southern Cameroon has 18 haploid chromosomes. The karyotype resembles those of the Bamkin and the Ndop populations of BUA. There is a smoothly graded series of chromosomes. There seems to be three metacentric elements, which are larger than the largest chromosome of the basic complement. If the karyotypes of these three strains are compared, it seems that EXI is closest to the 20-chromosomes strain (Bamkin), whereas the results of crossings indicated, that EXI is much closer to the Ndop strain with 19 chromosomes. The male hybrids from the crossing of EXI to both strains of BUA are characterized by the occurrence of uni-, bi- and tri-valents and the number of separate elements of meiotic divisions often is less than 18.

The hybrids from the EXI/BUA-Ndop crossing are quite fertile and in F_1 x F_1 spawnings, many individuals develop and mature. The F_2 individuals are fully viable. The hybrids from the EXI/BUA-Bamkin crossing are fully viable, but functionally almost sterile. Very few of the fertilized eggs develop viable fry, and most of these die before maturing or become dwarfs. The population of BUA last mentioned is the one which is in contact with EXI along the border between savannah and forest, thus it is to be expected that isolating mechanisms have evolved.

Morphologically and biologically, EXI and BUA are close relatives. The choice of biotope of EXI may indicate that this species has not fully acclimated to a life in the forest. The chromosome numbers indicate that EXI is a derivative of BUA. The form last mentioned seems to represent the sole species of its genus, which is restricted to the savannah, all other species being forest dwellers. Also both species (EXI and BUA) develop vertical

Adult hybrid male EXI/BUA-Bamkin. Notice the BUA pattern of anal fin.

bars in the caudal fin, which characterize savannah dwelling species (BIF and SPI) of *Rivulinae* of Atlantic Africa, when the species of *Fundulosoma* and of *Nothobranchius* are not considered.

Further information on the components of the EXI superspecies are given under BUA, (ELB, RUF, TES). JAU, JAC, LOL and LOB undoubtedly are synonyms.

FAL: *Aphyosemion (Fundulopanchax) fallax* Ahl 1935 (=SJO??)

The first information on this form was given by Schreitmüller (1933: 209) who said that he had been informed that an importation of "yellow *Fundulus*" (= GUL?) had arrived from the Gold Coast. Soon individuals from the reproduction of this new strain were sold, and he had some pairs from Härtel of Dresden. Schreitmüller and Härtel thought, that these individuals differed from the "true" GUL as important (?) differences in colors, color-patterns and shape of fins were observed. Schreitmüller sent some preserved individuals to Norman of the British Museum, who identified this material as SPU (= WAL) which had not been imported as an aquarium fish at that time. Schreitmüller also said that his strain differed substantially from individuals which were kept by Scholze and Potsche of Berlin. He said that his females were similar to those of AUS, and that Härtel agreed with him in this. He assumed that his individuals were hybrids.

Further information on this importation was given by Arnold (1934: 99) who had two preserved individuals from O. Scholze of Berlin for identification. Scholze told him that he thought this form was *Nothobranchius gularis*, but that some people in Berlin doubted this identification. Arnold himself compared the individuals to Boulenger's descriptions (1915) and found that they were closest to BAT. He also compared the individuals to Schreitmüller's description and figure and found that Scholze's individuals corresponded to those of Schreitmüller. He sent the individuals to Norman, who identified them as GUL.

Schreitmüller (1935: 339) realized that Norman's identifications were not in agreement and he asked Norman's opinion on this. Norman answered that the individuals sent by Arnold were from the Niger Delta and had more fin rays than Schreitmüller's individuals. Schreitmüller now sent two preserved individuals to Ahl in Berlin, and Ahl said that these individuals did not represent GUL or SPU, but belonged to a new species, closest to SPU and to SJO, but not close to GUL.

In 1935 (: 251) Ahl described FAL on the basis of Schreitmüller's material. He said that the types originated from the Gold Coast.

Three ink-drawings of the male and the female FAL have been published by Schreitmüller (1933: 209), Arnold and Ahl (1936: 296) and Mayer (in "*Die Aquarien Fische*" by Holly, Rachow and Meinken). The color patterns of these figures are in agreement, not the shape and position of fins. In

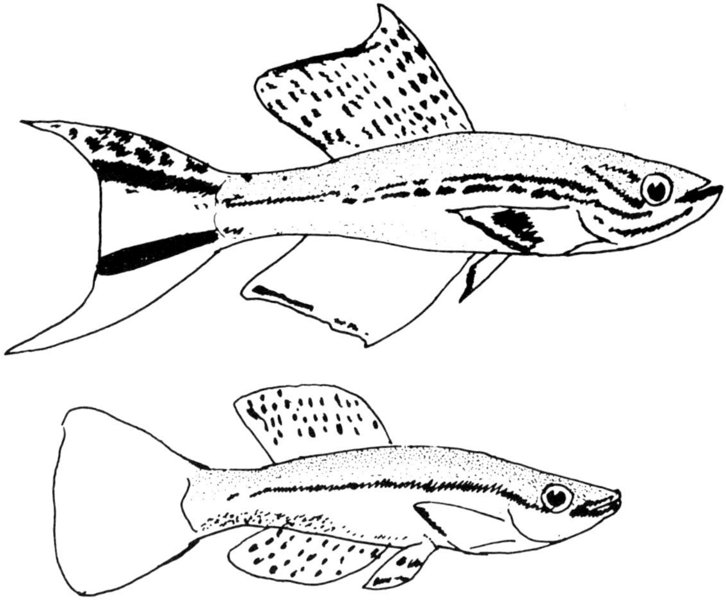

The male and female of FAL redrawn after Schreitmüller's illustration of 1933. The type for FAL originated from this particular aquarium strain.

Schreitmüller's and Arnold and Ahl's figures, the dorsal is in advance of the anal, whereas in Mayer's figure the dorsal is behind the anal, or almost above the anal. In Arnold and Ahl's figure some of the rays in the anterior third of the anal are greatly extended, as in SJO, whereas in Mayer's figure the anteriormost rays of this fin are extended. There are no elongated dorsal fin rays in these figures.

The color patterns of the sides and of the caudal fin of males as described by Schreitmüller and Ahl seems to characterize FAL. There are two (or one) narrow red median bands anteriorly on the side, these bands disappear or fade into rows of red spots posteriorly on the sides. The anterior part of (one of) these bands is present in the female, but not in all females. There is a series of faint "dark" crossbars posteriorly on sides in the male in Mayer's figure. The caudal fin has two submarginal bands of red color. The central part of the fin being dark yellow or orange, the upper edge is dotted with red and the lower edge is yellowish. The anal fin is yellowish, with or without red spots, and has a narrow red margin.

In 1962 Nieuwenhuizen sent me a black and white photo showing a pair of an unidentified species of *Fundulopanchax* which had been imported into Holland. Both individuals resembled FAL, but they also resembled SJO in some characters, in others not. In these two individuals, the dorsal fin is distinctively in advance of the anal fin, the anteriormost anal fin ray being inserted above the fifth dorsal fin ray. D. about 17, A. about 15, the base of the dorsal fin considerably longer than that of the anal fin, as in Arnold and Ahl's figure, but not in those of Schreitmüller's and Mayer's. There is a dark band from the gillcovers to above the root of the pelvics in the female and also there are dark crossbars posteriorly on the sides of the male. The pattern of the caudal fin correspond to that of FIL. Nieuwenhuizen's pair probably correspond to the phenotype described under the FAL name.

Although Ahl published the Gold Coast as the type locality for FAL, it is unlikely that the aquarium strain which was used by Ahl for the description of FAL originated from this part of West Africa. It is more likely that the types for FAL originated from the Niger Delta area. FAL is very close to SJO, but it seems to differ from SJO by the described details of the color patterns.

FAS: *Haplochilus fasciolatus* Günther 1866 **(Valid sp.)**

Günther (1866: 358) had four individuals from Sierra Leone and one individual from the Nile drainage. The individual last mentioned was later removed from FAS and placed in MAR (= SPI).

Günther (1866: 358) extended the range of FAS to Liberia. Nichols and La Monte (1933: 2) and Schultz (1942: 337) reported FAS from Western Liberia and Daget (1948: 42 and 1962: 127) and Lambert (1958: 51) reported FAS from Lower Guinée. The eastern limit of the range of FAS is unknown at the present time, and various reports of FAS from High Guinée (Mt. Nimba) by Daget (1948: 42) and from Ivory Coast by Pellegrin (1934: 101–120) need further investigation. The report last mentioned might concern OLB, which was referred to FAS as a subspecies by Daget and Iltis (1965: 190–92). (See also LAM and OLB.)

The history of the aquarium kept strains of FAS starts in 1908 when Grote (1910: 222) reported individuals which he thought were SEX from a small village near Conakry, Lower Guinée. It seems, that these individuals did not reach Hamburg alive. Arnold (1911a: 273 and 1911b: 576) reported live individuals of FAS which were imported into Germany from Sierra Leone in 1911. These strains were considered to be very feeble and juveniles were very slow-growing. Arnold himself identified these individuals (D. 12, A. 18) and this identification was verified by Boulenger. The aquarium society of Essen-Ruhr (1912: 59) reported that most of the juveniles of this species developed into males. Brieger (1913: 798) observed and described the very broad and dark longitudinal band which develops in the breeding female. These (?) strains of FAS survived in the aquarium in Germany during the First World War. FAS was kept by Ohlendorf (1917: 305), by Sachs (1919: 202) and Zindler (1922: 1), but apparently these post-war strains were soon lost. FAS was kept by German aquarists in 1935 according

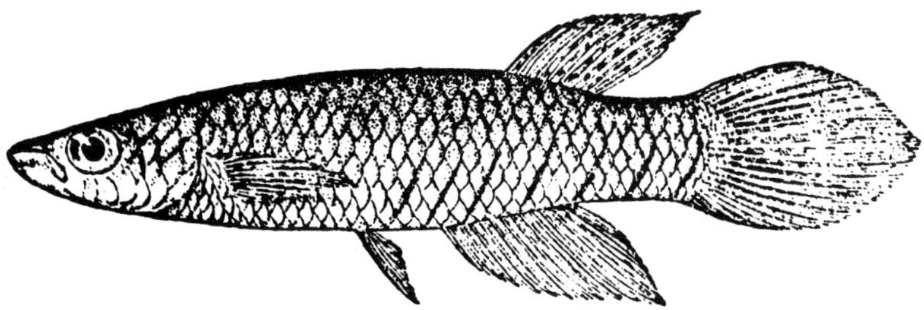

Haplochilus fasciolatus.
♂, type.

The type for FAS as figured by Boulenger 1915.

to Roloff (1936: 308–10), who studied a population of FAS near Kissy of the Freetown peninsula. Roloff brought to Germany preserved individuals from this population, and some of these were reported as FAS by Ahl in 1937, whereas one pair characterized by a broad dark longitudinal band was considered to represent a new sub-species of SEX! (See LEO.) If Ahl had studied Brieger's (1913: 798) report of the color pattern of FAS, the LEO name probably would not have been created. DOR and MAT represent two more synonyms, probably from Liberia.

In 1962, Roloff sent me additional material of FAS from the locality above mentioned, and LEO proved to be individuals of FAS, which developed the dark longitudinal band during preservation.

In 1962, 1963 and 1965, Roloff collected live individuals of FAS in many localities all over the southern part of Sierra Leone. He kindly placed most of these strains at my disposal for my biological studies.

Among these strains there were two strains from the Freetown peninsula. These strains had been caught in mountain brooks. The two strains were breeding true to their characteristics. The two strains differed by the size of adult individuals and by certain minor traits of the color patterns. In FAS, there is usually two thin red lines in the distant part of the anal fin

Large male of FAS from the Freetown area of Sierra Leone.

Adult male of FAS from western Sierra Leone.

Adult male of FAS from the Kenema area of eastern Sierra Leone.

Adult male of FAS from western Sierra Leone.

and in the lower part of the caudal fin. One line is marginal, the other submarginal. The submarginal line was absent in the strain which developed small individuals.

Hybrids from crossings of individuals of these two strains were fully viable and fully fertile. Both strains belong to the weakly colored, robust and deep-bodied Freetown area phenotype. Individuals of this phenotype never try to hide in the mud of the bottom of the aquarium probably because, in their natural biotope, the bottom is hard.

Two strains from the Barmoi district of western Sierra Leone were rather similar, but less deep-bodied. Also these two strains differed somewhat in the details of the color pattern, but their hybrids were fully fertile. The strains, which recently have been imported to Germany from the Conakry area of Lower Guinée, are quite similar but are more robust and deep bodied than the Barmoi strains.

Roloff also offered me live individuals of a strain from Kenema of eastern Sierra Leone. By size and shape of the body, the individuals of this strain resembled the Freetown strains, but the colors were much more saturated and beautiful. The color pattern, however, corresponded to that of other strains of FAS.

In 1962, Roloff also collected a few individuals of a very colorful and very

Mitosis of FAS. 2n = 38.

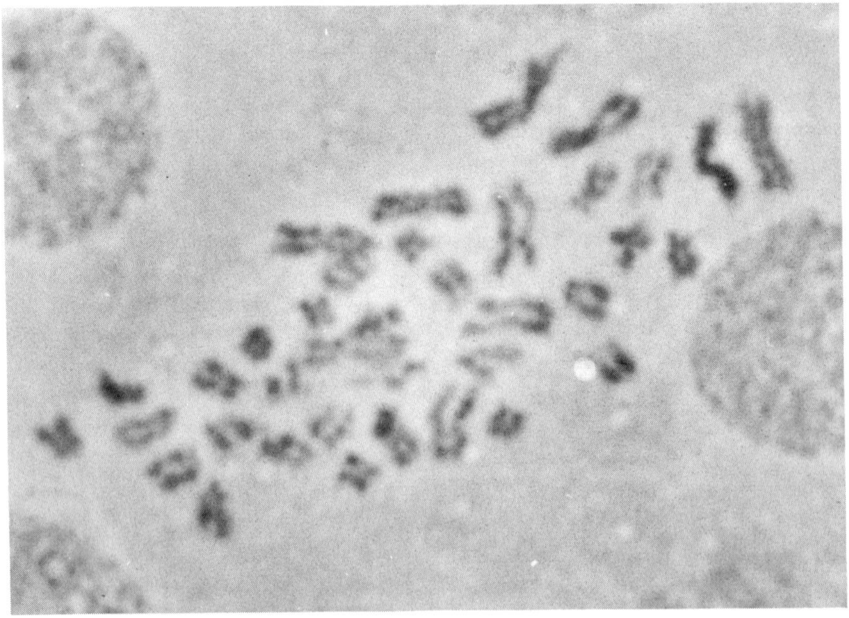

Adult female FAS. This female belongs to the slender and colorful phenotype in FAS from Sierra Leone.

slender strain from some inland locality. Unfortunately some accident during the transport of his live fishes obscured the locality of this strain. These individuals differed from other strains by the larger eggs (1.50–1.55 mm versus 1.30–1.40 mm in other strains), by the distinct hexagonal reticulation of the egg membrane (plain with small spots or tubercules in other strains), by their slender bodies and by behavior.

When afraid, these individuals dive deep into the mud of the aquarium, and they remained hidden in the mud for hours. This is the usual behavior of SPI individuals in the aquarium. There were one male and two females, probably originating from two different localities. D. 11–12, A. 17–18, Sq-l. 29–30. Their offspring (8 individuals : D. 10–13, A. 16–19, Sq-l 28–31) were fully viable and fertile, but the eggs from the aquarium-raised females differed markedly in size (1.35 to 1.55 mm). The offspring from this strain was not fully viable, and some of the embryos died in the eggs from "thrombus" and deformations of the body. A fairly stable strain could, however, be established based on the strongest individuals.

The intensity of the colors of the strain last mentioned and the Kenema strain was quite similar and for that reason these two strains were crossed. The offspring from this crossing proved to be fully viable. The fertility was poor, however. Development was studied in 43 eggs from a single pair. Twelve embryos died within the egg, mainly from "thrombus" and/or deformations. Thirty-one apparently normal fry hatched, but all except three died soon after. The three surviving individuals were very weak and difficult to raise. I do not consider FAS to be difficult to reproduce. Usually very few individuals are lost. This crossing indicates that some strains of FAS are almost completely reproductively isolated from other strains.

All males of FAS tend to develop narrow, oblique, brilliantly colored

Adult male of the slender and colorful phenotype in FAS from Sierra Leone.

Adult hybrid male FAS/SPI.

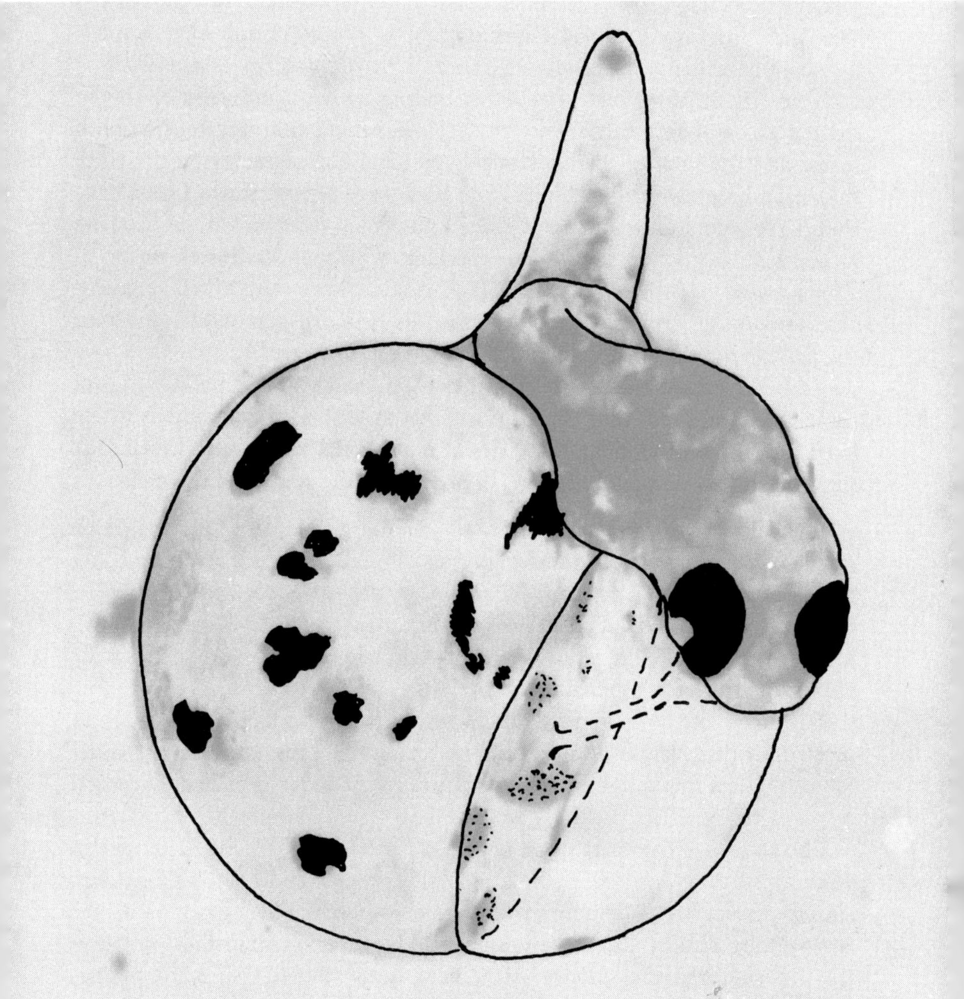

Embryo hybrid FAS/DAG with a large transparent "blister" below head. After the development of the blister, the egg-membrane burst and the embryo hatched out.

metallic crossbars or rows of spots, particularly on the posterior part of the side, but in the males of some strains there is also an overall metallic brilliance ranging between blue and green. The red pattern of the sides is variable. In the Freetown males this pattern is faint and diffuse. In the slender colorful form an almost perfect red reticulation develops in both sexes, and the females of this strain have more red pigments on the sides than have Freetown males. Males from Kasewe of central Sierra Leone have round red spots anteriorly and these spots may develop into a marking rather similar to the wound-like marking which characterizes males of *Aphyosemion* and *Roloffia.* These males also have much red pigment anteriorly on the throat. Here a broad red band is seen, quite different from the pattern of other males, in which the red pigment usually is covered by a dense black pigmentation. In all males a tendency for the building up of three black (and red) lines across the throat is seen. One such line is in the lower lip, the second along the series of neuromasts behind this lip and the third line, often very rudimentary, runs between the corners of the mouth.

The colors of the pectorals is variable. In most males this fin is yellow or orange, the color being rather evenly distributed all over the fin or being concentrated near the distant edge. In other males the pectorals are almost colorless, as in females of all strains, except for the slender and colorful form, in which the female has yellow pectorals and some orange and yellow color also in distant part of unpaired fins.

The shape of the caudal fin differs from most species of *Epiplatys,* as in most individuals the central rays do not extend and do not make a distinct lobe. In a few individuals, mainly females, a tendency for elongated central rays is seen.

The pattern of frontal scales is normal (G-type, no H-scales). Hemoglobin patterns of different strains, including the slender and colorful form, are similar and of the common rivuline type (4-lines). The karyotype of the two western strains and of various hybrid strains including the Freetown, the Kenema and the slender form, have been studied. All were similar. There are 19 haploid chromosomes and of these, five are large metacentric elements. There is also a metacentric element near maximum basic chromosome length.

Daget and Iltis (1965: 190–92) considered three subspecies of FAS. They considered LAM and OLB to represent subspecies of FAS. I do not agree with these authors in this, as LAM seems to represent a very distinct species, probably not closely related to FAS, whereas OLB appears to represent a subspecies of CHA, close to the SHE subspecies.

Embryos of FAS often hibernate in the egg; when they are ready to hatch, the circulation of blood stops completely. This finding and other peculiarities (the wound-like marking in males of some populations, the lack of elongate central caudal fin rays) seem to separate FAS from the main group of species of *Epiplatys*. FAS resembles GUI in general appearance, in chromosome number and apparently also in karyotype. These similarities may reflect close relationships and on the basis of the present evidence, I prefer to place FAS closer to GUI than to the different species of *Epiplatys*, except for SPI.

SPI and FAS undoubtedly are closely related. The hybrids are very viable but sterile. SPI probably represents a derivative of FAS, which has acclimated to savannah conditions. Various populations belonging to the FAS phenotype seem to replace SPI in savannah biotopes within Sierra Leone.

Benl and Foersch (1966: 265–66) deal with Guinée strains of FAS whereas Basch (1965: 203–04) deals with Liberian FAS. In males of the latter strain, the two red lines of the anal fin and of the lower part of the caudal fin seem to merge, producing a broad marginal band. (See MAT.)

FER: *Haplochilus ferranti* Boulenger 1910 **(Valid sp.?)**

Boulenger (1910: 285) had three specimens from Kondue, Kasai, southern Congo. These individuals were caught by M. F. Luja. Myers (1924: 2–4 and 1933: 183–84) placed FER in *Aphyosemion* after having seen the types. Poll (1952: 165) reported FER from Kondue, Sankuru and Kese, all of the Kasai drainage and in 1963 (: 96) Poll and Gosse said that FER had not been reported from the Central Congo drainage, excluding the Kasai drainage.

The color pattern of FER is interesting. There is a lightly colored band along the sides, from the gill covers to the root of caudal. This band might be carmine in life, according to Poll (1952).

The band might be continuous or interrupted. The dorsal and the caudal fins are brown with some yellow spots, and there is some orange color in pectorals and ventrals also.

The type for FER as figured by Boulenger 1915.

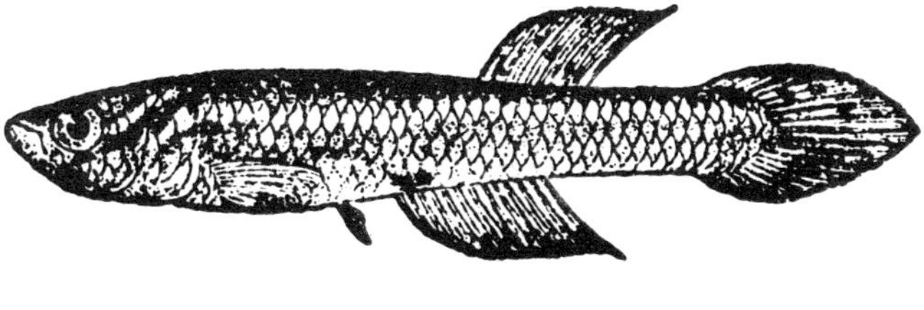

Haplochilus ferranti.
♂, type.

In the description, Boulenger said, that the body of FER was 6.0 to 6.5 times as long as deep, but these figures do not correspond to the drawing of FER published in 1915 (: 47). If this figure is used, the body (without caudal) is about 5.2 times as long as deep, still indicating that FER is a slender form.

On the basis of present evidence, FER cannot be attached to any known group of species of *Aphyosemion.*

FIL: *Fundulopanchax filamentosus* Meinken 1933 (Valid sp.)

The first information concerning FIL was published by Krüger (1913: 753), who described and pictured an aquarium fish which he called "The Steel-Blue *Fundulus.*" He was told that this strain originated from the mouth of the Congo River, but he also said that he considered this information to be rather unreliable. The picture, showing the male and the female was prepared by Unbehauen and it shows a form very close to the present common strain of FIL from SW Nigeria. Krüger said that the fins of the male were rounded, but Unbehauen pictured the male with a truncate, slightly trilobate caudal fin. The sides of his male were brilliant blue, and there was a broad band of brownish-red color in the ventrals, anal and caudal fins. These bands were submarginal. He considered his strain to be close to ARN.

The aquarium society "Nymphea" (1913: 809) corrected Krüger's description and said that this strain had been collected by one of their members and that the caudal fin was not rounded, but with long white streamers in old males. They said that the strain originated from Lomé, West Africa. This locality is on the coast of Togo.

Arnold (1914: 6) considered Krüger's strain to represent a local variety of ARN, but he was aware that scientific investigation might prove it to be distinct.

Dieterich (1915: 501–03) had a rather similar strain, which he described and pictured. In his male, the submarginal red band in the anal fin was not continuous, but broken into three large blotches. The pattern of the caudal

Large male of FIL from Ijebu-Ode of west Nigeria.

Two fighting males of FIL. Probably from the Badagri district of west Nigeria.

fin (lyre-pattern) resembles ARN more than FIL. He used the name ARN *var. blue* and pictured the male with short streamers in the caudal fin.

Sachs (1917: 213) bought a pair of a form which seems to be identical with Dieterich's strain, and he used the same name, but he said that this fish was also known under the name "*Fundulus* aus Togo." He considered this strain to be rare in Germany at that time.

Meinken (1930: 390) reconsidered the published information concerning this (?) form and said that Krüger introduced the name "The Steel-Blue *Fundulus*," and that this name had been attached to the present aquarium

Two fighting males of FIL.

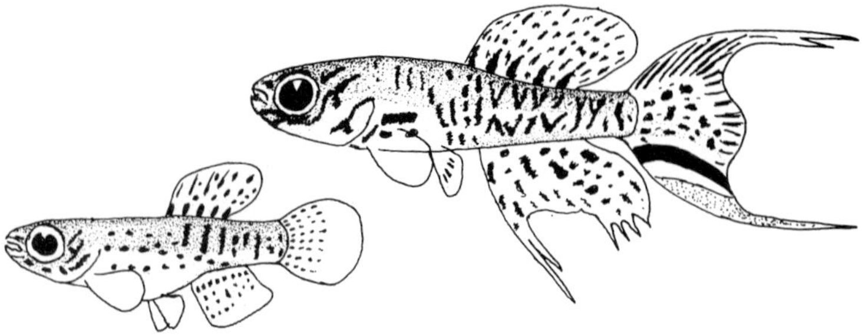

The types for FIL redrawn after Meinken 1933.

strain, which he himself recently identified as GAR (see GAR). He considered Krüger's strain to be different from the 1930 strain, because of the rounded caudal fin, and he thought that Krüger's strain represented a new species. He probably was not aware of "Nymphea's" correction of Krüger's information.

In 1933 (:249–51) Meinken apparently changed his mind concerning the identity of this form. He described and named FIL, which previously had been called "*Fundulus* aus Togo" and said that the origin of this strain was unknown, but that probably it originated from Tropical West Africa. He did not compare his new species to other species of the genus. In the *Die Aquarien Fische* (Holly, Meinken and Rachow) he wrote that the types were said to originate from Togo, but he indicated the mouth of the Niger as the true type locality.

A pair of FIL, probably belonging to the strain on which Meinken based his FIL, was photographed by Innes (1935: [48]–286). Innes used the GAR name for this strain. This photo also was published in *The Aquarium* 1954: 24. See also GAR and ARN.

Our present strains of FIL do not correspond to Meinken's FIL in all details. The pattern of the anal fin of the common FIL male is different and the development of filaments in this fin and in the caudal fin is less conspiciuous than in Meinken's strain. The first individuals belonging to the present strains of FIL were imported from Nigeria in 1952 according to Tusche. In 1959 Clausen caught live individuals near Ijebu-Ode of SW Nigeria. These were placed in my tanks and an aquarium strain was established. I crossed individuals of the 1952 and the 1959 strains and found the offspring to be fully viable and fertile. Clausen's strain differed from the 1952 strain by the pattern of the anal fin. In the strain first mentioned, the males often developed a narrow black marginal line or a row of black markings on fin edge. This pattern was absent in the 1952 males. Also in Clausen's strain, there were longer filaments in the anal fin.

In 1965, a somewhat different phenotype was imported into Germany, probably from the Badagri area of SW Nigeria. Males of this strain usually have no red submarginal band in the anal and the lower part of the caudal fin. Also eggs were smaller (1.0 mm versus 1.3 mm) corresponding in size to those of ARN. I produced a hybrid strain using the 1952 strain, and found the offspring to be fully viable and fertile. All males of this crossing had the red submarginal line above mentioned. In hybrid males the primary meiotic divisions were normal.

The geographical range of FIL was given by Clausen (1967: 26), who knows FIL very well from his investigations of the fish-fauna of SW Nigeria. It seems that FIL is restricted to a rather small geographical area of SW Nigeria, with populations up to Meko at the Nigeria-Dahomey border probably in the savannah. I have seen these individuals from the Meko district, and I do not agree with Clausen that they belong to FIL, although superficially they resemble FIL in color patterns, morphology and size. In nine specimens from the Meko area I found D.15–17, A.16–18, Sq-1. 30–32. All individuals had prominent H-scales. These individuals are referrable to GUL (D.15–18, A.17–19, Sq-1.29–35, with prominent H-scales) except for size, as the Meko individuals all measured less than 30 mm in standard length. The western and northern limits of the range of

Adult female FIL (above) and adult female ARN (below). Note the shape of the caudal fin of both.

FIL appear to be Badagri and Abeokuta of SW Nigeria. I have seen preserved material referrable to FIL from these localities and other western localities. The eastern limit of the range is unknown. FIL has not been reported east of Ijebu-Ode, and ARN has not been reported west of the Niger Delta.

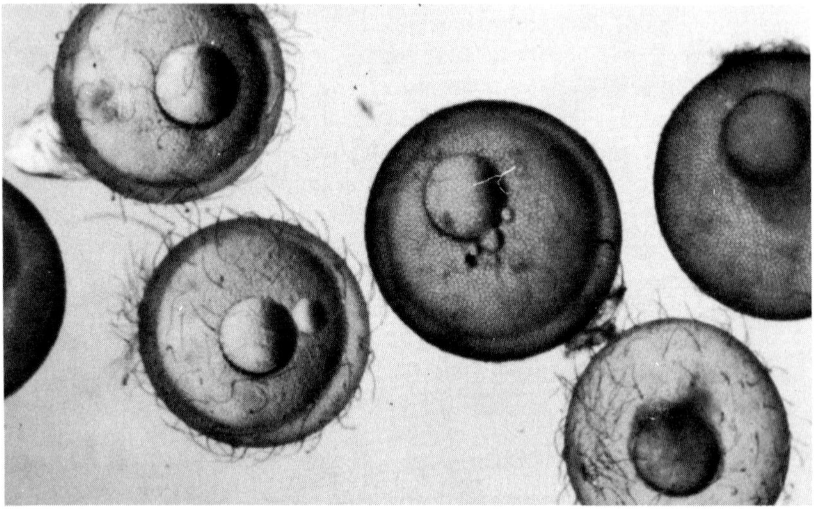

Eggs from FIL (large, without evenly distributed filaments) and from ARN (smaller). The three eggs in the center are alive, the others not.

ARN and FIL are closely related and seem to replace each other in their respective geographical areas. As has already been mentioned under ARN, their hybrids are rather feeble and sterile in both sexes. These two forms are difficult to separate by their meristic characters, but they differ markedly by egg type, by color patterns and by karyotypes.

The 1952 strain and the 1965 strain both develop 18 chromosomes and the karyotype is characterized by two very large, distinctly acrocentric elements, measuring about 3.5 times the mean length of the basic chromosome and more than 2 times the length of the largest acrocentric element of the basic complement. These two giant acrocentrics appear to be quite similar in morphology and size, but in meiotic divisions they always produce two big ring-bivalents, and no tendency for the development of tetravalents occur. One of these huge acrocentrics might have originated by aneuploidy and calculations based on the lengths of the individual chromosomes support this idea. There are also about five chromosomes exceeding the maxi-

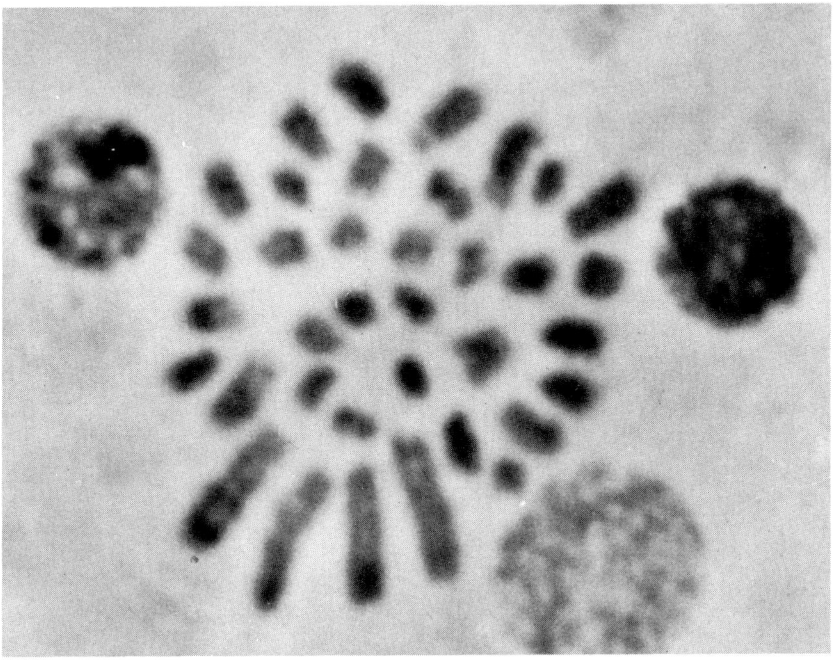

Mitosis of FIL. 2n = 36.

mum length of the (calculated) basic chromosomes, and two of these are metacentrics, the other three being strictly acrocentrics. In the hybrids, all 18 bivalents are present and they are symmetrical. No univalents have been observed.

Ctenoidy is strongly developed in some males and is almost absent in other males of similar size and age. The ctenoidy is mainly restricted to the median rows of scales, but in some males, ctenoid spines also are prominent in anal fin rays. The spines are often covered with tissues, producing long filaments. The frontal scalation is normal (G-type, no H-scales). The supraorbital lateral line groove usually is separated from the postorbital depression, but in a few individuals these two depressions fuse, almost the same as *Nothobranchius*.

Although the color photo of the male FIL does not show any yellow or orange pigment, such pigment sometimes occurs in young males, and the lower part of the sides and lower fins might be rather orange in these males, but not to the extent seen in young ARN males. In the hybrid strain (1952–1965) some males have distinct orange markings in the distant part of the anal.

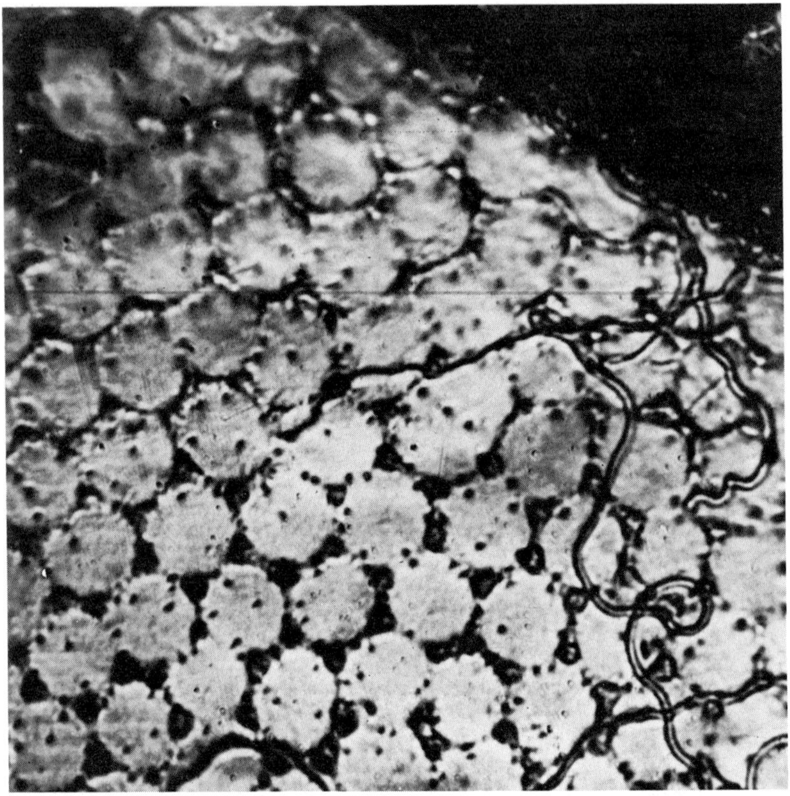

Pattern of egg membrane in FIL.

A submarginal red band very rarely occurs in the upper part of the caudal in males of FIL, whereas in ARN this line is always present and very close to the fin edge. Occasionally this band occurs in a male from a large brood. but I have not been able, on the basis of such individuals, to establish a strain breeding true to this character. The development of elongate anal fin rays (filaments) is highly variable, but in the 1965 Badagri strain this character usually is well developed.

FIL is an annual rivuline species and the eggs do not adhere to the mud. The hexagonal egg membrane pattern is very distinct. Eggs which become covered by fine mud remain transparent and without visible development for more than half a year, but after that time they usually are unable to develop an embryo.

Further information on FIL is given by Scheel (1963: 162–166, 277–280). Individuals of FIL have been called "*Aphyosemion ruwenzori*."

FLA: *Aplocheilichthys flavipinnis* Meinken 1932 **(Valid sp.)**

In 1929 Griem of Hamburg collected aquarium fishes in West Africa. Near Lagos of SW Nigeria, he collected some very small cyprinodonts, which were described as "*Panchax species nov.*, the Dwarf-Panchax" by Meinken (1929: 385–89). Meinken said that this form previously had been imported by the "Platy-Tischrunde," without information on its origin.

Meinken (1932: 54–56) changed his opinion on the generic status of this strain and described it as *Aplocheilichthys flavipinnis*, thus placing this form in *Procatopodinae* close to *A. normani* Ahl. Meinken did not explain the morphological reasons for which he removed this form from *Panchax* (= *Epiplatys* or *Aphyosemion s.s.*).

Since that time, there seems to have been no further reports of this dwarf species. The name FLA, however, has frequently been used for species of *Aplocheilichthys*, with (or without) yellow fins, but up to 1967 none of these strains represented Meinken's FLA.

In 1965, the author studied some unsorted samples of freshwater fishes which had been caught in southern Nigeria by various zoologists. When the large specimens had been sorted out and had been identified, some minute individuals usually remained in samples from swamp-like biotopes. Some of these small individuals were juveniles of *A. macrophthalmus* Meinken, but 38 individuals could not be identified. These individuals resembled rivulins

The types for FLA redrawn after Meinken's drawing of 1929.

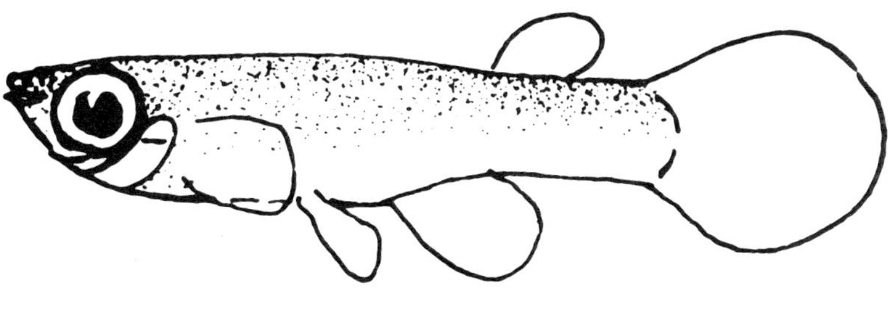

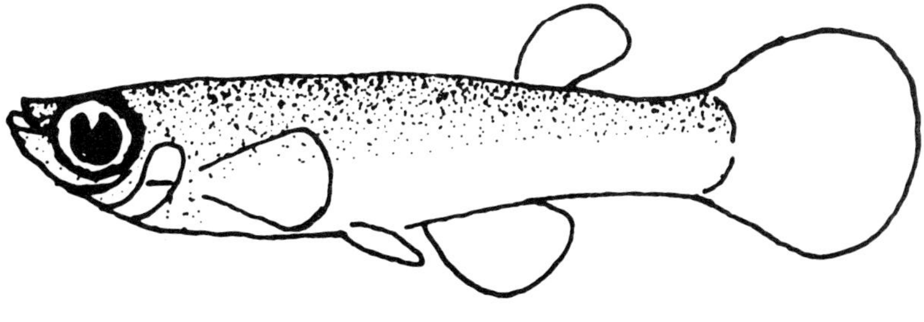

by some morphological traits and procatopodines by other such traits. They measured less than 15 mm standard length (without caudal) and the belly of a 14 mm female contained 26 almost fully developed eggs, so the individuals were not juveniles of some larger rivuline species. The individuals originated from Ibefun, south of Ijebu-Ode, from Ughelli, east of Warri of the Niger Delta, from Aduna, about 40 miles south of Isela-Ubu, from Auna between Ijebu-Ode and Shagamu, or from Aiyetoro or Badagri of SW Nigeria.

I place this minute form in *Rivulinae* because of the very narrow preorbital space, measuring about 10–15% of the diameter of the eye, of the rather low set pectorals, of the rounded rivuline-like scales, and of the red scale edges on the back of some males.

Head of a preserved FLA from Aduna, west of Asaba, Niger delta area. Notice the narrow pre-orbital space, the very big eye and the rather low set pectorals. The lateral line organs around the pre-opercle are situated in open grooves.

The procatopodine characters are: the social behavior (shoaling), the thin median black line on the sides, the corresponding line above the root of anal and the extremely large eye.

The dorsal fin is inserted behind the anal fin, the anteriormost fin ray of the former being behind the hindmost ray of the latter, or above this ray.

Preserved FLA from Aiyetoro, southwestern Nigeria.

The juxtaposition of these fins is quite unusual in *Rivulinae* and *Procatopodinae*. The caudal fin is of medium length for a rivuline species (30% of standard length) and lanceolate.

There is no "pouch" formed by large scales round the anus and anteriormost anal fin rays (often present in procatopodine females). The frontal scalation is normal for African rivulins (G-type), and the A-scale is covered distantly by the D-scales and not resting on the D-scales (as in many procatopodine species). The lateral line pattern of the head also resembles that of rivulins. The neuromasts are, however, situated in open, rather shallow grooves, as in *Nothobranchius*, ANN and HOM, except for the preorbital organs which are housed in a canal with two pores. The supraorbital and the postorbital grooves (or depressions) are not confluent and are distinctly separated by a row of scales. The pattern of the dorsal part of the snout is "open," corresponding to *Aphyosemion*.

In males from the Niger delta which have been preserved in alcohol, the scales above the median line are edged with red particularly anteriorly. There is also red pigment anteriorly on the throat in some males and in one male there is a red blotch in front of the root of caudal. The throat pattern is simple and corresponds to that of ANN and DUB, except for the red pigments in some males.

Live individuals were described by Meinken (1929 and 1933) and by Kaden (1967: 280). The sides are grey, with a faint greenish metallic brilliance, particularly on gill covers. The unpaired fins are yellow to orange, particularly in the male, which develop light-blue fin edges. This color description corresponds to males from SW Nigeria.

Meinken described the eggs as rather large, whereas Kaden for his strain, said that they were extremely small, considerably smaller than those of ANN.

FLA cannot be placed in any known genus of African cyprinodonts. It might represent an ancient form more closely related to ANN than to any further known species, but also quite different from ANN by important morphological characters. FLA seems to be closer to *Aphyosemion* than to *Epiplatys* and the name "*Aphyosemion*" *flavipinnis* (Meinken) might be used until a separate genus is created for this species.

FLV: *Panchax flavus* Ahl 1924

Ahl (1924: 311–12) had a single specimen which he thought originated from the Niger area. This individual was kept by the German aquarist Matte of Berlin, who is not known to be a collector of live fishes in Africa, so probably Matte had the information on the locality from another dealer or from a professional collector.

Ahl placed FLV closest to EXI and ELE. The description of FLV apparently has not been revised by any zoologist and it is so incomplete that it is impossible to determine to what genus FLV belongs. The meristic data published are very close to the average figures of the subfamily, and FLV cannot be separated from 69 of the other nominal species of rivulins from Atlantic Africa. FLV probably does not represent a valid species, but the type-individual should be seen if it is still in Berlin.

GAM: *Fundulus gambiensis* Svensson 1934 **(Valid sp.)**

Svensson (1934) had a single specimen, a female, from a rain water swamp on MacCarthy Island, situated in the Gambia River.

Johnels (1954: 327–411) revisited this area in 1950 and caught numerous individuals of GAM, but only five of these were at his disposal for his redescription of GAM. No further information concerning this most interesting species has been published.

Fins are rounded in both sexes, except for the caudal, which is more or less trilobate in the male. Dorsal is similar to anal and inserted slightly in advance of the latter.

The sides of the males are a light blue, with a greenish lustre, and sparsely spotted with dark brown. The head is dark brown with blue or bluish-green spots. The gill covers are yellowish-red. All fins are orange, the dorsal, anal and caudal in addition distinctly edged with deep red. The anal is dark at the base; the dorsal spotted along the lower half.

This species was exclusively found in the rainwater swamps, not influenced by the water of the river. These swamps contained GAM and *Protopterus* only. The temperature ranged between 24.5° and 31.0°C.

Johnels suggested, that GAM reproduced as an annual fish in isolated swamps, which dry up completely during the dry season.

Johnels, and Daget (1960: 615), placed GAM in *Nothobranchius* and considered GAM to be closest to WAL (see also THI). The present author prefers to place GAM in *Fundulosoma* because of the few brown spots on the sides, the trilobate caudal, the red fin edges and the geographical distribution.

It seems that THI and GAM cannot be separated by meristic characters but the color patterns are quite different. Importations of GAM are highly desirable.

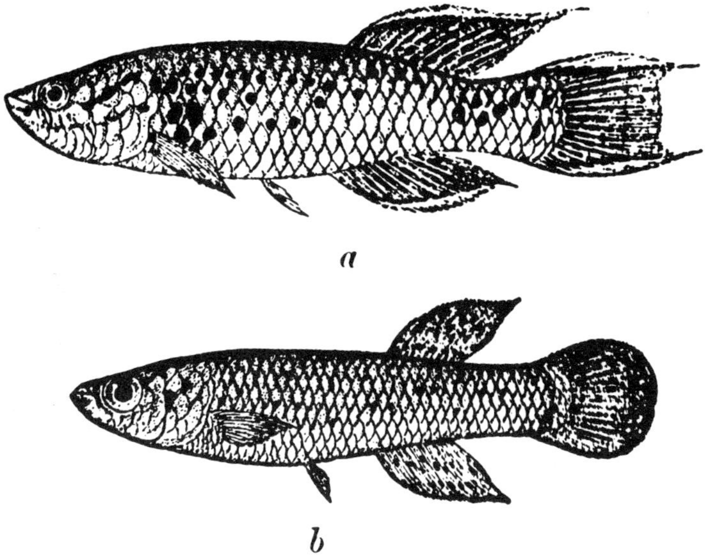

Fundulus gardneri.

a. ♂, type. *b.* ♀, type of *H. brucii* ($\frac{2}{1}$).

The types of GAR as figured by Boulenger 1915.

GAR: ***Fundulus gardneri*** **Boulenger 1911** **(Valid sp.)**

Boulenger (1911: 261) had three specimens caught by Capt. R. D. Gardner at Okwoga, the headwaters of the Cross River, in East Nigeria. In 1915 (:26) he concluded that the types for BRU belong to GAR and considered BRU to represent a synonym for GAR. (See BRU.) He also listed five individuals of GAR from the Bendi district of the Cross drainage; a young specimen from a swamp between Agwarba and Awka of the Lower Niger drainage; and a halfgrown individual from Old Calabar, which was presented by Arnold. No further reports of GAR from nature have been published, but the GAR name has frequently been used for individuals which apparently do not belong to this species but to the ARN superspecies. (See ARN and FIL.) On the other hand, there are numerous reports in literature concerning individuals of GAR, but these individuals and strains were reported under the CAL and the NIG names.

Adult male GAR from Eymojok of west Cameroon. 1966.

Large male GAR from Port Harcourt of east Nigeria. 1961.

Adult male of GAR from Akure. Yellow phenotype. W. Foersch photo.

Pair of GAR from Akure. Blue phenotype in male.

The first incorrect use of the GAR name was made by Rachow (1920: 294), who claimed that some aquarium fishes imported in 1908 from the Niger Delta area, were referrable to GAR, although these individuals probably belonged to BIV (see Scheel 1965: 273).

The confusion between the GAR and the ARN species and superspecies appears to begin with some identification of aquarium fishes prepared by Boulenger. Arnold (1925: 117) said that before the war Boulenger identified one preserved individual, presented by Arnold, as GAR without hesitation. He also said that before the war Boulenger also identified a pair of aquarium fishes to belong to GAR, although Arnold himself considered these individuals to represent a color variety of ARN. In this connection he referred to Dieterich's "Steel-Blue *Fundulus*" (of 1915, see FIL). Arnold probably is right, because the strain which Dieterich had belongs to the ARN superspecies. Meinken (1930: 553) however, said that the "Steel-Blue *Fundulus*" of Dieterich 1915 and of Krüger 1913 should be named GAR, and that this form also had been called "*Fundulus* aus Togo" or ARN *var. blue.*

Myers (1954: 138) said that he had had many letters concerning newly imported West African cyprinodonts. These letters dealt with a form called GAR. He also said that, in his opinion, no aquarium fishes had ever been imported from the type locality of GAR, nor had any aquarium strains been in full agreement with Boulenger's description in color. Myers referred to stocks which were imported into Germany during 1930 and considered these fishes to be extremely similar to the types for GAR in most anatomical features and he thought that they probably were referrable to GAR. The photo in Innes book (see also FIL) to which most of the readers refer, and which is labelled GAR, shows this form. This stock was also pictured by Arnold and Ahl (1936) and in some of Meinken's papers. Myers recommended that this form should be considered a race or a subspecies different from that to which the types for GAR belonged, and he added that the form described by Meinken as FIL might represent another subspecies of GAR.

The present author (1963: 162–166, 277–80) also considered such forms as ARN, FIL, GAR, SPU and WAL to represent a group of closely related forms and divided this group into two subgroups with ARN and FIL in one subgroup and the other species in the other subgroup. This separation was based on differences in the number of dorsal fin rays and of scales in a lateral series.

Further studies of these species, in addition to NIG, disclosed that the two subgroups were well separated. In all 68 individuals of the ARN superspecies (ARN and FIL) studied, the anteriormost anal fin ray is clearly behind the corresponding dorsal fin ray and usually below the 4th to 5th dorsal fin ray, and there are 24–28 scales in a longitudinal series.

In all 247 individuals of the GAR superspecies (GAR, NIG, SPU and WAL, with NIG = GAR and SPU = WAL) the anteriormost dorsal fin ray is clearly behind the corresponding anal fin ray and usually above the 4th to 5th anal fin ray. There are 29 to 34 scales in a longitudinal series.

Also, the eggs are different. Those of the GAR superspecies are sticky, whereas those of the ARN superspecies are not.

Meinken (1953: 86) identified some rivuline individuals caught by Sheljuzhko near Abidjan, Ivory Coast as GAR. The present author (1963: 166) disagreed with Meinken and referred these individuals to SPU (=WAL).

In 1957 Birket-Smith collected some live rivuline individuals near Akure of western Nigeria and brought this material to Denmark, where it was taken over by the author. This strain was first identified as CAL, but in 1963, Clausen described and named this form under NIG, as a new species.

Eggs from this strain were distributed to many friends all over the world, and this strain soon became a rather popular aquarium fish, which has often been described in aquarium magazines under the CAL and the NIG names. (See NIG.)

In 1961 Ulf Hannerz sent me two males of a similar form which he had taken in the Wokocha River, near Port Harcourt (see ARN) of East Nigeria. These males differed from the Akure males by the number of red spots on the sides. There were three times as many red spots in the Port Harcourt males as in the Akure males. Hybrids developed an intermediate number of red spots. These hybrids were very feeble and some could not be raised to sexual maturity. Both sexes were almost functionally sterile and the few individuals which could be raised from the F_1 **x** F_1 spawnings, also were very feeble and just as sterile as the F_1 individuals. The F_3 individuals were very feeble and almost inviable.

In 1962 I received additional live material of fishes referrable to NIG from Clausen who caught this material near Owo of western Nigeria. For further information about these strains see Scheel (1964: 510–15, 519, 550, 567–568, 596, 610–12).

In 1966 the author caught similar individuals in many localities within the Mamfe plains of West Cameroon and brought live individuals from a locality near Eyomojok, west of Mamfe to Denmark for further study.

The study of GAR and NIG also included many preserved individuals (including the types of NIG) from many localities of Nigeria and Cameroon. The individual variation of meristic characters is considerable, but this variation cannot be used to produce any grouping of different populations into different species or subspecies. D.12–16, A.14–18, Sq-1.29–34 was found in more than 247 individuals referrable to GAR (including NIG).

Adult male showing abnormal color pattern. This male belongs to the second hybrid generation of the Port Harcourt/Akure crossing.

Adult male of GAR from the Eymojok/Owo crossing. This hybrid has 20 + 18 chromosomes.

Adult yellow male GAR from crossing a Port Harcourt male to an Akure female.

Adult blue male GAR from crossing a Port Harcourt male to an Akure female.

The karyotypes of two Nigerian strains (Akure: n = 20, 2n = 40 and Owo: n = 18, 2n = 36) are rather similar, and are characterized by a very large sub-metacentric element. In the Owo karyotype, there is also one very long acrocentric element, not much shorter than the large submetacentric and there are also three smaller metacentric elements. The Akure karyotype does not have the large acrocentric element above mentioned. There seems to be two smaller submetacentrics and two metacentrics in this karyotype. Hybrids from crossing of these two strains have not been produced so far.

The karyotype of the Cross River strain (Eyomojok) has twenty haploid chromosomes. There is no exceptionally long submetacentric element. The largest acrocentric seems to correspond to the largest in the Owo

Primary meiotic division of GAR from Eymojok, west Cameroon. n = 20.

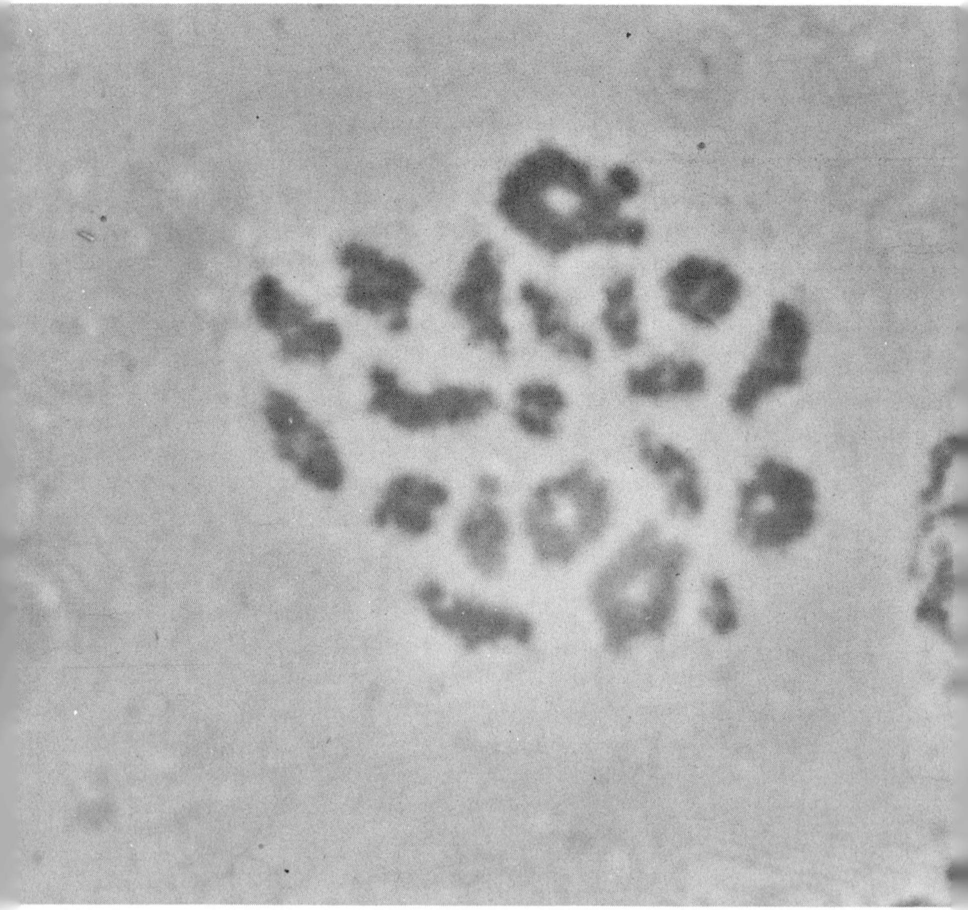

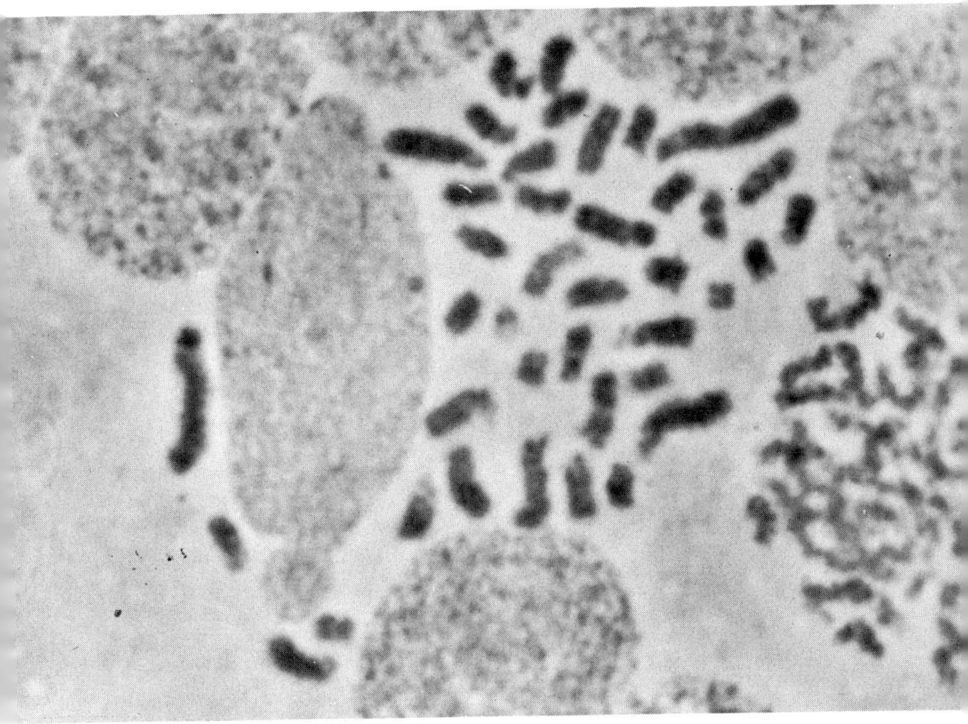

Mitosis of GAR male from Eymojok, west Cameroon. 2n = 40.

karyotype. There seems to be two metacentric elements only. The Eyomojok and the Owo strains have been crossed. The primary meiotic divisions of hybrid males have one large and one somewhat smaller trivalent. Univalents and other abnormalities were rare, and there are many very viable spermatozoa. Also, both sexes of these hybrid strains are almost fully viable and fertile and the F_1 **x** F_1 crossings produced similar individuals. Some F_2 individuals, however, died in the eggs.

Ctenoidy occurs in many males, but this character was found to be highly variable, even within the single population in nature and within aquarium raised broods. Ctenoid spines usually occur on median and lower rows of scales along the sides, in particular above the root of the anal fin. Except for the pectorals, ctenoid spines seems to be absent on fin rays.

The frontal scalation was variable, although all individuals studied developed the usual G-pattern. In some populations, almost all individuals had prominent H-scales situated on or below the G-scales, whereas in other populations H-scales appear to be very rare. Individuals from the Jos-Plateau (type-area for NIG) and from Akure had many H-scales, whereas

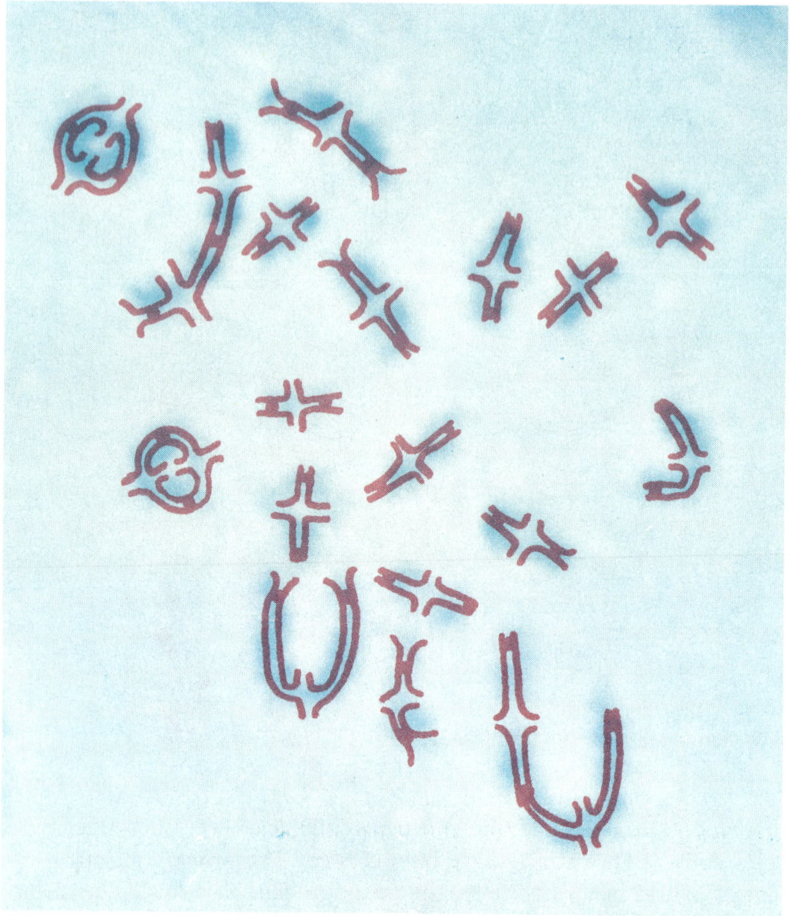

Primary meiotic division in hybrid male GAR-Owo/Eyomojok. 18 separate bodies are visible. Two of these are trivalents containing one (metacentric) chromosome from the Owo strain and two (acrocentric) chromosomes from the Eumojok strain. This male was quite fertile.

in individuals from Ondo (very near Akure, but within a different drainage) H-scales were very rare. There are two, rarely three rows of scales below the center of the eye.

The color patterns of GAR males from different populations are not very different and no separation of GAR into sub-units seems to be possible on the basis of color patterns. Males have from 30 to 90 (more or less) red spots on the sides. The spots are rather round, at least anteriorly. The number of red spots seems to be rather constant within the local population in

Adult hybrid male GAR/CIN. The GAR male belonged to the Port Harcourt population. Hybrids with the Eymojok population are similar, but more variable in color.

Adult hybrid male from GAR/OBS. The GAR male used belonged to the Eymojok population.

some geographical areas (Nigeria), whereas in other areas (Cameroon) this character is highly variable.

The red pattern of the throat of males also is variable, ranging from an almost uniform faint red pigmentation (Jos Plateau) to a well developed and marked pattern of red lines.

Most populations develop two rather different phenotypes in males. In the yellow males, the unpaired fins and usually the ventrals are broadly edged with gold and there is usually an unbroken broad red line in the median part of the anal (Nigerian populations, not in all Cameroon populations). This line is rather close to the root of the fin in other Cameroon populations, whereas in other populations or individual males from a certain population, there is no red line at all. In blue males, the younger males may develop bright lemon fin edges (these males have the recessive gene for yellow) but in older males, this color changes into a bluish white, although blotches of yellow or even gold may occur in Cameroon males. Nigerian "blue" males never seem to develop an unbroken red line in the anal, instead there is a row of large red spots. This row is median in Nigerian material and closer to the root and often doubled in Cameroon blue males. Males from Cameroon and the Nigerian part of the Cross drainage may develop a deep black narrow edge in their anal fin.

Adult male GAR from Eymojok, west Cameroon. Blue phenotype. The anal fin has an incomplete dark border.

Adult hybrid male GAR-Port Harcourt/CIN. Notice pattern of the sides and of caudal fin.

In other such males, the edge has a number of black spots or there are no black markings at all.

The lyre-pattern of the caudal fin is similar in blue and in yellow males. In Nigerian males the lyre-pattern is strictly submarginal, and the golden or bluish-white upper and lower edges are broad. In Cameroon males, the lyre-pattern is closer to the upper and lower edges and rather often the red lines of the lyre are not complete.

Preserved and live specimens from the Nigerian and Cameroon parts of the Cross drainage correspond well to Boulenger's GAR, and individuals from Ogoja, not far from the type locality for GAR, are very close or quite similar to GAR. It seems that, generally, the populations of GAR from the Cross drainage are much more variable than populations from the Niger drainage and the coastal rivers of SW Nigeria.

Based on the study of live and preserved material NIG is considered to represent a synonym for GAR. The reproductive isolation observed between the two populations of nominal NIG (Port Harcourt and Akure) indicates that GAR is a superspecies of a nature similar to the BIV superspecies. The closest known relative of the GAR superspecies is WAL from the humid forests of SW Ghana and Ivory Coast. WAL seems to be so close to GAR, that it might be considered a component of the GAR superspecies.

The relationships between GAR and other groups of species in *Fundulopanchax* and in *Aphyosemion* are obscure. Generally, individuals of GAR

Young hybrid male GAR-Owo/WAL and adult hybrid female WAL/GAR-Akure. Notice anal fin pattern of male.

produce fully viable hybrids with individuals of the subgenus first mentioned and rather often with individuals of the subgenus last mentioned. These findings indicate that GAR is a rather generalized form within the 20-chromosome group of *Aphyosemion.* There is, however, no evidence that GAR have given rise to any of the species of this genus living south of the Cameroon Mountain Chain, at least not directly.

GAR is very easy to keep and to breed. It reproduces well in hard, alkaline water. GAR is a semi-annual species, and eggs usually are handled as eggs from annual rivulins.

GER: *Aphyosemion guineense geryi* Lambert 1958 **(Valid sp.)**

Lambert (1958: 52–53) had twenty specimens from the Conakry-Dubreska district of Lower Guinée. There seems to be no further records of this species under the GER name, but it is very likely, that several reports of ROL and LIB in Lower Guinée concern GER, and not these rather similar species.

In 1961 Lambert informed me that he considered GER to represent a subspecies of ROL.

In 1962 Roloff caught a single individual of a ROL-like form in Sierra Leone. He kindly offered me that individual, a female. The individual resembled ROL, but as Roloff indicated, it probably did not belong to ROL but to a species unknown to us. It differed from ROL by the more slender body, by the broad, irregular, dark brown longitudinal band and by behavior. I tried in vain to cross this female to males of different strains of ROL and of BER. No sexual response was observed. The males ignored the GER female, and on the other hand this female ignored the males. The female usually preferred to stay among plants and was seldom seen in open water.

In 1965 Roloff sent me additional live and preserved material from Sierra Leone. The live material originated from Kenema, of eastern Sierra Leone. The females of this strain differed from the 1962-females by the absence of the broad, dark, longitudinal band, but they corresponded in other characters, including behavior. Roloff also supplied preserved material from Rotamp of the Barmoi district of western Sierra Leone and Foersch in 1967 sent me preserved individuals from Conakry, Lower Guinée.

The males of the Kenema strain have 20 haploid chromosomes. The karyotype seems to contain no metacentric elements.

There are two large double-sized acrocentrics which seem to be similar, or almost similar, and there are also a number of other chromosomes larger than the (estimated) largest acrocentrics of the basic complement. The two largest acrocentrics measure about 2.2 times the mean length of the basic chromosomes.

The frontal scalation is normal (G-type, no H-scales). In the male some of the scales develop the type of ctenoidy which I have called the "wrinkled

Adult male GER from the Kenema district of western Sierra Leone.

type" but no fin papillae have been found. The pattern of lateral line organs of the dorsal part of the snout is not as stable as in other nominal species of *Roloffia*, and some individuals are rather intermediate between *Aphyosemion* and *Roloffia*, whereas others are distinctly of the *Roloffia* type. (See MAE.)

Adult female GER from Sierra Leone. Roloff 1963.

The color pattern of the throat seems to differ markedly and constantly from those of ROL, LIB, BER and other species of the LIB group or superspecies. In the GER male there is a rather prominent pattern of red lines on the throat, resembling the usual pattern in *Aphyosemion* and quite different from the general pattern of *Roloffia*, where red lines never occur and where red pigments on the throat almost are absent. In the GER male, the throat of the active males develop a dark color, as in all other species of *Roloffia*, but this pattern also occurs in a number of species of *Epiplatys* and in CIN, AHL and AUS within *Aphyosemion*. In thc female GER, no rounded dark marking has been observed on the throat. In the female GER there is often a dark vertical blotch above the root of the pectorals, resembling the wound-like marking in males of *Aphyosemion* and *Roloffia*. So, generally speaking, the female GER resembles a male of *Roloffia*, and these differences in throat patterns might be one of the reasons for the absence of sexual response, when individuals of GER and other species of *Roloffia* are put together for crossing.

The marked, dark, zig-zag formed band along the side of the GER female is not present in females from some populations. It seems that generally individuals of GER are more slender than individuals of the LIB group. In the males from Kenema the lower two-thirds of the anal sometimes developed a uniform red color. The caudal fin is edged above and below with gold in all males I have seen.

GER probably has a rather wide range in West Africa. The individuals referred to MAE in this book, seem to be rather close to GER. These individuals originated from N'Zérégoré, High Guinée.

Although GER resembles the species of the LIB group superficially, I hesitate to incorporate GER in this group because of the difference in throat pattern, lateral line pattern, chromosome number, karyotype, and in behavior. GER seems to represent a form, which is closer to some groups in *Aphyosemion* (the CAL superspecies) than any species of nominal *Roloffia*. Individuals of GER have not been used in crossings to this group of *Aphyosemion* so far. GER may represent an important intergradient species between *Roloffia* and *Aphyosemion.*

GRA: *Haplochilus grahami* Boulenger 1911

Boulenger (1911: 267) had several specimens from swamps near Lagos, Nigeria.

GRA was introduced as an aquarium fish in 1905. It was represented in the very first shipment of aquarium fishes from Atlantic Africa, together with individuals of SEX, SJO and ARN. Arnold (1906: (26) and (27)) identified this strain of GRA as SPI. The aquarium society *Wasserrose* of Dresden (1906: 90) reported that their member Thumm had already bred this species. Köhler (1906: 398) said that the individuals of this species made about 12% of the (SEX and GRA) importations from the Niger Delta.

The type for GRA as figured by Boulenger 1915.

Haplochilus grahami.

♂, type.

Soon after the first importations of GRA and SEX, some discussion concerning the relationships of these two very similar species arose in Germany. Some aquarists considered SEX and GRA (still called SPI) to be conspecific, but Zimmermann (1908: 202), who crossed individuals of SEX and GRA, showed that the hybrids were not viable, as they all died in the eggs. In 1962 the present author confirmed Zimmermann's finding and had quite similar results from crossing of the sympatric populations of SEX and GRA from the western part of the Niger Delta.

Thumm (1909: 241) doubted that SPI was the correct name for the aquarium strains and found that the name INF was better, but his idea was not accepted. GRA was bred successfully by several German aquarists, but it seems that the different strains became extinct before 1911.

Arnold sent two preserved specimens to Boulenger for identification and Boulenger attached the LON name to these individuals. I have seen both and they belong to the NIM phenotype in GRA. (See NIM.)

In 1920 (:290) Rachow said that the strains from 1905, which had been called SPI, were not this species, but LON, and that Boulenger had made that identification. A photo (Unger's), which was published in Rachow's article, shows two males of GRA, labelled as LON. In 1924 (:677), however,

Adult male GRA from the Mboumboula River of east Cameroon.

Adult female GRA from swamps near Lagos, west Nigeria.

Rachow identified quite similar individuals of a new aquarium strain correctly as GRA, but it seems, that he still considered LON to be very similar. The 1924 strain originated from the Niger Delta area.

Pellegrin (1925: 104) reported GRA from Ouesso on the Sangha River of the formerly French Congo. These individuals are probably not GRA for zoogeographical reasons. In 1934 (: 118) he described DEF as a subspecies of GRA. DEF = SPI. (See DEF.)

Mayer of Hamburg had individuals of GRA in 1933, and these individuals, now in the British Museum, were identified as MAC, although Mayer said that they originated from Lagos. They correspond to GRA and not to MAC as I have seen these.

I have seen many preserved specimens of GRA from southern Nigeria, ranging from Badagri eastwards, in coastal swamps. Almost all these specimens are juveniles which occur in samples containing individuals of SEX of all sizes. It seems that adult individuals of GRA are not found in these localities, but that the juveniles of GRA "shoal" with the juveniles of SEX. Juveniles of these two forms are much more similar than are the adults. From a single habitat there are adults (and no juveniles) of GRA. This habitat is a swamp-forest.

The six live individuals of GRA which I have studied all originated from collections of juveniles of SEX, and they were not recognized as different from SEX by the collectors. I had two females from swamps near Lagos, Nigeria, one female from Benin City of the Western part of the Niger Delta and three males from the Mboumboula River, north of Douala of East Cameroon.

Although live individuals of GRA and SEX are very similar, they can always be separated quickly by the trained observer. Individuals of GRA have very brilliant grass-green eyes and individuals of SEX do not have this particular marking.

In preserved material it is easy to separate GRA and SEX by the pattern of lateral line organs on the dorsal part of the snout. In individuals of GRA, this pattern is the same as *Roloffia*, whereas in SEX the pattern resembles that of *Aphysemion*. LON which also develops the "*Aphyosemion*-pattern" is also sympatric with SEX and GRA within the Niger Delta.

The color pattern is useful for identification. Females of GRA develop a thin, but distinct black line just below the root of the anal fin and males develop a marbled dark pattern just above the root of this fin. These characteristic markings are absent in SEX and in LON. The dark crossbars are more narrow, more temporary, more irregular and more oblique in GRA than in SEX.

Mitosis of GRA from the Mboumboula River, east Cameroon. 2n = 46.

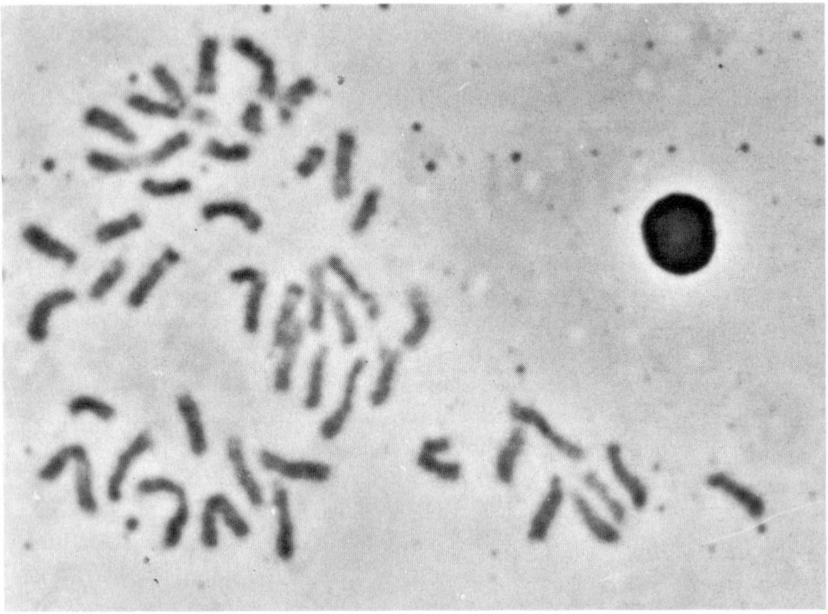

Adult female GRA from swamps near Lagos, west Nigeria.

Males of GRA from the Lagos area have two distinct bright red lines across the throat, corresponding to the usual *Epiplatys* pattern (which usually is dark brown or black). This particular red pattern also occurs in some males from the Niger Delta area, but in other males (populations?) from this area the red lines are completely absent. The populations last mentioned are NIM, which may or may not represent a distinct species within the GRA-ORN-MAC group or superspecies.

The dark crossbars on sides usually disappear completely in adult individuals of both sexes, these individuals are then similar to individuals of ORN and MAC. The black bars reappear, however, when the individuals are frightened.

A male from East Cameroon had 23 haploid chromosomes in meiotic divisions, whereas in mitotic spermatogonial divisions it appeared that there were five large metacentrics. Calculations on the lengths of these and the basic chromosomes indicated that GRA originated from a species having 25 haploid chromosomes. The frontal scalation is of the usual type (G-type, no H-scales), the hemoglobin pattern is of the common rivuline type (four lines) similar to that of ANN, DAG and MAC.

GRA does not differ from ORN and MAC by meristic characters, but the color patterns are different. It differs from individuals of MAC by the development of dark crossbars in adults and juveniles, and from ORN by the development of these dark markings in adults.

GRA probably is closely related to DAG. Hybrids are rather viable, but they are sterile female-like "intersexes." It seems to replace DAG in coastal swamps of the forest, east of the Dahomey Gap.

The metallic brilliance (blue) of the GRA male is very faint, and sometimes almost invisible. Fins are yellowish. This species is characterized by the very round red spots on the sides and on the back.

GUE: ***Fundulus guentheri*** **Pfeffer 1893** **(Distinct sp.)**

Pfeffer (1893: 167) had many specimens collected in Zanzibar by F. Stuhlmann in 1888 and in 1889. He considered these individuals to be conspecific to those reported from Zanzibar by Günther and Playfair (1867: 118) and which were referred by these authors to ORT. He also considered all these individuals to differ from Peter's ORT from Mozambique. He separated GUE from ORT on the basis of the pattern of teeth and by a few other details in morphology. Pfeffer (1895: 47) supplied further details on GUE and described the colors of the live male. The sides are brilliant green with a perfect red reticulation. Dorsal with rows of red spots and with a narrow white edge (?) and a submarginal (?) dark band, the dark color extending deep into the fin in some males. The anal is similar. The caudal is a uniform red with a dark distant edge or with two dark spots on the distant edge. Females are grey without dark markings. Pfeffer noticed that there were (ctenoid) spines on the fin rays in the anal and he also noted that the rays of this fin extended beyond the membrane of the fin. He figured GUE with 2–3 rows of scales on the preopercle below the eyes. He also referred some individuals from Longo Bay (see MEP) to GUE.

Hilgendorf (1905: 416) reported other specimens from Zanzibar and said that these individuals corresponded to Pfeffer's description, but not to his figure in which the dorsal fin is not correctly inserted, and by having less scales (3 and not 4 rows of scales on hindmost part of preopercle). See also MEP.

Boulenger (1915: 35–36) referred Günther's ORT from Zanzibar to GUE and reported other material from Mombassa of Kenya and from the Pangani River of Tanganyika.

There are two specimens of GUE (labelled ORT) in the Copenhagen Museum. These were presented by the British Museum in 1867 and they probably belong to the samples from Zanzibar which were reported by Günther and Playfair. They are both labelled Zanzibar. Both are males, the

The type for GUL as figured by Pfeffer 1893.

Adult male GUE from the aquarium trade. n = 19.

largest measuring 57 mm standard length, 65 mm total length. Fins, teeth and the few scales left are very dark, sometimes almost jet-black. There are three rows of scales below the center of eyes D.15–16, A.15–16, ctenoid anal fin papillae are very prominent and filamentous. Teeth as described by Pfeffer. Red pigments have been lost.

The GUE name frequently has been used for aquarium strains of *Nothobranchius*. It is, however, highly doubtful whether most of these strains are Pfeffer's GUE.

It seems that three distinct species of *Nothobranchius* have been known under the GUE name among aquarists. One of these is a small form with a single row of scales below the center of the eye (on the preopercle), with a broad black marginal band in distant part of caudal and with a broken red reticulation posterior on the sides. Thin red vertical lines are produced. This

Preserved male GUE from Zanzibar. Presented to the Copenhagen Museum by the British Museum on the 23rd of July 1867.

Head of preserved GUE male from Zanzibar. Presented to the Copenhagen Museum, by the British Museum on the 23rd of July 1867.

species has 18 haploid chromosomes. I refer this form to PAL because it produces fully viable and fully fertile hybrids with the strain which I think represent Lönnberg's PAL.

The other two species are larger and correspond to the two Zanzibar males above mentioned in size. These two forms are also rather similar in general appearance and in colors. They differ by the pattern of teeth and by the number of rows of scales on the preopercle and also to some extent by the shape of the anterior part of the head.

One of these forms has a single row of scales below the center of the eye on the preopercle. There are two rows of larger, curved, pointed teeth and between these rows there are rows with smaller teeth. The anterior part of the head is deeper than in the form mentioned below. I refer this form to ORT. It seems to have 18 haploid chromosomes.

The third form has up to 4 rows of rather small scales below the eyes on the preopercle. The teeth-pattern correspond to that described by Pfeffer for his GUE. The anterior row of teeth has long (longer than in the form mentioned just above) curved and pointed teeth, behind this row there are rows of minute teeth but none of these seem to be enlarged. This form has 19 haploid chromosomes. I refer this aquarium strain to GUE. The anterior part of the head is somewhat pointed and different from that of ORT.

In *Nothobranchius* it seems that large species produce large eggs (above 1.2 mm) and small species develop small eggs (less than about 1.0 mm). Although my material is limited, it seems that GUE differs from ORT by the number of scales in a longitudinal series. My GUE individuals have 29–32 scales, whereas my ORT individuals have 26–28. The PAL form, called GUE, cannot be separated from ORT by meristic characters. The number of fin rays is similar in all three forms, D. 15–18, A. 16–19, and in all three forms the foremost dorsal fin ray is inserted in front of the corresponding anal fin ray. All three forms were kept in the aquarium in Europe during the late 1960's.

The GUE-like aquarium form reaches 46 mm standard length in males, the females are smaller. The red reticulation on the sides of males is almost perfect, but it is absent on the lower part of the head and the belly. The head is rather pointed for a species of *Nothobranchius*. Males have a red caudal fin posteriorly edged with black; females have no dark spots on the sides.

Ctenoidy is well developed in the males and there are many short spines on the sides and even on the head, where they are very numerous on the preopercle. Anal fin papillae are well developed and filamentous, they are also rather colorless (see ORT). The hemoglobin pattern is complicated and has six lines. The pattern seems to correspond to that of BIF. The karyotype cannot be reconstructed at present, but there seems to be one large metacentric element among the 19 haploid chromosomes. There are also double-sized acrocentrics and some submetacentrics, the long arm being about four to five times the length of the short arm.

Ctenoid spines covered by tissue on the anterior fin ray of the anal. GUE from Zanzibar 1867.

GUE was imported from Zanzibar in 1913. The first report was given by Arnold (1914: 6–8) who considered the strain to represent ORT. He said that the individuals were taken in pure freshwater in Zanzibar. Kramp (1914: 209–11) reproduced this strain and said that the male reached 9 cm, the female 7 cm. He published two photos by C. Conn showing the male and the female. Arnold's strain was identified by Boulenger (Arnold 1914: 213–14) as GUE, and Kramp's strain was identified also as GUE by

Male probably related to **GUE** from the aquarium trade.

Rachow (1914: 211–12). Arnold's and Kramp's descriptions of the color patterns of this strain of GUE correspond well to that of the present strain, here identified as GUE. Kramp observed that the male held the female, using the large dorsal fin, during spawnings. Pugh (1914: 639–41) noticed that eggs, which remained in the sand after the spawning, did not develop and had to be removed from the sand in order to develop an embryo. The 1913-strain did not survive the War.

Haas (1965) reported the experimental introduction of a strain of GUE from Zanzibar into temporary pools in India (Delhi) and in the British Solomon Islands (Honiara) for mosquito control.

Adult male GUI from Lago of eastern Sierra Leone.

GUI: ***Aphyosemion guineense*** **Daget 1954** **(Valid sp.)**

The first report of GUI was given by Norman (1932: 185), who identified specimens from Bagbwe and Sandaru of Sierra Leone as CAM. I have seen these individuals and they are all typical GUI. In 1934 (466) Pellegrin reported specimens from Banamanan, Kissidougou, High Guinée as CAM. I have also seen these, and as Daget (1954: 300–325) pointed out, they are GUI. Daget (1954: 300–325) had 102 individuals from various localities of the Upper Niger drainage (Dabola, Tinkisso, Banian, Simandou and Pellegrin's individuals). I have seen 55 of these specimens. They are considerably smaller than Pellegrin's and Norman's specimens, but they correspond to these in all other morphological details.

In all individuals which have undisturbed scales on the front, the frontal pattern develops the rare E-pattern.

The color pattern of males is simple. Usually there are no red pigments visible to the naked eye, but under the microscope some red pigment cells might be found near the root of the pectorals and just behind the eyes.

Lambert (1958: 51–52) reported 13 specimens of GUI from High Guinée. I have been informed by Lambert that the 16 specimens reported from Macenta should be referred to ROL. In that paper he also described GER as a subspecies of GUI. He has informed me that this was an error

The largest male of the types for GUI. No. 60.475 in the Paris Museum. Chutes du Tinkisso, Dabola. 20–2–1952.

and that these individuals should be considered to represent a new subspecies of ROL. See GER.

In 1962 Roloff caught three individuals of GUI in Sierra Leone. These were females. He lost one and sent the remaining two to me for further study. These individuals developed the E-frontal pattern. In 1963 Schiötz caught many GUI individuals in the Loma Mountains of northern Sierra Leone and sent a lot of live individuals to me. Unfortunately, they died before they reached Copenhagen. All six corresponded well to GUI, and these and many previously preserved specimens showed the E-frontal scalation pattern.

In 1965 Roloff again collected rivuline species in Sierra Leone. Near Lago and Panguma of eastern Sierra Leone he caught many specimens of GUI. He kindly offered me a breeding pair of the Lago strain, and on the basis of this pair I raised an aquarium strain. The males from the Lago population correspond well to other males of GUI, except for the red pattern. There is a faint, but quite distinct, red reticulation on the sides of these males.

In the aquarium, GUI is a rather peaceful species in spite of its large size and robust appearance. Although it resembles the robust and large species of *Fundulopanchax* in general appearance, it is not at all an annual fish. The large eggs are placed close to the surface of the water within roots and the like. The eggs develop and hatch without delay. (See HOM.)

The throat pattern of the active GUI male is a uniform black all over the lower surfaces of the head. Indeed, the whole fish becomes a very dark color, and the milky-white marginal band in the lower part of the caudal contrasts very strongly with the other colors. In the female GUI a dark marking rather similar to those which develop in OCC and in the forms of the LIB group is present in pre-mating display. But in GUI the marking is oblong, not rounded, and similar to a short crossbar anteriorly on the throat, behind the lower lip. This marking resembles that of PLA, in *Pachypanchax*.

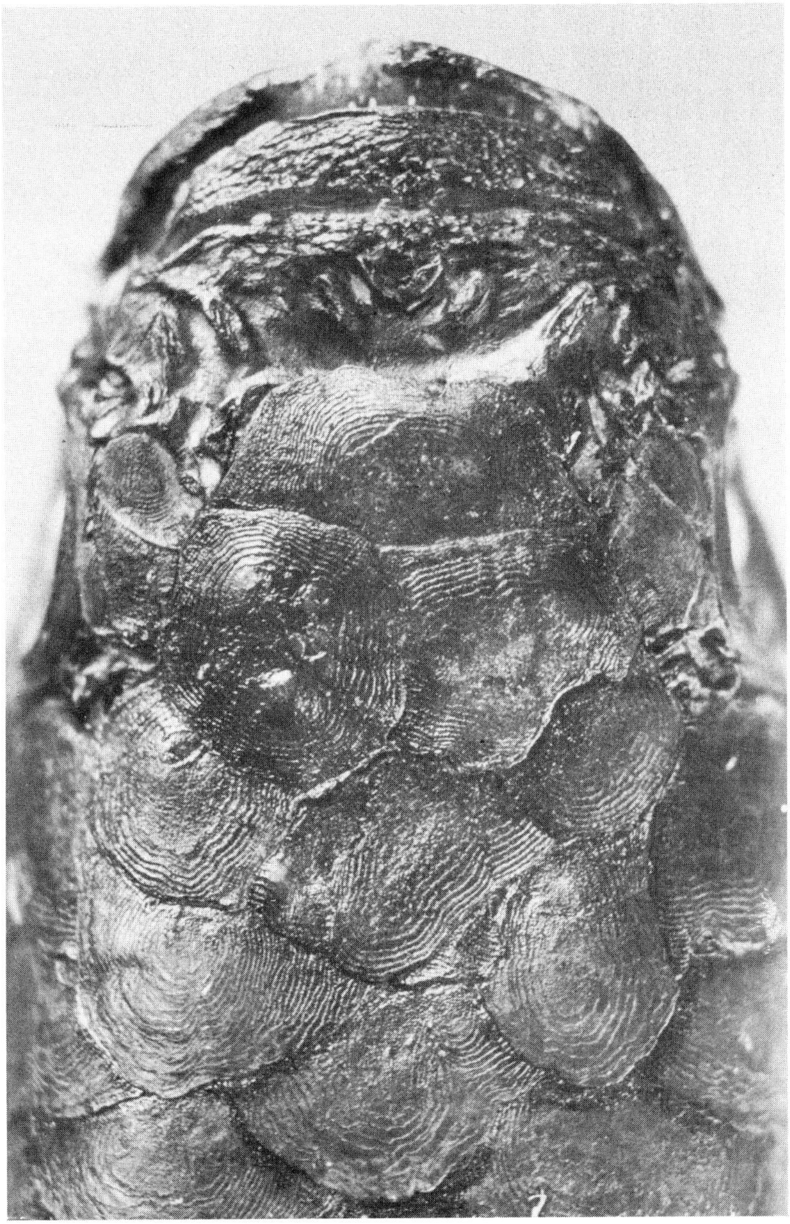

Front of preserved individual of GUI. No. 35.227 in the Paris Museum, from Kissidougou. Identified as CAM by Pellegrin 1934. One of the syntypes for GUI. Notice the pattern of frontal scales (E-type). The F scales above the eyes have reversed on the left side, not on the right side.

Hemoglobin patterns correspond to the usual rivuline type, with four bands. The frontal scalation does not contain H-scales. There are 19 haploid chromosomes in the Lago males. The karyotype seems to differ markedly from other species of nominal *Roloffia*. The two largest chromosomes are distinctly metacentric. There are also two more somewhat smaller metacentrics. One acrocentric element, however, is distinctly larger than the largest (calculated for n = 25) length of the basic acrocentrics, thus indicating that this acrocentric is a "double-sized" acrocentric and that GUI belongs to the *Roloffia-Aphyosemion-Nothobranchius* group and not to the *Aplocheilus-Epiplatys* group. But the karyotype also clearly resembles that of a species of the group last mentioned by the four large metacentrics.

Although GUI develops the *Roloffia* pattern ("closed") of lateral line organs on the dorsal part of the snout, its relation to this genus is obscure. It resembles the species of *Pachypanchax* more than those of *Roloffia* by the general appearance, by the frontal scalation, by the throat-pattern of females, by the lack of red pigments, and by fin pattern (HOM only). GUI certainly may represent a derivate of HOM now living as an ancient species in the mountains of the outmost West African area. GUI also resembles FAS in general appearance, in chromosome numbers and in karyotype, but probably it is not too closely related to FAS and SPI.

GUI appears to be restricted to mountains. Daget said that GUI was found only in places where BIF and SEN (=SPI) were absent. Roloff's Lago strain probably is not a mountain strain. It was taken together with BER and FAS, and not far from localities containing OCC.

Daget and Iltis (1965: 196, 200) doubted that GUI represented a distinct species, different from LIB, because individuals from the western part of Ivory Coast, which they thought were LIB (they probably were MAE; see MAE) were considered to be intermediate to LIB and GUI.

GUL: *Fundulus gularis* Boulenger 1901 **(Distinct sp.)**

Boulenger (1901: 623) had 16 specimens from Agberi of the western part of the Niger Delta. He placed GUL closest to SJO. Pellegrin (1907: 319) reported GUL from the Ntem River of the southernmost part of Cameroon and he, also, placed GUL closest to SJO. Pellegrin's individuals represent a real problem, as GUL undoubtedly does not occur in Cameroon, and GUL-like individuals were not found in coastal Cameroon during the 1966 expedition. There are, however, very large swampy areas near the mouth of the Ntem River, and in such biotopes the GUL-like form might be present.

Krüger (1913: 754) said that a friend in Cameroon named Schwab, who was a missionary in Ebolowa of southern Cameroon, had sent him preserved individuals of GUL taken in brackish water. These individuals probably belong to the same species as those reported by Pellegrin. It has

Adult male of GUL from southwestern Nigeria.

Adult male of GUL from southwestern Nigeria.

The holotype for GUL as figured by Boulenger 1901.

already been mentioned under BAT that Pellegrin (1929: 360) reported this species from Nyabessan and that a locality so named is situated in southern Cameroon, but there are other localities so named in Cameroon.

It is not quite clear when GUL first was introduced as an aquarium fish. It has already been mentioned under COE (and SJO) that this species (=SJO) was imported into Germany in 1905 and that this strain was identified as GUL by Boulenger, who maintained this opinion, even in 1915. (See COE.)

Köhler (1906: 380, 426) described a newly imported large rivuline species which resembles GUL (and FAL). He said that there was a violet longitudinal stripe on the throat (GUL pattern?). The indigo spots on the sides were confluent, producing a brilliant indigo stripe or band from the head to the root of caudal, where the band separated into two indigo lines extending into the caudal fin. The upper part of this fin was brownish and with large and small indigo spots and an indigo marginal band. The lower part of the fin had a median indigo band and was orange to vermilion and with a broad brilliant milky-white marginal band. The anal was a faint orange.

Arnold (1908: 568) said that GUL was first imported in 1907. Because individuals of SJO (see COE) which previously (1905) had been imported had been identified as GUL by Boulenger, Arnold was confused by the individuals of GUL, and he sent preserved material to London and asked for identification. Boulenger identified the 1907 individuals as GUL, and Arnold was still confused, because he thought that the two forms (occurring in the 1907 shipment) were distinct and different. Arnold proposed that the 1905-form (COE = SJO) should be called GUL *var. A* (blue) and that the 1907-form should be called GUL *var. B* (yellow).

The 1907 importation originated from the Niger Delta, according to Arnold (1911: 603) and Boulenger (1915: 623). The individuals first imported were considered to be very feeble, and they were not reproduced, but in 1911 German aquarists had improved their keeping of this species and it was reproduced by several aquarists. Arnold also reported that the two "varieties" (GUL and SJO) had been successfully crossed in Vienna. The report of this "crossing" was published by Fischer (1911), but I think that it is doubtful that any real crossing ever occurred.

Träber (1911: 233) reported a sex-reversal. His breeding female developed the colors of a male. He considered GUL to be rather variable in the male phenotype. All males had a zig-zag shaped red band along the side, and this band extended into the central part of the caudal fin, which was trilobate. Above and below this red band there were red spots tending to produce longitudinal rows. The sides were blue; the belly was white. The dorsal developed filaments and was reddish, with rows of red spots. The anal was yellowish-red, with some red spots posteriorly. The caudal was divided into two equal parts by the median red line, the upper part being reddish and spotted with red. The lower part was yellow to orange red. A red submarginal band was present in this part of the caudal fin. Fringes developed in the anterior part of the anal fin, but not to the extent seen in males of "*var. A*" (= SJO). He was not able to get a GUL male and a SJO female to cross.

Further information concerning the status of "*var. A*" and "*var. B*" is given under COE.

In 1955, some importations of GUL into Holland took place. I had a few individuals together with my first individuals of FIL, but the GUL males died before they reached Copenhagen. In 1961, GUL was again kept in Holland and was photographed by Nieuwenhuizen and de Looze. I had these photos sent to me. In 1959 Clausen collected live individuals of a very similar (or identical) strain at Ago-Iwoye, near Ijebu-Ode of SW Nigeria, and I had a single male from this collection. I crossed this male to an Akure GAR female and had one hybrid, a female. This hybrid was very viable and almost sterile in backcrossings.

As it has already been said under BEA, Klee and Turner (1962: 514–17) and Radda (1963: 70–73) identified individuals of SW Nigerian GUL as BEA. Since 1961, GUL has frequently been exported from Lagos.

These new strains of GUL are referrable to Boulenger's GUL, but they differ in color pattern from the first imported strains of GUL from the Niger Delta. But the SW GUL is also highly variable in color pattern, even within the local population. The strains described by Köhler and by Träber differ radically from the present strains.

The GUL male is characterized primarily by the very dark-red marking on the throat of males, which Boulenger used to name the fish, and which

Köhler also observed. It is further characterized by the fringes of the dorsal fin. The color pictures show two males from a single population of GUL of SW Nigeria. The collector, Kluge, said that at some time one of these phenotypes was dominant in the pool, whereas at other times the other type was dominant.

The color pattern of SW Nigerian GUL resembles that of the sympatric, but considerably smaller, FIL. A GUL-like population of FIL-sized rivulins from the Meko area of W Nigeria has been mentioned under FIL. FIL and GUL can easily be separated by the number of scales along the sides, by the frontal scalation and by size. The scales of the sides of individuals of GUL often are very irregular and difficult to count. Small scales occur together with large scales. In adult males the scales of median or lower rows usually develop many ctenoid spines, which are covered by tissues and produce fleshy filaments. Also, ctenoid papillae are present in the anal fin. The frontal scalation always has prominent H-scale, but in all specimens studied the usual G-type pattern was present.

Males from SW Nigeria have 16 haploid chromosomes. The largest element is a submetacentric. There is a smoothly graded series of acrocentric elements, the largest acrocentric being slightly less than five times the length of the smallest acrocentric element. Among these many acrocentrics there are two metacentrics of above mean basic length.

GUL and SJO (COE) usually are considered to represent closely related forms. I do not think that they are as close as generally assumed.

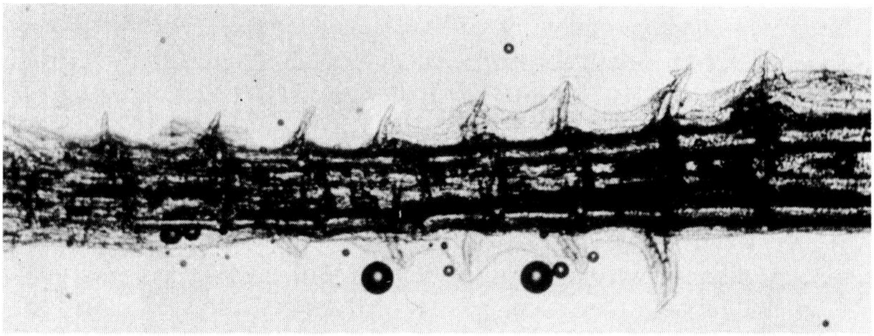

Ctenoid spines on tip of first anal fin ray of preserved male of GUL from southwestern Nigeria. These spines are covered by tissue.

Nieuwenhuizen (1964: 5–9) published many fine photos showing the variation in colors of males.

GUS: *Fundulus gustavi* Ahl 1924 **(=BEA)**

Ahl (1924b: 143–144) had two specimens from Sangmelima; they had been taken in the Upper Lobo River of South Cameroon. He placed GUS closest to BAT, which has been taken less than 50 km from Sangmelima.

Holly (1930: 209) studied the types and was in almost full agreement with Ahl. Radda (1963: 70–73) also studied the types and concluded that GUS (and SRE) were synonyms for BEA. No objections. (See also BEA and SRE.)

HJE: *Aphyosemion australe hjerreseni* Meinken 1953 **(Invalid ssp.)**

Meinken (1953: 123) created this name for a certain aquarium strain produced by Hjerresen of Flensburg. Hjerresen (1954: 73–75) said that he discovered an individual with a "white head" in a brood of AUS. He considered the individual to be diseased, but found that it was in reality very sound and the strongest of the brood. This ugly individual was a female. He spawned this female to one of its brothers and raised a hundred of their offspring. Among these there were seven which were white all over, but two of these were not able to swim and one more did not swim very well. A breeding pair developed from the remaining four. In the male, the amount of black pigment was much reduced; in the female there was a more yellowish color than in the head of the mother. Hjerresen also said that in the first generations of the new (golden) strain he had 7–10% of the individuals of the new phenotype, whereas from the fourth generation on the new strain was breeding true. Hjerresen sent live individuals to Meinken who based the HJE name on this material.

Hjerresen's strain does not represent any population of AUS in nature

Pair of orange aquarium variety AUS first raised by Hjerresen in 1953. G. Wolfsheimer photo.

and for that reason it does not deserve a zoological name. This strain represents the results of an accidental mutation and the experienced breeders' utilization of the changes for the creation of a strain of aquarium fishes. The activities of Hjerresen therefore should be honored by the name "Hjerresen's Golden Lyretail," and not by an invalid zoological name.

HOL: *Aphyosemion bivittatum hollyi* Myers 1933 **(=BIV)**

Meinken (1930: 177–79) had an aquarium strain of BIV which he thought to be different from Lönnberg's BIV (or different from other strains of BIV known to him), and he described and figured this strain under the name *Fundulopanchax bivittatus coerulea.* Myers (1933: 183–84) found the name "*coerulea*" to be preoccupied (see also AHL) and renamed Meinken's strain as indicated above.

The origin of Meinken's strain is not known; it does not differ from BIV by meristic characters, but by colors (?) only. In this connection it should be kept in mind that the color of the BIV male was never described. Lönnberg based BIV on a single specimen, a female. No other specimen of BIV from the Ndian drainage has been described, so one cannot separate any known form from an unknown one.

The types for HOL redrawn after Meinken's drawing of 1930. (*Fundulopanchax bivittatus var coerulea.*)

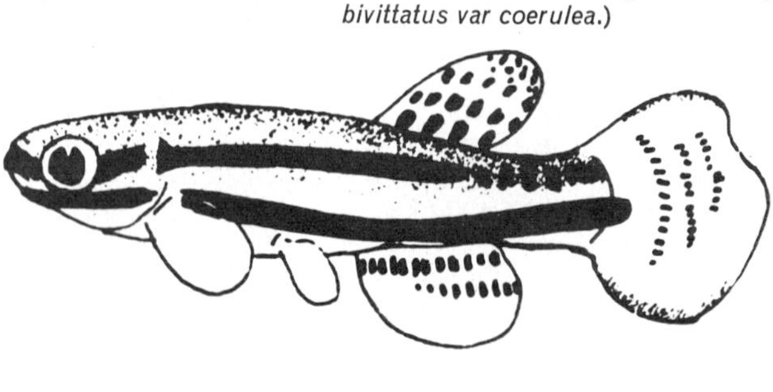

The male HOL had a faint blue metallic luster on its sides and in the anal and caudal fin. The dark violet-red marking behind the gillcover and the brown streamers of the dorsal and anal fin place HOL close to the populations of southwestern Nigeria, whereas the almost uncolored pectorals (with a faint red color) indicate a more easterly origin in Nigeria. The blue metallic color of the anal fin is a common feature in BIV, but the faint blue brilliance of the sides is rather extraordinary and probably is a result of a certain reduction of the overall red pigmentation of the BIV male. Other characteristics mentioned by Meinken seem to be generally present in various BIV strains.

When the total variation of colors of the BIV male in Nigeria and in Cameroon is considered, it appears unjustified to create a distinct subunit of BIV on the basis of the features mentioned for HOL. The distinguishing characters of HOL probably will not be present in preserved material.

HOM: *Poecilia omalonota* Dumeril 1861 **(Valid sp.)**

Dumeril (1861: 237) had numerous individuals from Nossi-Be, a small island just off the coast of northwest Madagascar. In 1953 Arnoult brought living specimens to France. These specimens originated in Maevatanana of western Madagascar, according to Nieuwenhuizen (1958: 73–76). This locality is situated about 100 km southeast of Majunga, within the Betsiboka drainage. The water of this locality measured 8 German degrees of hardness, pH 6.3–7.0 and 25–30° C., but HOM also has been taken in brackish water.

The color pattern of the HOM male is very simple. The sides are bluish, brownish above, white below. The unpaired fins are rather dark with a green hue; the ventrals, anal and lower part of caudal have a broad white to yellowish marginal band. In some males, the dorsal and the upper part of the caudal have a similarly colored narrow marginal band. In the HOM female the brilliance of the sides is much reduced, and the fins are less dark and without lightly colored marginal bands. In some individuals, some very

The type for HOM as figured by Dumeril 1861.

Adult male of HOM. Franz Werner photo.

ill-defined thin dark crossbars are seen posteriorly on the sides. No red pattern develops on the sides or in fins.

I kept Arnoult's strain from 1958 to 1960 and lost it because only females developed in the last brood. Like Franz (1958: 305–06) and Cahalan (1957: 358–59) I found HOM to be a form very easy to keep and to breed. Eggs are rather large (1.5 mm), and my females produced up to twenty eggs each day for months. HOM is not an annual fish, although it superficially resembles the robust species of *Fundulopanchax*. Eggs are deposited close to the surface of the water within roots and the like.

The pattern of lateral line organs of the dorsal part of the snout is interesting. From my last brood I have four preserved females left. Two of these developed the *Aphyosemion* pattern ("open"), whereas the other two developed the *Roloffia* pattern. It seems that individuals which develop the *Aphyosemion* pattern have more anal fin rays than those which develop the *Roloffia* pattern (A.17–19 and A.15–16 respectively.) It also seems that the neuromasts of the head are housed in open, rather shallow grooves, as in *Nothobranchius* and in ANN, and not in canals with pores on the sides of the head.

The pattern of frontal scalation seems to be very primitive for a rivuline species of the Old World. In the four individuals the E-pattern developed

Front of preserved individual of HOM. Notice E-pattern. Two of the H-scales are visible. The pattern of frontal neuromasts is "open" in this female.

Front of preserved individual of HOM. Notice E-pattern. The pattern of frontal neuromasts is "closed."

Preserved HOM.

and all have more than two prominent H-scales. Also, the pattern of scales above the eyes is extraordinary, as there is a scale behind the F-scale. The scalation is more "procatopodine-like" than in any other known rivuline species of the Old World. No color pattern is visible on the throat.

HOM and GUI are very close in general appearance and in color pattern. Both develop the rare E-scalation on the front. NUC appears to be a synonym for HOM.

INF: ***Haplochilus infrafasciatus*** **Günther 1866** **(=SEX)**

Günther (1866: 313) had a single male from Old Calabar of southern Nigeria and one male from the Upper Nile. Garman (1895: 131) placed INF and SEX as synonyms for SPI, whereas Keilhack (1910: 119) and Boulenger (1915: 65–66) considered INF to represent a synonym for SEX. Keilhack also placed the Nile individual under MAR (=SPI).

Lönnberg (1895: 188) considered INF to represent a distinct species and referred an individual from the Ndian River to this species. He also referred three specimens (1903: 39) from Sanye (Sunyi) of Cameroon to INF. He considered INF to be close to FAS. His individuals undoubtedly are SEX.

Steindachner (1895: 76) referred four specimens from Liberia to INF. From the information given these specimens appear to represent FAS, not DAG. Boulenger (1901: 7) reported INF from Sapelle of the Niger Delta and Günther (1902: 338) referred several specimens from the Bokitsa Mine of Ghana and from the Atesu River to INF. I have seen Günther's specimens. They belong to CHA, except for one individual of PET from the sample first mentioned.

JAC: ***Panchax (Aphyosemion) jacobi*** **Ahl 1928** **(=EXI)**

Ahl (1928: 121) had four specimens collected by Jacob at Lolodorf of southern Cameroon. This locality is situated on the Lokundje River, about 440 meters above the sea, and thus the locality probably belongs to the inland plateau although it seems to be isolated from the main part of the plateau. Ahl placed JAC closest to ELB (=BUA). Holly (1930: 195–281), who studied the types, seems to be in agreement with Ahl.

The holotype for JAC as figured by Ahl 1928.

This species is characterized by a dark brown band running from the lower jaw through the eye to the root of the caudal. The width of the band extends behind the gill covers and reaches the diameter of the eye. The upper edge of the band is just below the median line. In the male, the dorsal and the anal are pointed and dark-edged. The lower part of the head has red markings.

JAC corresponds to EXI by meristic characters, but my specimens of EXI in alcohol do not develop a dark lateral band along the sides. Live individuals of both sexes of EXI, however, may exhibit a broad longitudinal band from the head to the root of caudal occasionally and this marking was also described by Franz (1967: 340–43). As formalin usually develops temporary dark patterns in rivulins, I preserved individuals of EXI in 4% formalin and after preservation a broad, rather faint, dark longitudinal band appeared along the side. Similar markings did not, however, develop in specimens of BUA or in EXI/BUA hybrids after preservation in formalin. On the basis of this observation I do not hesitate to consider JAC to represent a synonym for EXI.

JAU: *Panchax jaundensis* Ahl 1924 **(=EXI)**

Ahl (1924: 305) had a single individual, probably a female, from Yaunde of East Cameroon. He placed JAU closest to CAL and STR. The individual was described as blackish-brown. Holly (1930: 216) studied the type material and changed some of the data given by Ahl. In 1963 Radda told me that the types for JAU probably were preserved in the same sample as OBS, because both individuals were stained with a greenish dye.

It seems that in the Jaunde area there are two species of *Aphyosemion* only: EXI and OBS. Morphologically JAU seems to be closer to EXI than to OBS, and in this particular area I have seen populations of EXI in which the females do not develop dark vertical lines on the sides. JAU cannot be separated from EXI and OBS by meristic characters.

KIY: ***Nothobranchius kiyawensis* Ahl 1928** **(Valid sp.?)**

Ahl (1928: 601) had two specimens, one male and one female, from the Kiyawa River. They were collected by Lt. Lloyd near Katagum of northern Nigeria.

The Kiyawa River is not on my maps or mentioned in the "Gazetteer of Place Names on Maps of Nigeria" (Lagos 1949), but there are two places so named near Kano, and in this area the Katagum River and a place named Katagum are situated. There is, however, also a place named Katagum within the Benue drainage, just east of the Jos Plateau. In the geographical area first mentioned, species of *Nothobranchius* (see RUR) or *Fundulosoma* are likely to occur, whereas in the area last mentioned populations of GAR (NIG) undoubtedly are present.

The male KIY was greyish-brown on the upper surfaces and bluish-white on the sides and lower surfaces, with a number of large, rounded, irregularly distributed carmine-red spots. The gill covers had some vermiform red stripes, and there was a longitudinal red line below the eyes. Gill membranes were red. The base of the dorsal was bluish, with large red spots, and above these there was a red line followed by a broad white submarginal band and a red margin. The anal had a carmine-red band, with a broad white area and a reddish-brown margin, near its base. The central part of the caudal was dark grey, with a broad submarginal white band above and a similar marginal band below, the upper edge being narrowly edged with dark color.

The female KIY was greyish, without markings.

The fins were rounded, except for the caudal of males, which is obliquely truncated, the upper lobe a little longer than the lower.

The description of KIY indicates that this species belongs to *Aphyosemion* or *Fundulosoma* (?), not to *Nothobranchius*.

KUH: ***Fundulus kuhntae* Ahl 1926** **(=ORT)**

Ahl (1926: 222) had several specimens from an importation of aquarium fishes from East Africa. This collection was said to originate in Mozambique. The teeth were arranged in broad bands, with the outer row somewhat enlarged. Each scale had a red edge. A dense pattern of red spots appeared on the head. The dorsal and anal were light red, with rows of darker spots and a dark-red distant edge. The caudal and ventrals were red. The females had some dark spots on the head. The description of the color pattern was, however, changed by Arnold and Ahl (1936: 331–332), who said that the male was reddish all over and densely spotted with red, with red fins, the female being spotted or unspotted, and with a red basic color on the body and in the fins. It was also said that KUH reached 7 cm.

Rachow (page 933 in *Die Aquarien Fische*, Holly, Rachow and Meinken) considered KUH to represent a synonym for ORT. Aquarium strains called

KUH were also kept in Germany during the thirties, according to Beyer (1937: 113) and others, and these were probably used by Arnold and Ahl for the revision of the color description.

Various aquarium strains referrable to KUH were collected near Beira, Mozambique by Roloff (1959. (6): 12–19), Hansen (1957: 129–132) and by Griem of Hamburg (unpublished). The males of these strains are characterized by their violet-red markings and by the absence of a red reticulation on sides. Instead, there are many crescent and square red spots anteriorly. These strains are also characterized by the rather dense pattern of red spots on the lower surfaces of the head.

Females usually have dark red spots irregularly distributed on the sides and many similar but smaller spots on unpaired fins.

In the male, the caudal fin often, or always, is not dark-edged, and the dorsal and anal fins have narrow bluish edges. Individuals of these strains appear to reach medium size for the species of the genus. Eggs measure 1.20–1.25 mm in Griem's strain.

These Mozambique strains probably represent ORT, but they may represent a different species.

LAB: *Aphyosemion (Fundulopanchax) labarrei* Poll 1952 (Valid sp.)

Poll (1952: 168–69) had 73 specimens from the Lower Congo area: Madimba (Inkisi) near Kiavo; Kisantu; Thysville; Kigemba near Kisantu. Some of the Kisantu material had previously been identified as LUJ by Pellegrin. It has already been mentioned under CAM that the individuals from Mayanga identified as CAM by Boulenger might represent LAB (or LOU, or a similar form) although this locality is situated north of the Congo River.

LAB resembles CON but seems to be differentiated from the latter by the number of dorsal fin rays.

I have found no report of aquarium-kept individuals of LAB prior to 1956, but in that year I received some from the United States. These fish had not been identified and were sold as "Blue Panchax," the usual name for PAN (Axelrod and Schultz 1955: 417). Poll kindly identified these individuals for me. Dvoskin (1956: 223) also used the name "Blue Panchax" for LAB.

The relationships of LAB are still obscure. It resembles the species of the NDI-CAM-LOU group of "flame-tailed" species of *Aphyosemion*, but it also resembles the species of the ELE-CHR group of Central Congo, and these two groups appear to represent two different lines of evolution in *Aphyosemion*. LAB appears to be very close to LOU in general appearance and color pattern, but it differs markedly from LOU by chromosome numbers and, in particular, by karyotype.

LAB has 14 haploid chromosomes with one very long acrocentric (or

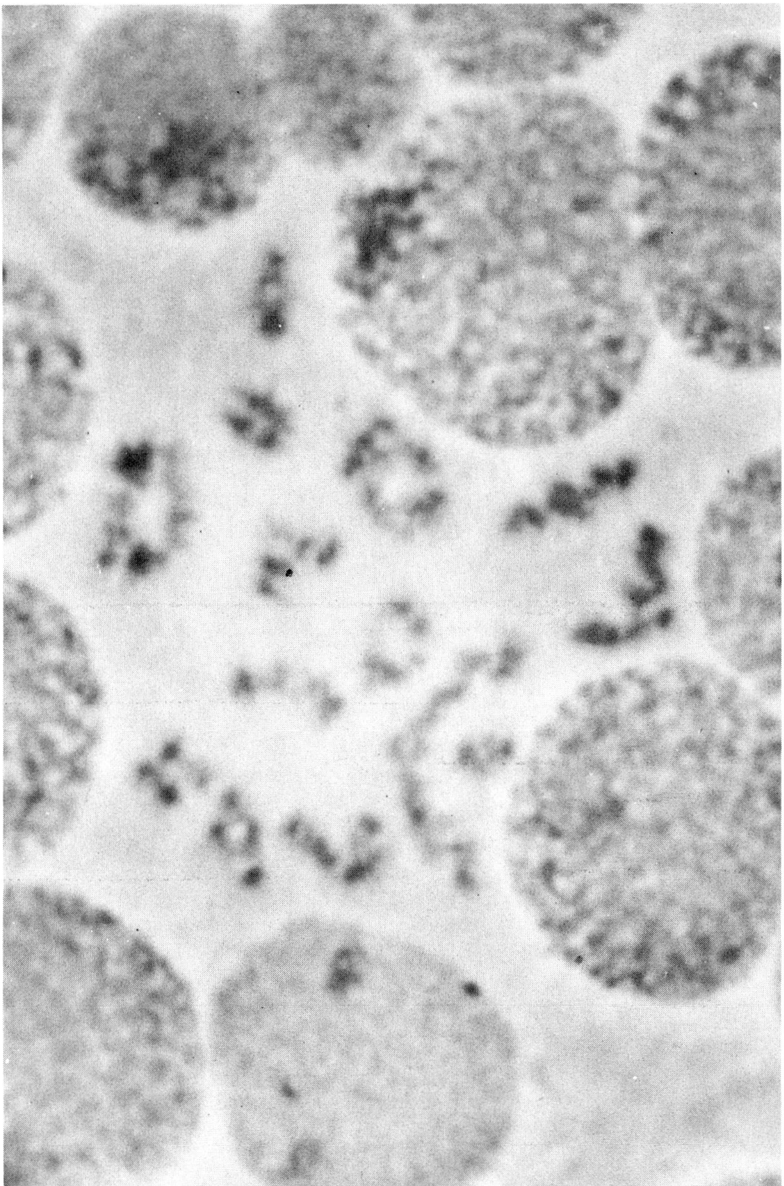

Primary meiotic division in LAB. n = 14.

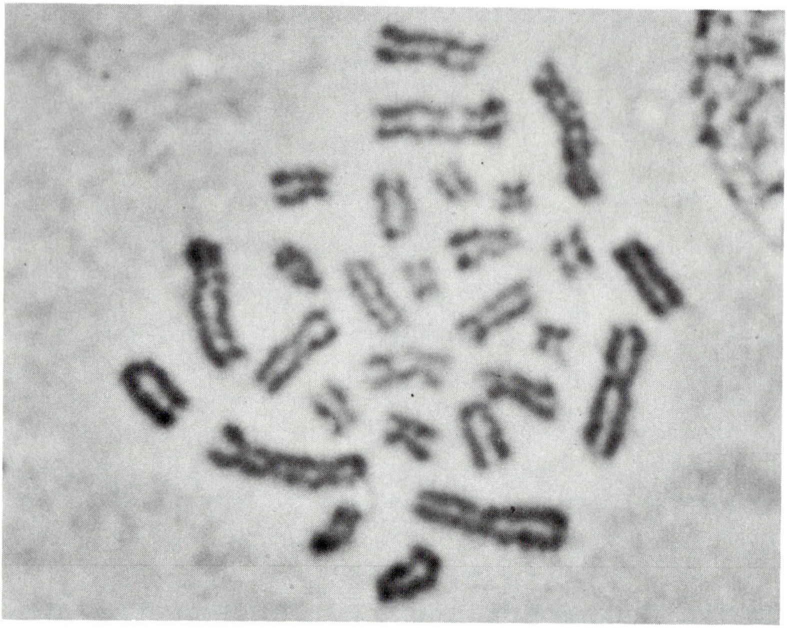

Mitosis in LAB. $2n = 28$.

GAR-Akure/ LAB from the Lower Congo drainage. Seligmann photo.

Adult male of LAB from the Lower Congo drainage.

almost submetacentric, the short arm being about one-fifth of the long arm). There are also two large metacentrics, which in morphology seem to correspond to two of the chromosomes of the n = 9 strain of CHR. One smaller metacentric and three or more acrocentrics of above basic length are also present.

The frontal scalation, hemoglobin and lateral line patterns are normal for the genus.

The LAB male is characterized by the broad very dark red margins of unpaired fins, by the cold, very strong blue metallic shine on the sides and in the basal part of fins and by the strong tendency for development of a rather broad, irregular, often almost zig-zag-shaped red band posterior on median part of the side. Yellow pigments seem to be absent in LAB.

Although the juxtaposition of the dorsal and anal fin indicates a relationship to *Fundulopanchax,* LAB probably is not a true member of that subgenus.

LAB is easily kept and bred, but the present strain seems to give a very large surplus of males, in my own tanks at least. Males usually are very aggressive and may kill females during reproduction. The eggs from LAB are very large compared to the size of the individuals. They measure 1.6–1.8 mm.

Young male LAB, common aquarium strain.

LAM : *Epiplatys fasciolatus lamottei* Daget 1954 **(Valid sp.)**

Daget (1954: 300–325) had 22 specimens from Simandou of the Upper Niger drainage. Daget kindly informed me that the type-locality is a range of small hills between the Milo and the Sankarani Rivers at about 9°W, 9°N. He placed LAM so close to FAS, that he considered LAM to represent a subspecies of the latter.

Each scale of the sides has a red spot. The fins of males are yellowish-red, except for pectorals, which are yellow. The anal and caudal are bordered with red. No dark crossbars were present in any individual. This species is also characterized by the long and pointed ventrals in the adult male.

In 1964 Herold Olsen of N'Zérégoré, High Guinée caught several individuals near N'Zérégoré, about 140 km from the type locality. There were seven males and six females. The largest individual measured 57 mm standard length; 68 mm total length. These individuals correspond so well to Daget's LAM that I do not hesitate to refer them to this species. There are about five rows of very rounded red spots along the sides, these spots are confluent in some males, producing almost uniform red longitudinal lines. There is no tendency for the development of a red reticulation or crescent shaped red spots (see FAS). The ventrals of males may reach the 10th anal fin ray. The throat pattern is simple, with slightly more dark pigments in the male than in the female and with no particular red or dark pattern. The branchiostegal membrane has much red pigment in most males, but not in all males. Pectorals are colorless, with a faint dark edge, faint yellow, orange with reddish edge, or uniformly dark. Other fins as in the type. The yellow and orange pigments in unpaired fins are strongly developed and the LAM male is probably a beauty in life.

Front of LAM. Notice the partial splitting of the central frontal fleshy lobe.

Preserved male of LAM from N'Zérégoré, Guinée. Notice the long ventrals.

The frontal scalation is normal (G-type, no H-scales). The pattern of lateral line organs on the dorsal part of the snout is somewhat variable. In ten individuals, the neuromasts are rather close as in *Roloffia*, but the central lobe is distinctly divided into two parts. In one individual (female) the lobe is undivided and in two more individuals the division is intermediate.

There are no indications of dark crossbars or of a dark longitudinal band, except for one male, which has a large dark blotch near the root of caudal.

LAM undoubtedly represents a very distinct species, not more closely related to FAS than to other known species of *Epiplatys*. Until live specimens have been studied it appears to be impossible to attach LAM to any known group of species of this genus.

LEO: *Epiplatys sexfasciatus leonensis* Ahl 1937 **(=FAS)**

It has already been mentioned under FAS that the types for LEO originated from a population of FAS in the Freetown area of Sierra Leone, and that LEO does not differ from FAS by meristic characters nor by color pattern. Ahl probably was not aware of the dark longitudinal band which develop in individuals of FAS during certain conditions (fear, pre-mating display). Roloff (1936: 308–10) supplied information on the type-locality for LEO. The description of LEO is in Ahl (1937: 240).

LIB: *Haplochilus liberiensis* Boulenger 1908 **(Valid sp.)**

According to Arnold (1908: 314) there were three small individuals of a rivuline species in the first recorded shipment of DAG from Monrovia of Liberia. These three individuals were females described by Arnold as yellowish-brown, with no pattern. Arnold considered these individuals to be rather "*Rivulus*-like." He preserved one individual and another died shortly after. The two individuals were sent to Boulenger who (1908: 30) based his description of LIB on these two females.

LIB was reported from Man of western Ivory Coast by Pellegrin (1934: 117). This locality is close to the type-locality for MAE, so probably Pellegrin's individuals are MAE. He also (1935: 466) reported LIB from Kindia of Guinée (455 meters above the sea). These individuals probably represent GER or (?) ROL. LIB also was reported from two localities near Monrovia (Bromley and Du River) by Schultz (1942: 330).

Daget (1948: 41) reported LIB from Mount Nimba and (1960: 326) from Banfora (S.W. of Bobo-Dioulasso of the Upper Volta drainage). The Mount Nimba individuals probably belong to the same populations which are reported under MAE in this book. The Banfora individuals are very interesting because this report extends the range of the LIB-group (or GER-group) far eastwards.

Daget (1962: 141) also reported LIB from several localities of Fouta Dialon and Lower Guinée. These individuals probably are GER, ROL, or similar forms. There are no individuals referrable to LIB in Roloff's and Schiötz's collections of rivulins from Sierra Leone.

The present author prefers to consider LIB as a form restricted to the odd "coastal swamps" near Monrovia. We know that a population of BER lives in the forests north of Monrovia. (See BER.) The different usually rather ill-defined species of the LIB-group or superspecies are difficult to identify and for that reason it is very likely that mistakes occur. Also, the color pattern of the LIB male was never published because Boulenger had females only.

Old male of LIB. This male belongs to the old aquarium strain previously called CAB.

Haplochilus liberiensis.

♀, type.

The type for LIB as figured by Boulenger 1913.

It has already been mentioned under CAB, that this form undoubtedly represents LIB. (See CAB.)

The frontal scalation of specimens of the "old aquarium strain" (since 1935) is often irregular and H-scales may occur. These H-scales are situated on and not below the G-scale as in GAR, whereas usually such scales should be situated below the G-scale. The four individuals collected by Clausen in 1965, which died before they reached Copenhagen, have normal frontal scalation without any H-scales. By meristic characters and by color pattern these individuals correspond very well to the aquarium strains of CAB. Ctenoid scales of the "wrinkled" type occur frequently in males of the 1935-strain and in particular such scales are present in median scale rows of the sides. Fin papillae have not been seen.

It is a remarkable fact, that LIB (CAB) has been able to survive in the aquarium since 1935. Breeders have observed that eggs which were left in the aquarium often did not hatch on time (after 10–12 days) and were delayed by several weeks. If kept in clean water, without peat or mud, the eggs of LIB do not exhibit "annual characters."

LIB is very close to such forms as ROL and BER and with these LIB probably represents a superspecies. GER and MAE also appear to be close, but I hesitate to incorporate GER because of the important differences given under GER.

LIN: *Panchax lineatus* Cuvier & Valenciennes 1846 **(Valid sp.)**

The types (1846: 381) for LIN originated from the Bombay area of India. According to Job (1940: 67–69) LIN is very common in the rice-fields, ponds, canals and estuaries at Bombay, the Malabar District, Corg, Wynaad and in the states of Cochin and Travancore. He also said that LIN has been

Front of LIN. Notice the wrinkled appearance of the G-scale and the two pairs of neuromasts in front of that scale.

reported from Ceylon. In India, LIN is called "piku" or "poochaatie," whereas in Ceylon it is called "Irri Nalla Handhya." It resembles the northern PAN in habits and it is considered to be superior to PAN in the mosquito larvae control. He also said that, generally, the broods contain more males than females, and that up to 90% of males have been reported in some broods. Eggs measure 1.4–1.7 mm and have more filaments on the membrane than have eggs from PAN.

Various synonyms for the morphologically defined LIN have not been taken into separate consideration in this text because they have not been used for aquarium-kept strains of LIN:

VIT: *Aplocheilus vittatus* Jerdon (1849: 330–31)

AFF: *Aplocheilus affinis* Jerdon (1849: 330–31)

Some misspellings of the LIN and the RUM (=LIN) names have occurred *H. lineolatus* Freund 1913 (: 469) and *H. rubropictus* Stansch 1910 (: 469).

Pair of LIN.

The present common aquarium strain of LIN probably has 25 haploid chromosomes, and the karyotype seems to contain no enlarged metacentric elements, and probably correspond to those of ANN and DAG. The hemoglobin pattern is very simple and has two almost confluent lines only. The lateral line pattern of the dorsal part of the snout resembles the pattern found in ESE, LON and SEX, but in LIN there are a number of neuromasts in each series, resembling the condition found in BLO and in most South American rivulin species. The frontal scalation is normal (G-type, no H-scales).

Female of LIN.

The dark crossbars on the sides are longer and more regular than in DAY, and they are usually lost in adult males. I observed a very broad and distinct black longitudinal band from the gill covers to the caudal and extending somewhat into that fin in two females. These females were used in crossings and the *Epiplatys* males refused to serve them. Soon after this "signal" had been observed, eggs were deposited, and the eggs were fertilized. I have never observed this marked color pattern in individuals of LIN under normal conditions, and it seems that similar patterns have not been reported by others, although LIN is an old and rather common aquarium fish which often has been described in literature. I never observed such bands in preserved material.

The throat pattern of LIN is weakly developed in both sexes, but when observed it is seen to be of the common *Epiplatys* type with two dark crossbars anteriorly on the throat.

LIN was introduced as an aquarium fish in Germany 1909 (Stansch 1910: 469) under the RUM name. It soon established itself as a common, robust aquarium fish, rather popular, when not fully grown. The ventrals are pointed in both sexes.

LOB: *Panchax loboanus* Ahl 1924 **(=EXI)**

Ahl (1924: 306–07) had a single individual from Sangmelima, southern Cameroon. The individual had been caught by Schreiner at the upper reaches of the Lobo River. Some information concerning the Lobo River is given under BEA. Ahl placed LOB closest to EXI and in this, he is right.

Holly (1930: 215–16) re-examined the type and changed some of the data for meristic characters given by Ahl. He also considered LOB to be closest to EXI.

LOB corresponds to EXI in meristic characters and also in color pattern. It has ten to twelve dark brown crossbars and three broad red similar bars in the caudal, which is rounded. The dorsal and anal are pointed, the latter with a broad white edge. The author caught many individuals of EXI near Sangmelima and also at the upper reaches of the Lobo River. The present aquarium strain was taken 30 km north of Sangmelima, in 1966.

LOE: *Fundulus loennbergii* Boulenger 1903 **(=BIV)**

Boulenger (1903: 440–41) had two individuals caught by Bates in the Kribi (now Kienke) River of southern coastal Cameroon. The description corresponds well to that of BIV, but Boulenger said: *Yellowish olive, most of the scales with a crimson spot or vertical bar, head with crimson wavy lines, fins dark, all except the pectorals with small crimson spots.*

In 1915 (: 25–26) he also reported four individuals of LOE from the Zima County and four more from Efulen, all of southern Cameroon and from Bates' collections. He figured the type, a male, without any trace of dark longitudinal bands on sides, but in all other respects the phenotype of

The type for LOE as figured by Boulenger 1915.

Fundulus loennbergii.

♂, type.

the type corresponds to that of the BIV male. It should, however, be kept in mind that he also reported individuals of BIV from the Kribi area.

Fowler (1930: 318) described and pictured three males identified as BIV from the Kribi area, and only in one of these are there two dark lateral bands, whereas in one more there are traces of such bands and in the third, there are not even traces of lateral dark bands.

In 1966, I caught numerous individuals of BIV in several localities within the Kribi area. There were no indications of the occurrence of two similar BIV-like forms, but in many populations of BIV in East Cameroon, the dark lateral bands, so prominent in Nigerian material, are less developed and less likely to occur in live and preserved material. It seems, that in a few populations, no such bands develop, not even in juveniles and in females. (See RIG.)

The two Cameroon strains which have now been kept in my tanks for almost two years are rather different in this respect. In the Buea strain from the Cameroon Volcano, the dark lateral bands are very rarely seen even in females, whereas in the individuals of the Mboumboula strain from East Cameroon, these bands are frequently seen, although they are not very marked if compared to my Ijebu-Ode and Meko strains.

Thus LOE cannot be separated from BIV by external morphological characters, but it is highly probable that the Kribi strains of BIV are strictly reproductively isolated from the population of BIV (Ndian River) from where the types for BIV originated. (See BIV.)

According to Arnold (1908: 585–86) one individual of a rivulin species which had been imported into Germany from Warri of the Niger Delta was identified as LOE by Boulenger, but Boulenger did not list this individual in 1915, at least not under LOE. There are no reliable reports of importations of aquarium fishes from Kribi, and LOE probably has not been kept in the aquarium so far.

LOL: *Panchax (Aphyosemion) loloensis* Ahl 1928 **(=EXI)**

Ahl (1928b: 121) had four specimens caught by Jacobs at Lolodorf of East Cameroon. He placed LOL closest to EXI. The town Lolodorf is situated on the Lokundje River, a coastal river, north of Kribi, south of the Nyong River. Lolodorf is situated about 440 meters above the sea, and the locality probably belongs to the inland plateau, although it appears to be isolated from the main part of the plateau by some wedges of lowlands. The types for JAC also originated from this locality and JAC undoubtedly represents a synonym for EXI, a very common species of the inland plateau.

LOL correspond to EXI (and JAC) by meristic characters. It has a lightly colored band from the gill covers to the root of the caudal. This band is situated above the median line according to Ahl. Individuals of EXI which have been preserved in formalin (see JAC) may show a dark median band along the side. In case this band is faint, it seems then that there is a broad lightly colored band between this band and the dark color of the back. Thus LOL undoubtedly also represents a synonym for EXI.

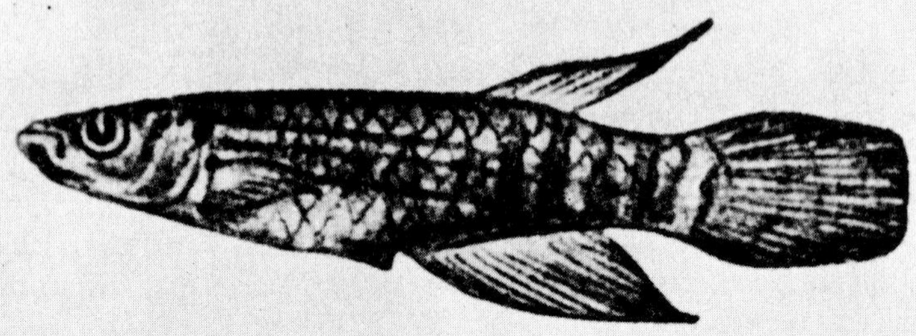

The holotype for LOL as figured by Ahl 1928.

LON: *Haplochilus longiventralis* Boulenger 1911 **(Valid sp.)**

The first reports of this form were given by Arnold (1908: 575–76) who said that a large collection had been imported from the Niger Delta. These individuals resembled SEX, which was well-known to him since 1905, and they differed from SEX by the more slender body and by the larger number of dark crossbars. Boulenger identified preserved material to belong to SEN (=SPI), but German aquarists used the PET name for this strain. This form was re-imported in 1911 and again Boulenger identified preserved material as SEN, but that year (1911: 266) he described LON on the basis of numerous individuals taken by Gardner in swamps between Agwarda and Awka, just north of the delta zone. All individuals of LON known at that time originated from the Niger Delta area. It seems that LON is endemic to this area. In the first description Boulenger said, however, that he had received preserved material referrable to LON from Arnold and that he previously had identified this material as SEN.

The LON name has also been used after 1911, but it seems that after that year no further importations of LON took place prior to 1961. That year Ulf Hannerz of Stockholm sent me live individuals of rivulins which he had caught during his Nigerian expedition. From the Benin City area there was a collection of *Epiplatys* which later on turned out to contain three different species: GRA, LON and SEX.

LON resembles SEX so much that untrained people probably will not be able to distinguish between these closely related forms. There are, however, small differences in meristic characters, color pattern and in general appearance. To make things more complicated, SEX from the Niger Delta area resembles LON more than do any other known populations of SEX. Also hybrids between SEX and LON are fully viable and in the hybrid males at least (females not tested) they are fully virile.

All this evidence indicates that LON is not a distinct species but represents a variation of the SEX phenotype within the Niger Delta. Having kept live individuals of SEX from many localities for years and also having kept three different strains of LON for some years I believe LON to be distinct.

An analysis of 44 individuals of LON and 22 individuals of SEX from the Niger Delta disclosed minor, but constant differences. The individuals of LON had D. 7–10, mean figure D=8.9. SEX had D. 9–13, mean figure D. = 10.5. LON had 25–29 scales in a longitudinal series (mean 26.8) and SEX had 27–31 scales (mean 28.8).

Individuals of SEX-like rivulins from the Niger Delta, which have D. 7–8, Sq-1. 25–26 probably will be LON, whereas others having D. 11–13, Sq-1. 30–31 probably will be SEX.

These two lots of preserved specimens also differed by color patterns and general appearance.

Old male of LON, from the Ologbo River of the western part of the Niger Delta.

My different strains of LON and of SEX from the Niger Delta were breeding true to the characteristics of both forms.

The color slide published in this book unfortunately does not give the true colors of the adult LON male. The general basic color is a brilliant grass green, which anteriorly on the sides grades into a marbled golden, lemon and grass green pattern. Rather similar colors develop in males of SEX from the Niger Delta, but not in other males of SEX. Compared to males of SEX from the Delta, the red pattern of the sides of LON males usually is different, because in the latter form, the red spots are completely absent in some populations and much reduced in other populations. Indeed these spots also are smaller than usual in SEX from the Delta.

Fully grown males of LON usually have six dark crossbars on the sides, but younger males have more such bars than similarly aged and sized males of SEX. Whereas in SEX, the adult males usually lose the dark bars, except during a fight, the dark pattern is more stable in LON, and the bars usually are quite distinct. Females of LON always develop considerably more dark crossbars than females of SEX. In the LON female there is often up to 13 such bars, whereas in females of SEX and in particular in those from the Delta area, the broad, dark longitudinal band which develops during fear and during pre-mating display, is inconspicuous or absent. This signal is very marked in the LON female. When this band is shown by the SEX female (all populations) the dark crossbars usually remain visible or they are somewhat reduced in size and occur below the band only. In the active LON female the dark crossbars disappear completely when the lateral band is visible.

The observed differences of the pre-mating color pattern of the SEX and

Juvenile males of LON from the Ologbo River, western Niger Delta area.

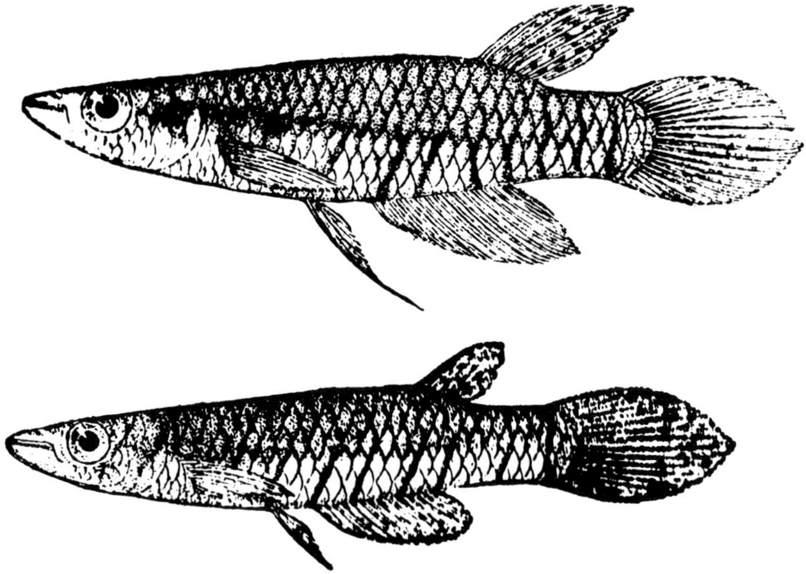

Haplochilus longiventralis.

♂ ♀, types, Old Calabar.

The types for LON as figured by Boulenger 1915.

the LON females might be important. It is a well-known fact that within rivulins, the females look for the males when they are ready to spawn. Probably the females distinguish between different males by the color patterns and from my crossings I know that males of most species will spawn all willing females, whereas females usually discriminate if the males differ from males of their own species. From my crossing experiments I know that the SEX male responses very actively to the pre-mating color pattern of the previously isolated LON female. It seems that the response of the SEX male is less prompt to females of its own species. This probably is the result of the very distinct lateral band of the LON female (over-stimuli). Under natural conditions the LON female probably rejects the flirt of the SEX male because to that female this male looks (or behaves) differently than the LON male. The LON male, on the other hand, probably is not stimulated to sexual activities by the weaker stimuli which is produced by the pre-mating color pattern of the SEX-female. Perhaps these deficiencies of pre-mating color "signals" prevent crossings between SEX and LON under natural conditions.

The similarities of the SEX and the LON phenotypes within the Niger Delta area indicate that the reproductive isolation between these two forms is not complete, or has not always been complete. In this connection it is worth noting that males of SEX living in the Niger Delta and west of the Delta right up to the Togo Hills of Ghana, have long and pointed ventrals, similar to those of the adult LON male, whereas males living east of the Delta in Nigeria, Cameroon and Gaboon have short ventrals in both sexes. Even in the Umudike (near Umuahia) population of SEX, which lives almost in the watershed between the Niger and the Cross drainage, close to the Delta area, the ventrals of adult males are not long and pointed to the extent seen in Delta males.

This division of SEX into two major groups of populations on the basis of the length of ventrals of males again indicates that crossing between SEX and LON once took place within the Niger Delta area. See also SEX.

Although LON is not so easily kept and bred as SEX, it is not a difficult rivuline species. It should be reproduced like SEX, but soft and slightly acid water is recommended. If kept together with individuals of SEX of similar size, the individuals of LON are rarely seen. They seek cover within plants and the like. Individuals of SEX are more robust and more aggressive. LON probably has become extinct in the aquarium.

Male probably belonging to LOU from the Niari-Kouilou drainage of the former French Congo. J. Lambert photo.

Adult male of LOU from Mindouli of the former French Congo. Upper Niari drainage.

Compared to most populations of SEX, the throat pattern of LON is very faint, but in some males the two crossbars of the throat are rather marked, because they are red, not dark brown or black.

The frontal scalation and pattern of lateral line organs of the head and of hemoglobin correspond to SEX. The eggs also are similar.

LOU: ***Haplochilus lujae louessensis*** **Pellegrin 1931** **(Valid sp.)**

It seems that Pellegrin (1931 : 221) had only one specimen, which had been taken in the Louesse River, a tributary of the Kouilou River of what was formerly the French Congo. Pellegrin considered this individual to represent a simple color variety of LUJ.

There were about ten carmine spots anteriorly on the sides; these spots sometimes were confluent. There was also a rather broad orange longitudinal band posteriorly on the side. The dorsal had five or six red lines parallel to the fin rays, whereas in the anal fin there was a red bar, similar to that in OGO. The caudal had a number of red lines and a large red line in its lower half.

In 1964 Brichard and Roberts collected live specimens of various cyprinodont species in the Congo. Some of these were brought to me by the Danish United Nations Military Forces serving in the Congo at that time. Among them there was a strain of *Aphyosemion* collected near Mindouli on the upper

Adult male LON from Benin City, western Niger Delta area.

reaches of the Niari River, an affluent of the Kouilou River. I identified these individuals as LOU. A quite similar strain also was kept in Scotland in 1964, of which I had a color photo from Seymour. Lambert also has kept rather similar individuals, originating from other parts of the Kouilou drainage, but from this drainage he also has had other strains, which probably are not LOU although they have some traits in common with my LOU strain.

LOU resembles LAB superficially, but it differs from LAB in several respects. The haploid chromosome number is different (LAB has n = 14). My LOU males had 10 haploid chromosomes, the karyotype being quite unique among rivuline species of the Old World. There are nine large metacentric elements and one submetacentric, which is the smallest of the chromosomes. The submetacentric element measures about 1.8 times the average length of the basic chromosomes (on the base of n = 25), and thus there are none of the ancient and basic chromosomes of the cyprinodont karyotype left unchanged. The differences in size of the ten chromosomes is not marked compared with other species of *Aphyosemion*, the largest metacentric element measuring about 1.7 times the length of the smallest chromosome. In LAB the longest element measures 5.5 times the length of the shortest, in CHR (n = 9) 4.2 times or (n = 15) 4.5 times.

Apparently, three of the LOU chromosomes correspond to chromosomes found in the LAB karyotype. It also seems that two chromosomes correspond in morphology to chromosomes found in CHR (n = 9), but apparently no chromosome found in the LOU karyotype corresponds to a chromosome of the CHR (n = 15) karyotype. Probably very large and complicated recombinations of the chromosome material took place during the evolution of these four forms.

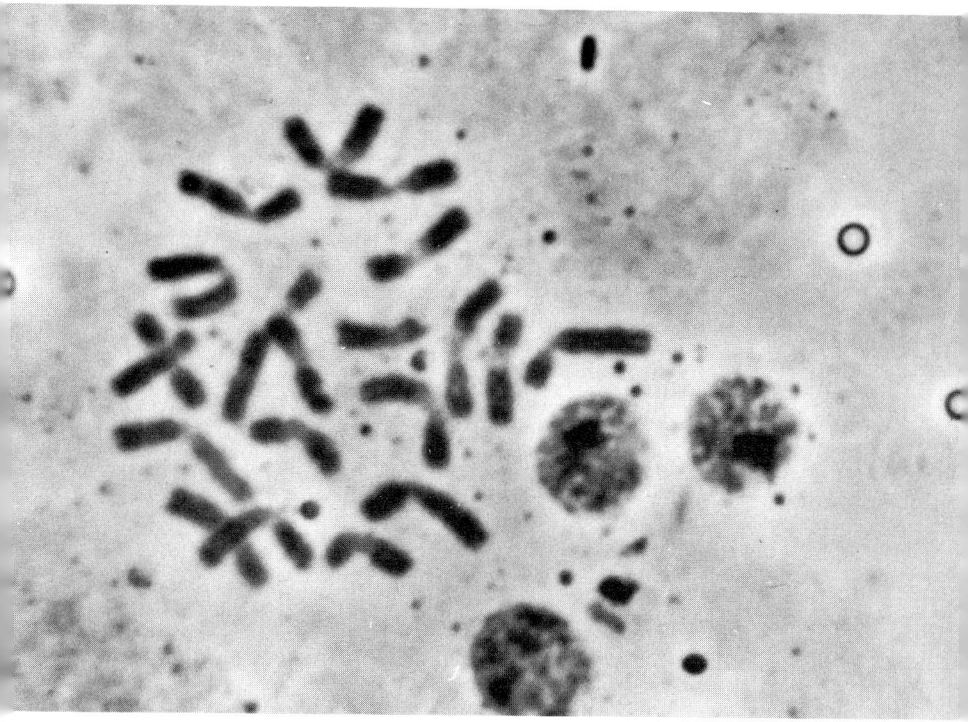

Mitosis of LOU from Mindouli, upper Niari River, the former French Congo. $2n = 20$.

It has been a serious problem to keep the LOU strain. The individuals are very sensitive towards attack from mycobacteria and *Oodinium*, and the aquarium-raised individuals were just as feeble as imported individuals. Some breeding pairs were distributed, but no new strain established. In 1965 the last male was lost; females survived until 1967. Although I have tried to cross individuals of LOU to different species of *Aphyosemion* and to LAB and NDI in particular, no sexual response was observed and no eggs were laid.

If color patterns reflect relationships, and it is very likely that this is true, then LOU might be close to CAM and, through CAM, to NDI. Indeed, males of NDI and LOU are very similar in general appearance, but they differ markedly in the number of dorsal fin rays and by scales along the sides. Also, NDI has 20 haploid chromosomes, and of these apparently no element is metacentric. (See LUJ also.)

LUJ: *Haplochilus lujae* Boulenger 1911 **(Valid sp.?)**

Boulenger (1911: 263) had numerous specimens taken by M. E. Luja at Kondue, Sankuru River, Kasai, Congo. See also ELE.

Male probably related to LUJ. This male originated from the former French Congo. J. Lambert photo.

Male of LUJ from locality about 200 km north of Kinshasa (Leopoldville), Congo. R. la Corte photo.

Male of LUJ. La Corte's strain of 1961.

Pair of LUJ. This aquarium strain also often has been called STR.

Holly (1930: 213–14) considered SAN to represent a synonym for LUJ, but this probably is not correct. See SAN. Poll (1941a: 17–18) reported LUJ from the Uele River of the northern part of the Central Congo drainage and noted that LUJ also had been taken within the central part of the drainage. In 1952 (: 165) Poll restricted the range of LUJ to the Kasai drainage and to gallery forests. In 1930 and 1931 Pellegrin described two subspecies of LUJ; see LOU and OGO. In 1961 Weitzman identified a strain of aquarium-kept Congo *Aphyosemion* as LUJ, according to La Corte (1961: 64–67), who kept this strain and sent me eggs. La Corte's strain or other similar strains had

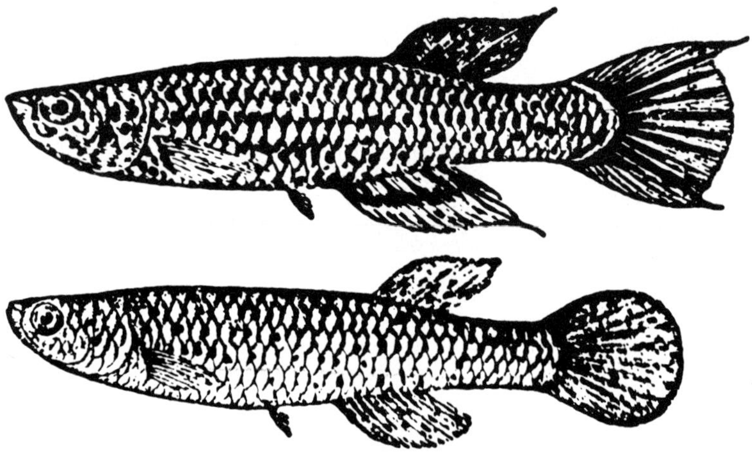

Haplochilus lujæ.

♂ & ♀, types.

The types for LUJ as figured by Boulenger 1915.

been kept in the aquarium since 1961, but usually the STR name has been used for these strains. See Scheel (1967: 6–8). The identity of this strain is obscure; perhaps it represents an undescribed and unnamed species.

This strain originated in a locality about 200 km north of Stanley Pool, about near the mouth of the Kasai River. The meristic characters and some traits of the color pattern, and in particular the pattern of the caudal fin, place this strain close to Boulenger's LUJ, and for that reason I think that Weitzman's identification should not be changed on the base of present evidence.

Boulenger said: ". . . with numerous rather large carmine spots dispersed irregularly, dorsal fin with small round purple spots, anal and caudal with large purple markings which may form streaks in the directjon of the rays." This indicates that LUJ belong to the difficult "group," of "flame-tailed" species of *Aphyosemion*. This group has CAM, LAB, LOU, LUJ, NDI and OBS and probably some more species. In these, the red pattern of the caudal fin generally is a number of red lines on the membrane in between the fin rays.

From the eggs from La Corte I raised an aquarium strain, which became extinct mostly because many individuals became dwarfed and could not be raised to sexual maturity. Hybrids from crossings of this species to some other species of *Aphyosemion* were very difficult to raise. I had a new strain in 1967.

The present strain has been described and pictured by Elek (1965: 296–97), Hesse (1967: 1) and Häfelin (1963: 332–34).

LUJ was cytologically studied after the manuscript had been finished. It has 20 haploid chromosomes and thus LUJ is the sole species of *Aphyosemion* of the n = 20 group which lives south of Cameroon. The karyotype is characterized by a graded series of rather short superacrocentrics, the longest of these measuring about 1.4 times the length of the largest element of the basic complement for the family (for n = 25). Also the lengths of the superacrocentrics of LUJ appear to correspond well to the supermetacentrics of LOU and the latter might represent a derivate of the former, the twenty acrocentrics having fused into ten metacentrics. Compared to all species of *Aphyosemion* and *Roloffia* so far studied, LUJ and LOU have the shortest superacrocentric elements and probably these two Congo species represent rather ancient rivuline forms in Africa.

Young male of LUJ from locality about 200 km north of Kinshasa (Leopoldville), Congo.

LUT: ***Haplochilus panchax lutescens*** **Köhler 1906** (=PAN?)

Köhler (1906: 387–91) based this name on a strain of aquarium-kept PAN, which he thought originated from New Goa in India. He defined the new subspecies on the basis of color patterns only. The sides of the male were olive-brown, darker above, lighter below. Each scale had an indigo spot edged with gold. The ventrals were lemon; the anal also was lemon, or greenish with violet spots, and posteriorly with an orange margin. The caudal was marked like the sides, with a broad sulphorous to yellowish-white submarginal band and a black edge. The basal part of the dorsal was black, shading into sulphorous to yellowish-white, with a broad black edge. The colors of the female were not described because he had no females. In 1907 (: 16–19) Köhler reported that he had had the "true" female of LUT, and that the strain was breeding true to colors in three generations (apparently within less than one year!!). Unfortunately, he did not describe the colors of the LUT female. Further information is given under PAN. As I said under PAN, Köhler's LUT may represent a distinct species, but until living specimens from many localities in India, including New Goa and the Ganges River, have been carefully studied, LUT probably should be considered a subspecies or an accidental color variety of PAN. See also MAI, MAN, DOM, and SIA.

Individual of MAC. H. Hansen photo.

Pair of one of the southern phenotypes of the GRA-MAC group. Photo by Dr. Herbert R. Axelrod.

MAC: ***Haplochilus macrostigma*** **Boulenger 1911** **(Valid sp.?)**

Boulenger (1911: 268) had a single individual caught by Ansorge at Cabinda, Portuguese Congo. In 1912b (: 22) he had numerous individuals from Lundo on the Luali River and from the Luculla River, both rivers being affluents of the Chiloango River. He also said that the holotype originated from the Lucola (River) near Cabinda. In 1915 (: 73) he also had four individuals from Boma of the Lower Congo area.

Pellegrin (1924: 290) reported MAC from Loango (probably the town situated on the coast between the Loeme and the Kouilou Rivers in the former French Congo, although Pellegrin placed the locality in Gaboon.

Fowler (1930: 79) reported MAC from the Chiloango drainage. He said that his ten individuals were paratypes. Poll (1952: 161) said that the Congo Museum had numerous individuals of MAC from Portuguese Congo, from the Mayumbe district of the Lower Congo (Chiloango drainage) and from the Lower Congo drainage.

All these reports seem to restrict MAC to the coastal parts of the Loeme, Kouilou, Chiloango and Congo drainages. But see also GRA.

I have seen some of the specimens reported by Boulenger and by Poll, ten specimens from the Luali River at Lundo, one from the Luculla River, five from Boma (coll. Nicolay 1942), and fifteen individuals from the Lubi River (coll. Csikay 1948).

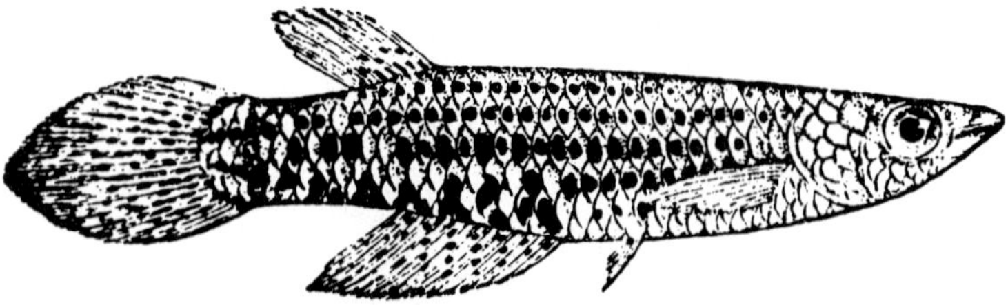

Haplochilus macrostigma.
Type.

The type for MAC as figured by Boulenger 1915.

The Luculla, the Luali and the Luibi individuals are rather similar.

The red spots on the sides are rather large and perfectly rounded. The spots produce irregular rows along the sides, and generally there is one such row just above the root of the anal and along the lower outline of the side, just as in GRA. On the throat of males, the anterior-most band of the usual *Epiplatys* pattern is brilliant red and usually complete. The second band, between the corners of the mouth, is rudimentary or absent. In the females, the anterior-most red band is faint or absent. The anterior part of the throat is faintly dark and distinctly separated from the posterior part, which is white. Dark pigments also are present on the throat of males, but in these the dark color extends all over the throat.

The Lubi individuals of both sexes have a faint dark band from the head to the root of the caudal, this band being $1\frac{1}{2}$ scales wide. This band is absent in all other individuals. These three lots had: D. 9–10, A. 16–17, Sq-l. 27–29.

Preserved male of "MAC" from Boma, Lower Congo. No. 1942, 10.6.10–14 of the British Museum. Nicolay collection.

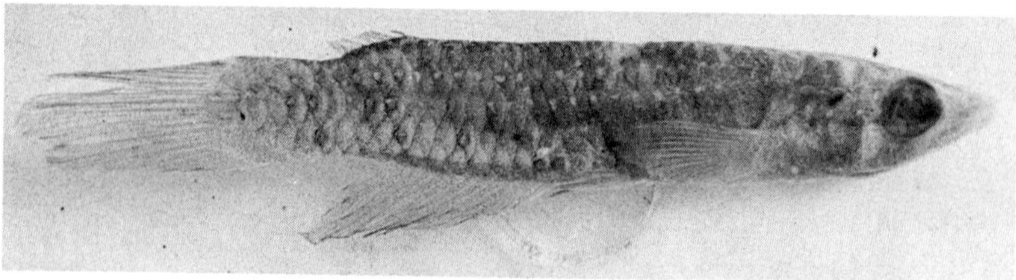

The single Luculla individual is more slender and has fewer red spots on the sides, and these spots are irregularly arranged.

The Boma specimens resemble the Luculla specimen in general appearance and in color pattern. They have, however, fewer fin rays and scales: D. 8–10, A. 15–17, Sq-l. 24–28, but probably they are referable to MAC. They are, at least, referable to SIN, see SIN.

Preserved male MAC from Lundo on the Luali River, Portuguese Congo. No. 1912, 4.1, 465–74 of the British Museum. Ansorge collection.

The British Museum has interesting lots of MAC-like fishes. Two individuals, identified as SEX, originated from Gaboon and were collected by Higgins in 1872. They do not belong to SEX. The pattern of lateral line organs on the dorsal part of the snout is "closed." The specimens resemble MAC in general appearance and in color pattern. The red spots are, however, crescent-shaped and irregularly distributed, tending to develop thin red cross-lines posteriorly on the sides of the largest specimen. The throat pattern does not correspond to MAC, although the two cross-lines of the usual *Epiplatys* pattern are present, but in these individuals the lines are dark, although they contain red pigments in the small specimens (but not in the largest). In the small individuals, the throat-pattern resembles that of GRA from Lagos: the two lines are fullly developed. D. 9–11, A. 15, scales lost in these two individuals.

There are also six individuals, all males, collected at the Benito River of NW Gaboon in 1907. These individuals first were identified as SEX, which they are not; later they were identified (by Radda 1963) as MAC. The general appearance is that of very large specimens of MAC and the color patterns are similar. Faint dark crossbars are, however, visible in some specimens, these bars being of the GRA type. The row of red spots above the root of the anal fin is present. The red spots on the sides are perfectly rounded and large. The anterior-most red band of the throat is fully developed in most. The second band is absent or represented by a very faint red pigmentation.

Adult female of MAC from the aquarium trade.

D. 9–10, A. 16–17, Sq-l. 28–30 in these six individuals. Although Bates did not give the exact locality for this sample, these specimens can not possibly originate from the coastal part of the Benito drainage, which is in Spanish Guinea, not in Gaboon.

The first live representatives of MAC were imported into Germany in 1910, according to Brünning (1911: 650–51), who said that specimens of this strain had been identified as SEN (=SPI) by Boulenger. These individuals were said to originate from Congo. His figure, showing the male and the female, is very bad and not suited for identification, but his description corresponds rather well to that of MAC. Träber (1912: 17), referring to Brünning's description, described and figured (Thumm) other aquarium-kept individuals said to originate from the Congo. His description and his figures probably are referable to MAC. He described the caudal fin of the male to be similar to that of the DAG male, and his figure shows a short lower streamer. He used the SEN name for his material, but Wolterstorff (in a footnote) expressed doubt that SEN was the correct name for this strain. Wolterstorff (1913: 356–58) said that the strain called SEN by aquarists should be named MAC, and that this identification was made by Boulenger on the basis of new preserved material from Wolterstorff. Arnold (1913: 230) said that the collector of this strain first said that the individuals originated from the Lower Cross River, but later it was made clear that they originated from Boma of the Lower Congo. The preserved specimens, which were first sent to Boulenger, were labelled "Lower Cross River." Later Arnold sent more preserved material, and this was identified as MAC by Boulenger. The error of the first identifications might have been caused by an exchange of labels, because at the same time, Arnold also sent two individuals of SEN labelled PET to Boulenger. See SPI.

Sachs (1919: 202) said that MAC survived in Germany. Beyer (1930a: 169) reported MAC in the aquarium, and Arnold (1933: 649) reported MAC in a shipment of aquarium fish from Spanish Guinea (Rio Muni). His figure, showing this strain, indicates a broad dark longitudinal band in the female.

This pattern appears to be very rare in GRA and MAC.

Since World War II, MAC has frequently been imported, and Meinken (*Die Aquarien Fische*, pages 1191–93) considered MAC and SHE to represent subspecies of CHE, a point of view which cannot be accepted. MAC and CHE are probably not closely related, and SHE appears to be a subspecies of CHA not related to MAC.

The sides of the MAC male are brilliant bluish-green; the fins usually are yellow. Some of the rays of the lower part of the caudal sometimes extend and build up a short streamer or a pointed edge, as in DAG, CHA (including SHE) and sometimes in SEX and CHE.

MAC undoubtedly is closely related to GRA and ORN. These three forms correspond very well in meristic characters, in color pattern (red pigments) and in general appearance. MAC and GRA also correspond in the size of their eggs and in hemoglobin pattern (4 lines). These three forms probably represent a superspecies ranging from southwest Nigeria to the mouth of the Congo River, being restricted to coastal biotopes, probably usually in swamps. See also GRA and ORN.

MAE: *Aphyosemion maeseni* Poll 1941 **(Valid sp.)**

Poll (1941: 141) had a single specimen, probably a male, from the Maeseni River, an affluent of the Nuon River of western Ivory Coast. The preserved specimen was brownish, darker on the back with the dorsal, anal and caudal very dark, the former with some faint dark spots. Poll placed MAE closest to CAM. No further information concerning this species has been published, but reports of LIB from western Ivory Coast (Pellegrin 1934: 117), from Mount Nimba (Daget 1948: 41) and from Banfora of the Upper Volta drainage (Daget 1960: 326) might concern MAE and not LIB.

I have studied a large collection of preserved specimens of a nominal *Roloffia* species caught by Herold Olsen near N'Zérégoré of High Guinée in 1963–64. The number of dorsal finrays was counted in 476 individuals:

fin rays	8	9	10	11	12	13	14	
frequencies	1	2	177	210	42	37	7	D_{mean}: 10.9.

The variation observed, appears to be rather high for a single (?) population of a *Roloffia* species, and other meristic characters were also rather variable: A. 14–18, Sq-l. 27–34. The different samples taken at different places at different times did not indicate any clear separation of two nonsympatric forms, but for large samples of *Aplocheilichthys* individuals such separation into two distinct (partly sympatric) species was possible. The color patterns of the *Roloffia* individuals strongly indicated that two sibling species were present, but I was not able to make a clear diagnosis. Individuals having a low dorsal and anal fin count generally were more heavily built, were less dark on the sides in the males, had larger eyes (7.8–8.1% of standard length) and in the pattern of lateral line organs on the dorsal part of the snout, the fleshy lobe in front of the neuromast was undivided. Individuals with high dorsal and anal fin counts were slender, larger, and darker on the sides in males, with smaller eyes (5.7–6.4%); in some individuals, the frontal fleshy lobe was divided into two lobes, in part, as in the sympatric LAM.

Preserved large male of the smaller, heavily built phenotype, generally with low dorsal fin count (D.10 in this male). MAE from N'Zérégoré, Guinée.

The individuals last mentioned resembled GUI by the almost total lack of red spots on the sides and in the fins of males and by the shape of the head, when seen from above. The head of GUI resembles that of an *Epiplatys*, when seen from above.

The frontal scalation of all individuals was, however, normal (G-type). In some males of both "forms" the dorsal, anal and upper and lower part of the caudal were often broadly edged with white. In some males, the lightly colored marginal band was narrow or absent in some fins; in large males of both "forms," the caudal fin was edged all around by white, as in CIN. In other males, only the hindmost part of this fin was lightly colored, a very rare fin pattern among rivulins of the Old World.

In some of the robust males, always smaller than the slender males, there were red spots or crescent shaped markings anteriorly on the sides, especially above the median line. In a few males there was a dark vertical blotch above the root of pectorals, as in GER females, but in most males no trace of such

Preserved large male of MAE from N'Zérégoré, Guinée. H. Olsen 1964. Notice the lightly colored borders of unpaired fins are fully developed in this male. 14 dorsal finrays.

marking was visible. The throat pattern generally was a very dark over-all pigmentation, sometimes with a rounded darker marking anteriorly, as in most active individuals of *Roloffia*.

Females of both "forms" were brownish. The rounded dark spot anteriorly on the throat was present in some specimens; in others it was not. Some females had a very conspicuous dark marking just above the root of the pectorals, as in the GER female, but other females had no trace of such marking. The dorsal usually was marbled and with a broad white margin; the anal had a similar edge, but without the marbled pattern.

A separation based on the color patterns of males and females did not correspond to any separation based on meristic characters.

The development (or lack of development) of red pigments on the sides of males seems to separate the N'Zérégoré form (or forms) distinctly from LIB, ROL, BER and GER, and Herold Olsen's individuals are referred to MAE, an identification which Poll accepted. A study of live material from High Guinée is needed.

MAG: *Aphyosemion margaretae* Fowler 1936 **(Valid sp.??)**

Fowler (1936: 320) had two specimens from a locality near Saidi's Village, Irumu-Avakubi road, 10 miles west of Epulu River ferry, Kibali-Ituri districts, of the former Belgian Congo. Elevation 2800 feet. The village clearing (elevation 2800 feet) was in a dense and very tall rain forest of the typical Ituri-Congo character. Occasional small areas of scattered second growth forest indicated former clearings. The primeval forest was of the storied type, with the upper story reaching as much as 200 feet above the ground; this was in 3–16 September, 1934.

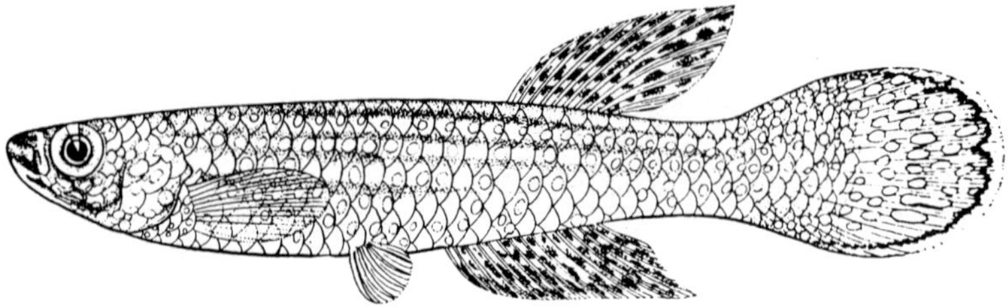

The type for MAG as figured by Fowler 1936.

Fowler's excellent ink drawing shows a male of the common Congo *Aphyosemion* phenotype, but differing from other species by the completely rounded caudal fin and by a series of faint dark stripes in between the rows of red spots. There are rather many red spots on the sides and many such spots in the unpaired fins.

Fowler placed his MAG closest to ORN, a point of view which is very difficult to understand. In the description he said that the depth of the two specimens was 5.75 to 6.66 times in length, indicating a very slender form, but the figure shows an individual not extraordinarily slender. If the figure of the type is correct, Fowler used the total length, not the standard length, as a unit of measurement, but still his MAG is rather slender, if full grown.

Poll (1952: 167) considered MAG to be based on juveniles or females and said that it might be close to FER on the basis of published data.

Lambert, who visited this part of the old Belgian Congo, has told me that individuals of rivuline species are very rare east of Stanleyville and are replaced by individuals of protatopodines ("lampeyes"), but he found a form rather similar to MAG much more farther south at Nyangwe, on the Lualaba River (northwest of Kasongo of Katanga), also in a geographical area where individuals of *Aphyosemion* are not expected to occur.

The dark lines on the sides, shown in the figure, but not mentioned in the description, are a character also present in males of STR and in OGO, and this pattern probably is important. The rounded caudal may or may not represent a distinguishing characteristic (see BIV), as various grades of development of streamers are known to exist among Congo *Aphyosemion* species. ELE and CHR have long streamers, whereas in COG the streamers are very short.

Although MAG may represent a distinct species, it is undoubtedly very closely related to the main group of Congo *Aphyosemion* species.

MAI: *Haplochilus panchax mattei* Köhler 1906 **(=PAN?)**

Köhler (1906: 387–91) had a strain of aquarium fish said to originate in Bombay, extending southwards to New Goa. He received this strain from

Matte of Berlin. Köhler said that the MAI strain bred true to color patterns, which he thought differed from those of PAN. He defined MAI on the basis of these differences only. It is difficult to understand how such differentiation of PAN could be prepared, as Köhler did not know the color pattern of Hamilton-Buchanan's PAN. Köhler compared MAI to another aquarium strain of PAN, and he probably thought that this strain represented the "typical" PAN by colors. See PAN.

In MAI there was a dark stripe anteriorly on the back. The basic color was olive-green to golden above, with a brilliant sky-blue on the median part of the sides and below the median line. The male had lemon ventrals; the anterior part of the anal also was lemon, the posterior part being green and with dark spots and a broad orange-red marginal band. The caudal was greenish-grey, with a broad orange margin and a narrow dark edge. The basal part of the dorsal was black, followed by a yellow zone which changed into orange in the outer part of the fin, which has a narrow dark edge. The colors of the sides of the female were similar to those of the male, but less brilliant. The ventrals were transparent, with yellow fin rays in the outer part of the fin. The anal also was transparent, with a broad cinnabar marginal band; the caudal was transparent, with a cinnabar margin. The dorsal was black at the base, cinnabar at the top. See also LUT, MAN, DOM, PAN and SIA.

MAN: *Haplochilus panchax marginatus* Köhler 1907 **(=PAN?)**

Köhler (1907: 16–19) attached the MAN name to an aquarium strain of PAN which he thought came from Ceylon. This strain had previously been named *Haplochilus panchax* var *dayi* by German aquarists. Köhler found that this strain was not DAY; he probably created the MAN name in order to separate this strain from the usual aquarium strain of PAN. He thought that MAN differed from other strains of PAN by the colors of the female, and that in MAN the anal and caudal were edged with black. The males of MAN were similar to those of the Calcutta strain, which he thought represented the typical PAN. In my opinion, the differences published by Köhler for his separation of MAN from PAN are not sufficient for the creation of a new subspecies. This aquarium strain also was described by Ruda (1908: 671–73) under the DAY name, as given above.

MAR: *Haplochilus marnoi* Steindachner 1881 **(=SPI)**

Steindachner (1881: 198) based the description of MAR on individuals from Bahr-el-Seraf and Bahr-el-Gebel of the Nile drainage. Garman (1895: 130) considered MAR to represent a synonym for SEN, and Boulenger (1907: 418–19) apparently had a similar opinion concerning the individual from a locality near Khartoum of Sudan, which was used by Günther for the description of FAS. Boulenger also referred an individual caught by

Loat at Lake No and individuals caught at Mongalla, near Gondokoro of Sudan, by Werner to MAR, but he did not join MAR and SEN (=SPI), because he had not seen Steindachner's type. Werner (1919: 447) said that he found MAR to be common at Gonga within the vegetation along the banks of the river, and that he also found this species at Renk. He considered MAR to be viviparous and said that one individual gave birth to a fry with yolk-sac during preservation. He observed up to ten dark crossbars in the individuals from Renk.

I have seen two specimens (British Museum) caught by King at Khor Barbry in 1913. They (D. 8–9, A. 15–16, Sq-l. 27–28) correspond to SPI in all details, including color patterns. There are three distinct dark brown crossbars in the caudal fin. I have also seen seventeen individuals caught at Lake No by Lewis in 1948. Although these individuals are small, they do not differ from juveniles and semi-adults of SPI. The largest individuals had: D. 8–9, A. 16, Sq-l. 28.

There seems to be no reason to consider MAR to differ from SPI, and the Nile populations of BIF do not seem to differ from West African BIF.

MAT: *Epiplatys matlocki* Fowler 1950 **(=FAS)**

Fowler (1950: 225) based his description and naming of MAT on twenty-five specimens from Robertsport of western Liberia. He placed MAT closest to FAS, but thought that it differed from FAS by minor details in morphology and by important differences in color patterns. In all his individuals of MAT there was a broad median dark band from the head to the root of the caudal, and he considered the dark crossbars on the sides to be much wider than those figured by Boulenger (1915: 52–53). I agree with Fowler in this. The bars of Boulenger's FAS are very narrow for an individual of Sierra Leone FAS. Fowler also said that the MAT individuals first were in formalin, then in alcohol. This explains the presence of the dark lateral band and the full development of the dark crossbars. Rivuline individuals preserved in formalin always develop a maximum of black pigments; usually the temporary dark patterns, which were invisible before preservation, develop maximum intensity. At the same time, the red and yellow pigments are completely destroyed if the individuals are not removed to large amounts of alcohol soon after the first preservation.

The broad, dark lateral band which Fowler considered to characterize MAT is a pattern found in all specimens of FAS, but because of the temporary nature of this marking, it usually is not present in individuals preserved in alcohol. Also Ahl used this lateral band for his description of LEO. Fowler did not describe any red pattern for MAT, probably because the individuals were kept far too long in formalin before they were transferred to alcohol.

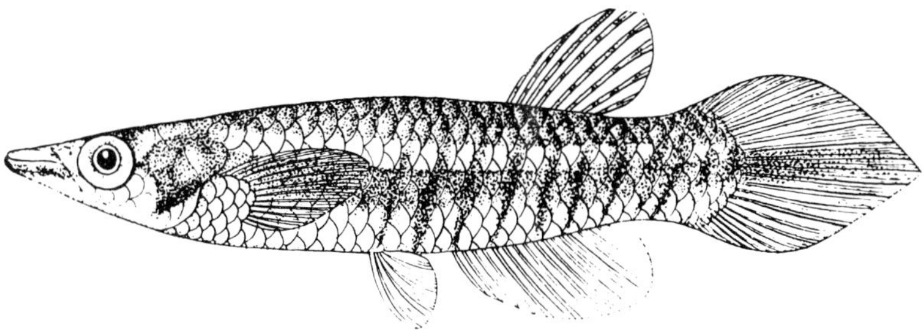

The type for MAT as figured by Fowler 1950.

MAT does not seem to differ from FAS by any external character, and FAS is likely to occur at Robertsport.

Live specimens of Liberian FAS were imported into Germany from Totota, north of Monrovia, in 1965, according to Basch (1965: 203–04). J. Franz sent me photos showing individuals from this strain. The males correspond to Sierra Leone FAS, except for the red pattern of the anal fin. In Sierra Leone FAS males there are two thin red lines in the bottom part of this fin, whereas in the Totota males, the two lines merge, producing a rather broad band. In 1965 Clausen sent me individuals from a population of FAS, which he found about 20 miles north of Monrovia. The individuals unfortunately died just before they reached Copenhagen. In the two adult males of this lot, the anal fin and the lower part of the caudal fin have a broad almost black marginal band, with no indications of two red lines. The upper part of the side has an almost perfect red reticulation, whereas in the middle of the side there is a strong tendency for the building up of rounded red spots, the red pigments being heaviest at the junctions of the reticulation. On the throat of both sexes, the usual *Epiplatys* pattern of two dark lines is visible. D. 11–12, A. 16–18, Sq-l. 28–30 for six individuals.

It has already been said (under FAS) that different populations of this morpho-species appear to be reproductively isolated and for that reason any constant difference in morphology or in color pattern which will be maintained in preserved material might be important for the future separation of various bio-species. The color pattern of the anal fin of certain Liberian FAS seems to separate these populations sufficiently from Sierra Leone FAS, but an intensive study in biology of many populations of FAS is needed before any differentiation of sub-units of the FAS superspecies can be made.

MAY: ***Nothobranchius mayeri*** **Ahl 1935** **(=ORT?)**

Ahl (1935c: 126) based his description of MAY on aquarium-kept individuals which he received from F. Mayer, who (1936: 258) said that he had received them from a friend in Beira, Mozambique. Ahl separated MAY from the very similar KUH by the number of scales around the body (40 in MAY and 27 in KUH) and placed MAY closest to NEU and MIL. The color pattern of MAY is very similar to that of KUH (=ORT?) and the sole difference seems to be the number of scales. Ahl's type for MAY probably needs re-examination, because in many cases his data for meristic characters are not too reliable.

Rachow (*Die Aquarienfische*, page 909) suggested that MAY may be a variety of ORT. No further reports of MAY have been published, and Mayer's strain probably soon became extinct. Adult males measured 7 cm total length.

MEI: ***Aphyosemion meinkeni*** **Myers 1933** **(Valid sp.??)**

This species was imported into Germany by the "Platy-Tischrunde" of Hamburg in 1930. Meinken (1932: 56–57) was aware that it resembled CAM (as defined by German aquarists of that time), but he thought that it differed from CAM by important characters. He sent two preserved specimens to Norman of the British Museum and asked Norman to compare these individuals to the types for CAM. Norman answered that the two individuals of "*Panchax* from Cameroon" had been compared to the types for CAM and that they differed by being shorter and deeper, also by their considerably shorter caudal fin and by the very different color pattern. Meinken now described and named the strain under "*Panchax* (*Aphyosemion*) *pictus* sp. nov.". He placed the new species closest to BUA, although he was aware that Holly considered BUA to represent a synonym for CAM. Meinken, however, considered BUA to represent a valid species, differing from CAM by the shorter caudal fin.

He gave "Tropical West Africa" as the type locality.

Myers (1933: 181) considered the name "*pictus*" to be preoccupied by *Panchax pictum* Cuvier and Valenciennes 1846 (=*Betta picta*) and renamed Meinken's PIC as MEI.

When Meinken considered his new species to differ from CAM, he probably had in mind the strain of aquarium fishes imported from Gaboon(?) in 1913, which Arnold identified as CAM. The 1913 strain identified by Arnold as CAM definitely was not CAM, but probably a form close to OBS and STR. MEI also seems to be rather close to OBS and STR, but on the basis of published information it is impossible to determine whether MEI is a component of STR or similar forms.

The male MEI had brilliant bluish-green sides with four rows of red spots, which sometimes merged, producing red lines along the sides. The

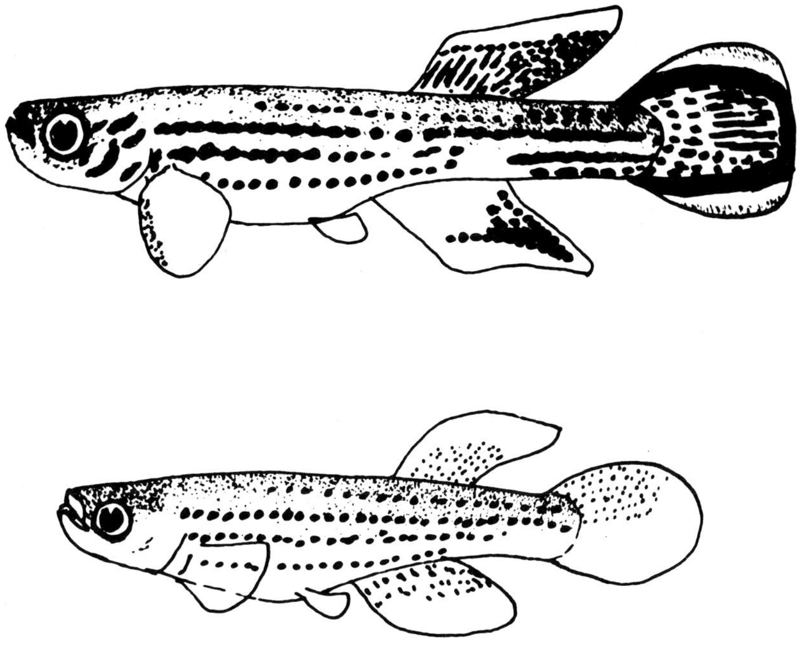

The types for MEI redrawn after Meinken's drawing of 1931. (*Panchax* (*Aphyosemion*) *pictus* Meinken).

fins were yellowish, changing into gold at the tips and with many red spots and streaks in the dorsal and central part of the caudal. The red spots of the anal produced a triangular pattern in the posterior part of fin. The caudal had an upper and a lower submarginal red band, which became marginal in the posterior part of the fin.

MEL: *Epiplatys melanteron* Fowler 1950 **(Valid sp.??)**

Fowler (1950: 225) had three individuals from Robertsport in western Liberia. The description and the figure of MEL cannot be referred to a species of *Epiplatys*, but to a species of nominal *Roloffia*. Colors in alcohol: dull brown, with more or less uniform appearance, most scales with traces of small pale spots, perhaps crimson in life. Fins pale greyish to whitish. A black blotch on front of chin. Dark transverse bar at caudal base. Some specimens show traces of a dark axial band. Fowler said that this species did not appear to be allied with any of the known species of its genus, and certainly he was right in this. He also said that it should not be identified with the Liberian LIB and BIF species. His individuals were small, being only 20–25 mm.

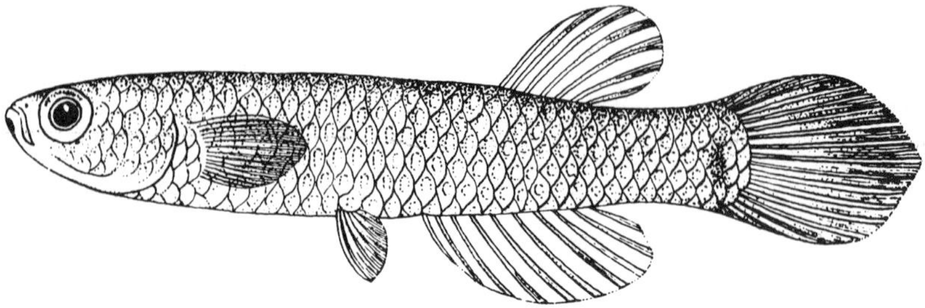

The type for MEL as figured by Fowler 1950.

It has already been mentioned under MAT that Fowler said that the Robertsport sample first was in formalin, then in alcohol. The red pigments of the MAT specimens have been lost completely. The pale spots in the MEL individuals probably were red in life, as indicated by Fowler.

The published number of fin rays and of scales in a transverse series separate MEL distinctly from such forms as LIB, ROL, BER and GER. The black blotch on the front of the chin indicates that MEL belongs to *Roloffia* (or *Aphyosemion*), and the general appearance of MEL strongly supports this idea. In an analysis of meristic characters MEL comes closest to PET within nominal *Roloffia*. PET has, however, not been reported west of the Ivory Coast, although Bruce Turner of New York wrote me some years ago that he had received live individuals of a PET-like species from a friend in Monrovia.

MEP: *Fundulus melanospilus* Pfeffer 1896 **(Distinct sp.??)**

Pfeffer (1895: 48) based his description of MEP on four specimens from Longo Bay and referred some individuals reported by Günther from the Seychelles Islands to MEP. According to Klee (1965: 11–16) the Longo Bay is in reality a small stream running past the modern town of Frehami, Tanganyika (on the western edge of the coastal zone). Under GUE (see GUE) Pfeffer also refers to the Seychelles individuals, which Günther thought were the females (of ORT), but which Pfeffer thought belonged to a different species, on the basis of the position of the dorsal fin.

In GUE, the first ray of the dorsal fin is inserted midway between the posterior border or center of the eye and the root of the caudal, whereas in MEP (according to Pfeffer) this ray is midway between the middle of the opercle and the root of the caudal. In Pfeffer's figure showing the type (?) for GUE, the first dorsal fin ray is placed as in MEP, not as in GUE. Hilgendorf (1905: 416) said that Pfeffer's figure of GUE had been copied from a figure published by Günther and Playfair.

The colors of MEP were given by Pfeffer. The basic color was a light greenish-grey with irregularly distributed dark markings almost of the size of the pupil. These spots occurred in particular on the posterior part of the

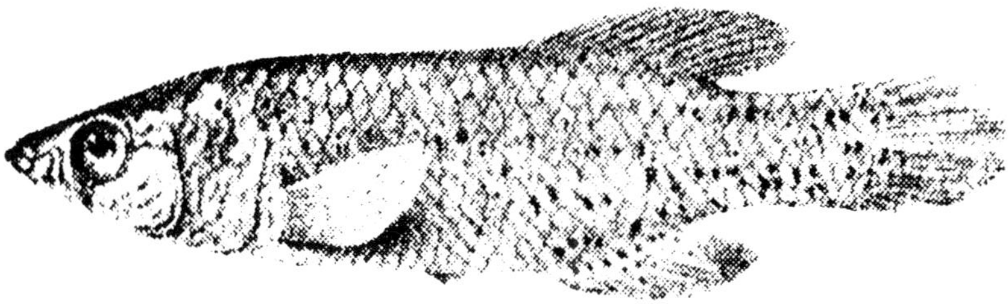

The type for MEP as figured by Pfeffer 1896.

side. There were smaller dark spots on the back and on the caudal peduncle and in the lower part of the dorsal. He said that he had only females of MEP. He did not publish any information concerning the teeth, but as he separated MEP from GUE on the character above mentioned, it appears that the tooth patterns are similar.

Boulenger (1915: 34) considered Playfair's (and Günther's) specimens from the Seychelles to be MEP, but he also said that the dorsal originated midway between the occiput and the root of the caudal. Males of MEP were unknown to him. See also SEY.

A strain of MEP-like individuals was brought to the United States by Hansen in 1955. This strain originated from a locality near Dar-es-Salaam in Tanganyika. The males of this strain were very similar to those of GUE and, like these, they had a rather pointed head. The fins were dark; the red pattern of the sides was very regular, with a full reticulation. There were no red spots on the lower surfaces of the head or on the belly. The females of Hansen's strain had a variable number of dark markings on the sides, in particular posteriorly and near the median line. The fins often had small dark spots near base. This form belongs to the larger species of the genus, measuring at least 8 cm in the male. Eggs measured 1.3 mm. The strain soon became extinct. I kept it for some years, but I did not preserve any individuals.

Pair of MEP (?). W. Foersch photo.

MIC: ***Panchax microstomus*** **Ahl 1924** **(=CAM)**

Ahl (1924: 303–04) had four specimens from Cameroon. No further information concerning the type locality was published. The types had been collected by Zenker. Holly (1930: 212–13) placed MIC as a simple synonym for CAM, because he found that the sole distinguishing morphological character used by Ahl to separate MIC from CAM was not correctly given by Ahl. Ahl thought that there were 26–28 scales around the body in front of the ventrals, but Holly found only 22–24 such scales.

This means that MIC cannot be separated from CAM (and 47 other nominal species of rivulins from Atlantic Africa). Radda has informed me that the types are no longer kept in the Berlin Museum, and they probably have been lost.

The length of the caudal fin rays and the proportions of the caudal peduncle indicate that MIC belongs to *Aphyosemion*, not to *Epiplatys*.

MIL: ***Fundulus microlepis*** **Vinciguerra 1897** **(Valid sp.)**

Vinciguerra (1897: 356) had several specimens, caught by Capt. Bottago in temporary pools at Saha, at the foot of the Egherta Mountains, between Brava and Lugh of formerly Italian Somaliland. No further report of MIL has been published.

The largest specimen measured 52 mm standard length. MIL differs markedly from all other species of Rivulinae of the Old World by two features. It has 40–42 scales along the sides, whereas in other species, 37 such scales are reached in SJO and SPL only. MIL has a distinct black band from the corners of the mouth, through the eye and across the nape. This particular marking is absent in all other Old World rivulins, but it occurs in some species of *Cynolebias* and *Austrofundulus* of South America, and in the genus last mentioned, *A. dolichopterus* spawns eggs which correspond to those of *Nothobranchius*. The dark band through the eye is marked in the species last mentioned.

Vinciguerra noted that the MIL individuals had been preserved in strong alcohol and that they had been deformed by the preservation. He did not, however, observe any red pattern, which usually preserves well in strong alcohol. Apart from the black band through the eye, he found brown scale edges in some specimens which he thought were males. He did not disclose any ctenoidy in fins of males although he looked for such structures.

MIL may indeed represent a very interesting species which may reduce the gap between the rivulins of the Old World and those of the New World.

MKU: *Fundulus mkuziensis* Fowler 1934

Fowler's (1934: 411) individuals originated from the Mkuzi River drainage of Natal, South Africa. No further information concerning MKU appears to have been published. I have not seen the description.

Adult male of DAG MON. W. Foersch photo.

MON: *Epiplatys dageti monroviae* Daget & Arnoult 1964 **(Valid ssp.)**

The early history of this subspecies of DAG was given under DAG. The first individuals of MON were reported by Peters (1863: 396), who described it under the SEF name, which, however, was preoccupied by Gill's SEX. SEF (=MON) also is the type-species for *Lycocyprinus*, a genus created for SEF by Peters in 1868. This is important to keep in mind in a future division of *Epiplatys* into subgenera.

Live individuals of MON were imported as aquarium fishes in 1908 and were known under the CHA name up to 1964, when Daget and Arnoult (1964: 596–98) realized the true identity of this strain and named it by the MON name. They separated MON from DAG on the basis of the color pattern of the throat of males and by the size of the dark crossbars. Further information concerning the differences between DAG and MON has been given under DAG.

MUC: *Fundulopanchax multicolor* Meinken 1930 **(=BIV)**

Meinken (1930: 18) based MUC on two specimens, a male and a female. He also had one live male for his study of MUC. He said that these individuals came from Cameroon ("Zwei neue *Fundulopanchax*-arten aus Kamerun"), but in a subsequent article (1930: 306) he said that the type-

Mitosis of DAG from Monrovia, Liberia. 1965 strain. 2n = 50.

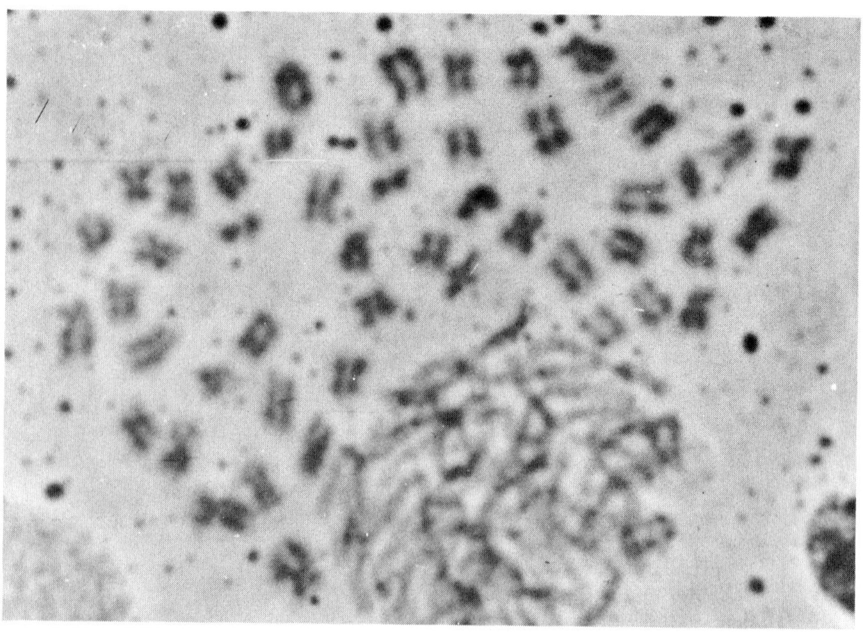

Adult male of DAG MON representing the 1965 strain from Monrovia, Liberia.

individuals were caught by Griem of Hamburg at Ajakapulka, near Lagos of southwestern Nigeria. Ladiges (1951: 97–98), however, said that Griem caught these individuals at Apapa, near Lagos. It is known that Griem (the owner of Aquarium Hamburg) also collected aquarium fishes within the Niger Delta and in West Cameroon, and for that reason we cannot be quite sure that the types for MUC originated from the Lagos area.

Meinken's description of the color patterns of MUC indicates, however, that the types originated from southwestern Nigeria and not from the Niger Delta or from West Cameroon. He said that the area below the dorsal fin was cupreous. This is a character of southwestern Nigerian males of BIV, whereas in other populations of BIV, this area is grassy green, except for the Buea strain from the Cameroon Volcano, which develops a bronze shine below the root of the dorsal. The streamers of the unpaired fins of the MUC individuals were light-brown to congo-red. This also is a character of Nigerian males, whereas in Cameroon males these streamers usually are lemon. The pectorals of the MUC males were light yellow, with a green margin. Males from southwestern Nigeria usually develop in the pectorals a more yellow-orange-red color than described by Meinken. Meinken also said that the anterior-most dorsal fin ray was above the center of the anal fin, also indicating a Nigerian origin of the types.

Adult male of DAG. Aquarium strain. This male belongs to the MON subspecies.

Adult male of DAG. Bruce Turner's strain of 1962. This male belongs to the MON subspecies.

Two hybrids DAG/GRA. These female-like hybrids were completely sterile. Notice the red pattern of the sides.

Adult hybrid MON/SEX.

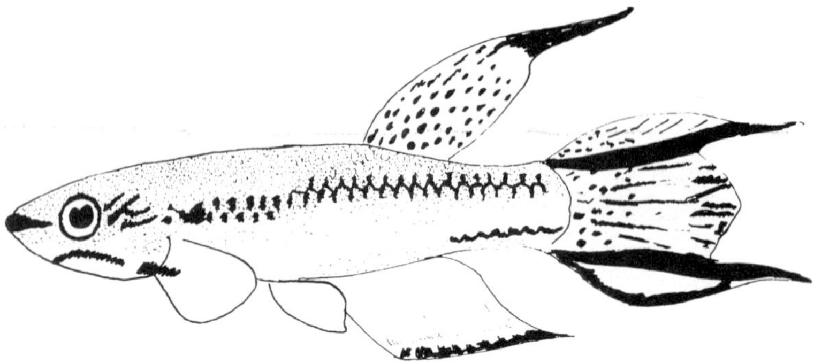

The type for MUC redrawn after Meinken's drawing of 1929.

For some unknown reason, German aquarists soon exchanged the names for Meinken's new species: MUC and SPP. And the aquarium strains of SPP were named MUC, whereas the aquarium strains of MUC were named BIV. Wöhlert (1937: 81) described and figured (two photos) males of SPP under the MUC name.

When colors are not described or pictured, males of MUC and the 1930-strain of SPP (Tiko strain) might be separated by the (red) pattern of pigments of the caudal fin. In the MUC male, there usually are many rounded spots in the basal part of the fin, whereas in SPP there are lines between the fin rays. My Buea males also differ from the almost sympatric nominal SPP by having rounded spots in the lower part of the caudal. The juxtaposition of dorsal and anal fins also might be used to distinguish males of MUC and SPP in black and white photos.

The 1930-strain of SPP seems to be kept by a few aquarists at the present time. Why this strain has been able to survive in the aquarium for all these years probably is a result of the fact that MUC and SPP are completely reproductively isolated.

The usual aquarium strain called BIV by aquarists all over the world actually is MUC, but as preserved individuals of this strain cannot probably be separated from other Nigerian populations of BIV, including the Ndian River males (type-area), I think that the BIV name still should be used for this strain. See also BIV and SPP.

MUF: *Haplochilus multifasciatus* Boulenger 1913 **(Valid sp.)**

Boulenger (1913a: 43) had numerous individuals from Kondue, Kasai, Congo. He placed MUF closest to NIC and GRA, and separated MUF from these by color pattern and by the more elongated body, which he considered to be at least five times as long as deep.

In 1915 (: 75) he figured the type, with six dark crossbars and with two black bars across the throat.

Nichols and Griscom (1917: 725) reported MUF from Stanleyville. These individuals were taken in pools in a clear running forest river, an affluent of the Tschopo River. Fowler (1936: 316) reported MUF from Batangafo of Ubangi-Shari (Uam-Fara district, near the upper reaches of Bahr-Sara) and from a locality 30 km east of Kribi of southern Cameroon. He also caught many individuals of SEX in the locality last mentioned, and probably his individuals of MUF are actually SEX. As for the Batangafo individuals, these might not be MUF, but a different species. See SAN and BAU.

Poll (1952: 162) made BOU a synonym for MUF (see BOU) and reported numerous individuals of MUF from many rivers within the Central Congo drainage, except for Katanga and the upper affluents of the Congo River. In 1959 (: 141) he reported many individuals from the Stanley Pool area.

Lambert (1961: 30–33) reported numerous individuals of MUF from the eastern part of the Central Congo drainage (Banalia, Yangambi, Stanleyville, Opala, Ikela and Lomela districts) and considered, on the basis of morphological similarities, MUF to represent a subspecies of SEX. Clausen (1967: 27) separated MUF distinctly from SEX by the pattern of lateral line organs on the dorsal part of the snout.

Gosse (1963: 214) reported MUF from the Yangambi, where it was found to be very common in brooks and in swamps of the forest. He said that MUF prefers the shallow parts of the biotopes and was found close to the surface of the water. It feeds on terrestrial insects (often rather large) and in particular on ants. It reproduces throughout the year. Water: 23–24°C., pH 3.6–4.0, redoxpotential below ÷100 millivolts. Matthes (1964: 126) reported MUF from the Lake Tumba district and from Ikela,

The type for MUF as figured by Boulenger 1915.

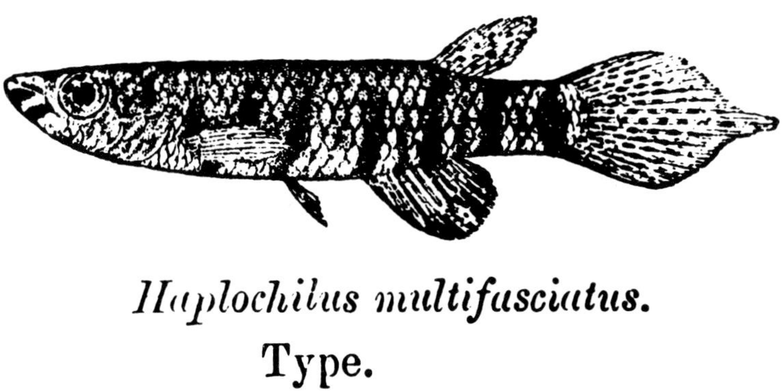

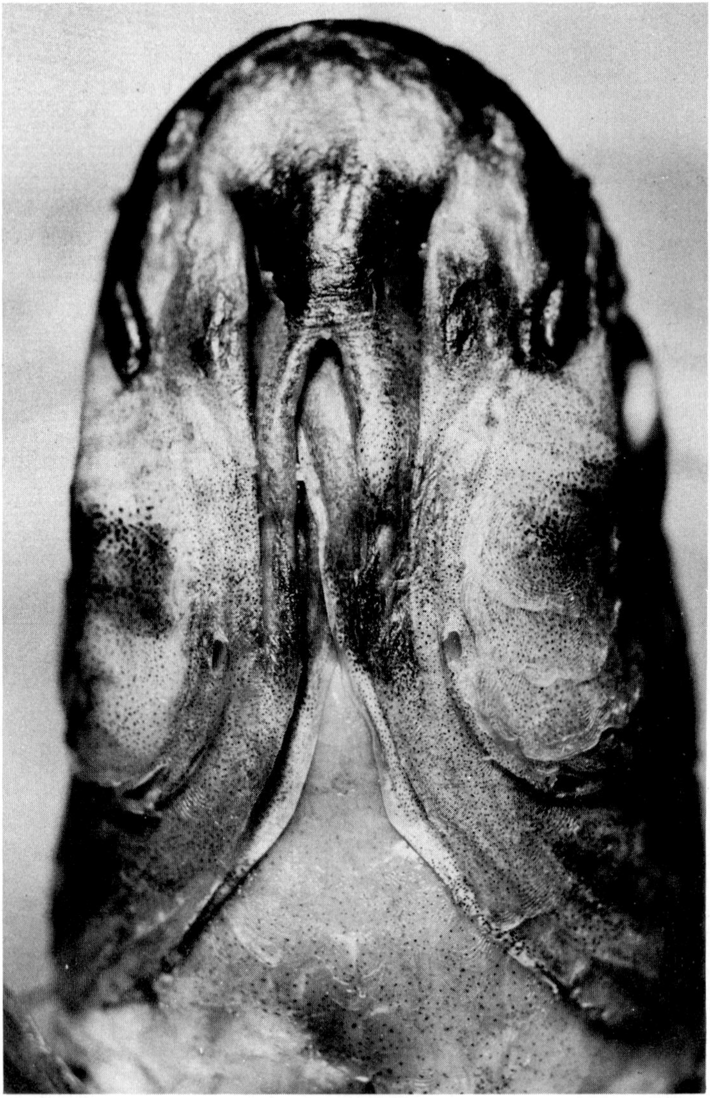

Throat (above) and body of preserved male of MUF from Lundo, on the Luali River, Portuguese Congo. No. 1912, 4.1, 450–55 of the British Museum. Ansorge collection. (SEX by Boulenger.)

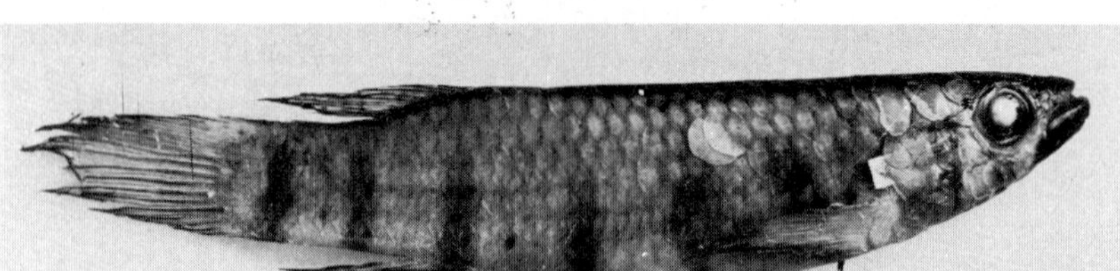

both localities within the Central Congo drainage. He described the colors of the live male. There are six dark crossbars on the sides, the bars often separated by narrow dark to reddish additional bars. The two bars above the root of the anal fin extend into that fin, and the posteriormost bars on the side extend into the lower part of the caudal fin. The back is yellowish-brown, the belly being lightly colored. Each scale has a violet-red spot. The anterior part of the side is bluish-green, the posterior part being orange, more or less. The throat is reddish-violet or violet, with two marked dark crossbars. The dorsal is orange near the base and also has dark red spots in its basal half. The caudal is reddish-orange or salmon, sometimes olive, and it has a broad dark lower and distant marginal band.

Lambert (1961: 30–33) described the sides of the male as bluish, with six to eleven dark crossbars. The fins were yellow, with red spots.

I have seen preserved specimens only. In eighteen individuals from Ibembo (northern part of the Central Congo drainage, west of Buta) of the collections of the Congo Museum, the throat pattern is interesting. It is very marked in the males and often almost invisible in females. In the male, the lower lip is dark brown. Just behind the lip, but clearly separated by a narrow lightly colored band, is another dark band. There is also a third band between the corners of the mouth, often with much red pigment and sometimes not quite complete. The central part of this band is always present. On the edges of the gill covers (interopercles) just below the eyes, there is a dark marking, similar to that found in BIF and in CHE. The branchiostegal membrane has a great deal of red pigmentation. The faint throat-pattern of the female is similar to that of the male, but generally it seems to lack red pigments. The throat pattern of MUF resembles that of MAC and Lagos males of GRA, by the way, and a reduction of the black pigments in MUF would produce the MAC and GRA patterns. A red pattern of two lines across the throat and no pattern in the female would result.

The red spots on the sides of the MUF male are perfectly rounded, not crescent shaped, not even on the back. There is a characteristic pattern of dark red spots on the belly of males. This pattern is not present in the females. The pattern is situated in front of the ventrals.

The dark crossbars are situated low on the sides compared to the condition in SEX and are often rounded dark blotches rather than real crossbars.

Six large individuals from the Luali River at Lundo, Portuguese Congo (British Museum collections) are rather similar, and probably they are referable to MUF. They were caught in 1912 by Ansorge. Two small individuals from Banzyville, Ubangi, Congo also from the British Museum are similar. D. 9–11, A. 14–16, Sq-l. 28–31 in these three lots.

I have not been able to find any notes on MUF in the aquarium literature, except for a photo by Foersch in Sterba (1959: 320) showing young individuals, undoubtedly referrable to MUF. It is very likely that MUF have been imported into Europe and the United States in the many collections of live fishes which have been shipped from Leopoldville, where MUF seems to be rather common.

NDE: *Epiplatys ndelensis* Fowler 1949 **(=BIF)**

Fowler (1949: 265–67) had three small specimens from a locality near Ndele, Ubangui-Shari. This locality is within the Chad drainage, situated at the western foot of the Bongos Mountains of the Central African Republic, at the Bangoran River, an affluent of the Shari River. The largest individual measured 25 mm. The description and the figure, which shows the type, correspond very well to juveniles of BIF. In these, the median dark longitudinal band usually is not complete, but is broken and is made up of a series of black vertical short blotches, just as in Fowler's figure of NDE. BIF has been recorded from the Chad drainage, the Niger drainage and the Nile drainage and it is to be expected to find BIF at Ndele.

Fowler placed NDE closest to MUF and did not even mention BIF (or TAE, which is a synonym for BIF).

NDI: *Aphyosemion ndianum* sp. nov. **(New species)**

The data for the holotype are given in the appendix.

The first specimens of NDI were caught by Clausen in February, 1959, and in April, 1959, and in June, 1965, Schiötz caught more. All were taken in localities near Osomba on the southern part of the Calabar-Eyomojok-Mamfe road. Clausen brought live specimens of NDI to Denmark in 1959. These were placed in my tanks; an aquarium strain was established, but not distributed to aquarists, because we were not able to attach any zoological name to this form. This strain became extinct in 1965.

The NDI name is proposed for this species, because it was said to originate from the Ndian River drainage, situated east of the Cross drainage. Similar individuals apparently have not been taken elsewhere. In 1966 I caught fishes in very many localities east of the Ndian drainage from Mbonge west of the Volcano to Kribi in the southern part of East Cameroon, without finding any individuals resembling NDI north of the Nyong drainage. South of the Nyong drainage and in particular east of Kribi some populations resembling CAM were found, but the individuals were so small and their color patterns so faint that they usually could not be identified. As I have already mentioned under CAM, I have seen nine specimens of CAM from the Nyong River. This British Museum sample was collected by Bates in 1904, and these nine males correspond very well to Boulenger's description of CAM.

Adult male NDI from the type locality.

NDI resembles CAM by the general appearance and by color pattern. It differs from CAM by the higher number of fin rays and of scales. NDI: D. 14–16, A. 15–18. Sq-l. 31–35. CAM: D. 11–12, A. 14–15, Sq-l. 30–33, according to the original description. I found D. 11–13, A. 15–17, Sq-l. 31–33 in the nine Nyong individuals of CAM. The Nyong individuals of CAM also differ from the twenty-two individuals of NDI, which I have seen, by throat pattern (a very marked red pattern in NDI males and almost no red throat-pigments in CAM). In spite of the morphological differences between NDI and CAM, these two forms are without doubt very closely related.

The NDI male has twenty haploid chromosomes. The karyotype is characterized by two superacrocentric elements, of different length. The largest seems to measure about $3\frac{1}{2}$ times the mean length of the basic chromosomes (for $n = 25$), the second measures about 2.7 times the length. There are no indications of metacentric elements larger than the largest acrocentric element of the basic complement, but in primary meiotic divisions, one larger and one smaller bivalent tend to produce ring-elements, indicating that submetacentrics are present. The number of "double-sized" acrocentrics cannot be estimated, because these chromosomes grade smoothly into the basic chromosomes by length.

The sides and fins of the adult NDI male are brilliant blue. For some unknown reason, some color-slides and all color-prints do not reproduce this basic color with the natural very cold blue shine, which often tend towards violet (as in FIL), and the color-pictures show NDI with a greenish-

Adult male of NDI.

Adult hybrid "male" NDI/CIN.

Small hybrid male NDI/GAR.

Adult hybrid male NDI/COG.

blue color, which does not occur in the live NDI. The color and the intensity of the brilliance correspond to that of the LAB male. The red spots on the sides are more numerous and also larger than in CAM. Anteriorly on the sides, the spots often merge, producing short longitudinal lines above the median line, whereas below that line there is a tendency for the development of short vertical bars. All males develop an irregular, often rather broad, longitudinal band along the keel. Whereas in CAM this band usually begins just below the root of the pectorals, this band in NDI usually begins above the anterior part of the anal fin, and is more irregular. Also the upper red longitudinal band is more regular in CAM than in NDI, where it usually does not develop at all and is represented by a row of large red spots or markings. In both forms, there are usually no red markings between these two posterior bands.

The dorsal has many small red spots and streaks, which tend to build up a narrow marginal red line. The margin is blue. The anal usually has almost no red markings except for a more or less complete marginal band, rather narrow and often also rather dark. In some individuals the edge of this fin is blue. The red pattern of the caudal fin is very irregular and no distinct pattern develops. There are red lines between the fin rays ("flame-tailed pattern") and a tendency for the building up of a very irregular lyre-pattern.

The pectorals have a red submarginal band. The throat pattern is well developed and has two complete red lines crossing the throat as in BIV and a number of red spots and streaks. The branchiostegal membrane has much red pigment.

The female does not develop any distinguishing color patterns. There are a number of small brown spots anteriorly on the sides and many such spots in the unpaired fins. The belly is pinkish.

Adult male NDI from the type locality.

Adult hybrid male NDI/CIN.

Ctenoid spines often are present on the scales of median rows, but usually these spines are short and without filamentous fleshy protuberances. Similar small spines occur on the anal fin rays of some males, but no fin-papillae develop. On the upper rays of the pectorals of some males, bud-like papillae supported by ctenoid spines occur. The frontal scalation is normal (G-type, no H-scales) and also the hemoglobin pattern is normal (four-lines). There are two or even three rows of scales found on the opercle below the center of the eye. The anterior-most dorsal fin ray is inserted above the 5th to 7th anal fin ray.

The reproduction of NDI corresponds to that of the species of *Aphyosemion* s.s. No annual characters have been observed. The eggs measure 1.3 mm; the egg membrane has a marked and complete hexagonal pattern. Individuals of NDI are easily kept and bred in the aquarium.

Individuals of NDI have been crossed to GAR, CIN and COG. With GAR, ten hybrids were produced. Two of these died before they reached full size. The eight hybrids all were males, which were functionally sterile and rather viable. With CIN, twenty hybrids were produced, but eleven of these died before reaching full size, and all were feeble and very difficult to raise. Eight individuals developed a male-like phenotype and the ninth resembled a female. With COG, eight hybrids were produced, but the development was not normal, not even while still in the eggs. After hatching, these hybrids were extremely difficult to keep alive. They never matured sexually. In all these hybrids, the NDI phenotype was not expressed, and the hybrids resembled the three species used in crossings, although they clearly differed by details of the color pattern.

The haploid chromosome number of NDI indicates that this species belongs to the large group of species within Nigeria and West Cameroon from which other species of *Aphyosemion* probably originated. It is likely that NDI gave rise to CAM and the CAM-like species living in coastal areas of Central Africa. LOU resembles NDI, but I was unable to get any eggs from crossings. The individuals of NDI and LOU did not exhibit any sexual response at all; they simply ignored each other. Such behavior is very rare in *Aphyosemion*, but individuals of LOU generally seem to be difficult to cross to other species.

Although NDI appears to stand rather isolated within the group of twenty-chromosome species, it probably is closer to GAR than to other species.

The different samples of NDI did not originate from the Ndian River of West Cameroon, but from a somewhat smaller river-drainage situated in East Nigeria (Akpa-Yafe River), about 10 miles west of the Ndian River. Both rivers are short, rising in the Oban Mountains and joining in the Rio del Rey, east of Calabar.

NEU: *Fundulus neumanni* Hilgendorf 1905 **(Distinct sp.?)**

Hilgendorf (1905: 417) had many specimens from a pool between Tisso of Nord-Ugogo and Usandawi, probably within Tanganyika. These localities are not on my maps. Hilgendorf observed (ctenoid) spines on the anal fin rays in males. The type individuals were preserved in sublimate, which may have destroyed red and yellow pigments. He found faint red markings on the head above the opercle and at the tip of the dorsal and transverse irregular red lines in the lower third of the anal. The distant edge of the caudal also had a faint red color. In small males only, red pigments occurred in the anal fin. The dorsal fin of males was darker than other fins. The females had no dark markings.

Preserved male NEU from Malindi, Kenya. 27 mm standard length. British Museum.

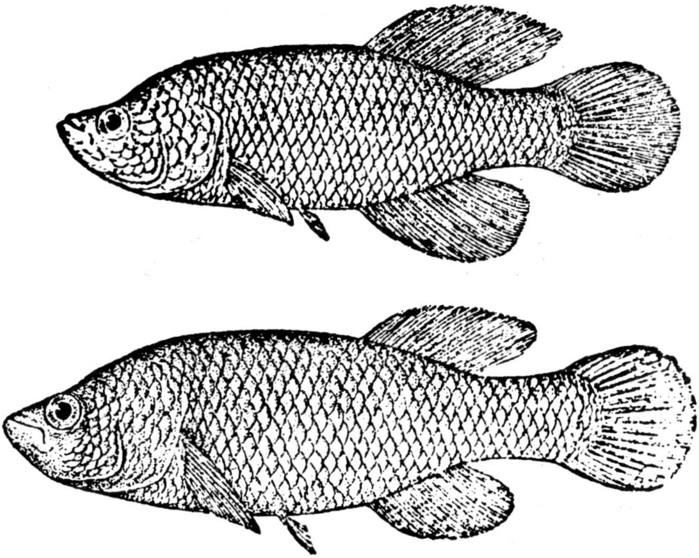

Fundulus neumanni.

♂ ♀, **types.**

Two "paratypes" for NEU as figured by Boulenger 1915.

Anal region of female NEU from the Lower Tano River, Kenya. British Museum, not identified, Whitehead collection. The structure around the anus resembles that of some *Fundulus species.*

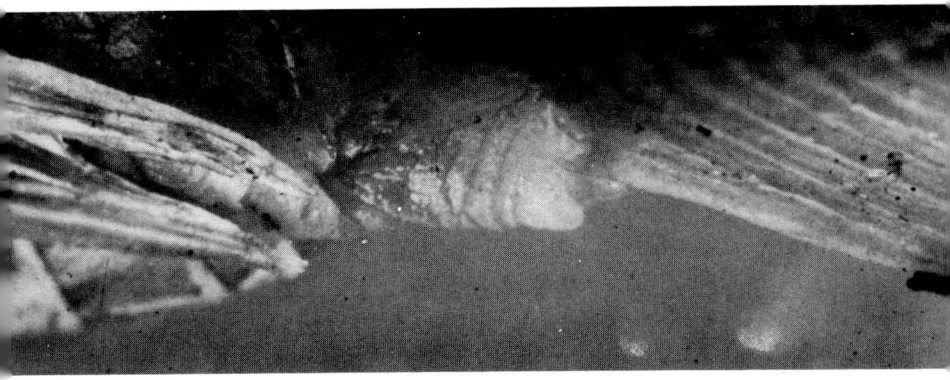

I have seen three males from a locality north of Malindi of Kenya and one female, probably referable to NEU, from a locality within the Lower Tano drainage in Kenya. These individuals are in the British Museum. The males had been identified as NEU by the museum. All these individuals unfortunately have been in formalin, and all red pigments had been completely destroyed. In the female, there is a fleshy anal pouch, rather similar to that of females of *Fundulus* and of some females of *Cynolebias*. These rather small specimens, the largest male of which measured 27 mm in standard length and 33 mm in total length, are rather similar to individuals which in this book have been referred to PAL. I found D. 16–18, A. 18, Sq-l. 27–31.

NIC: *Haplochilus nigricans* Boulenger 1913 **(Valid sp.)**

Boulenger (1913: 160) had three specimens from Dungu on the Uelle River of the northern part of the Central Congo drainage. He placed NIC close to GRA.

Pellegrin (1925: 99) reported NIC from M'Bouamo, on the Ebaka River, an affluent of the Djoua River, which empties into the Ivindo River, flowing into the Ogowe River of Gaboon. I hesitate to place these individuals in NIC. See SAN, and under SAN, the individuals which Lambert had from the northern part of Gaboon.

Pellegrin (1930: 210) also reported NIC from the Sangha River of the Congo drainage. Poll (1952: 162) reported NIC from various localities, all of the Central Congo drainage between Stanleyville and Leopoldville, and he said that the geographical distribution of NIC corresponded to that of MUF. NIC was found in the High Forest and in the Gallery Forest. Lambert (1961: 29–30) reported NIC from many localities of the eastern part of the Central Congo drainage (Banalia, Opala and Ikela districts). Gosse (1963: 214) reported NIC from Yangambi and considered NIC to be much more rare than MUF in this geographical area.

Matthes (1964: 125–26) reported NIC from Lake Tumba and the Ikela district of the Central Congo drainage and said that NIC fed on terrestrial insects and on aquatic larvae of insects. He also found filamentous algae in the stomachs of NIC. He caught NIC in swamps and in small brooks, where he observed small "shoals" of NIC near the surface of the water, in particular in almost stagnant water.

Matthes described the life colors of NIC as olive, darker on the back and lighter on the belly. Scales with a red (males) or rust-like (females) spot. A dark violet-brown band low on the side from the eye to the root of caudal. Males with reddish metallic brilliance on the sides of the head and on the gill covers. Belly and lower part of head faint yellow. Two dark crossbars on the throat. Fins yellow, unpaired fins with red or rust spots and rather dark. Lambert's description is rather similar. The male has a sky-blue metallic sheen on the lower part of the sides and a strong bluish-green sheen on the gill covers and the area just behind.

Poll (1952) said that NIC probably represented a subspecies of CHE, and Lambert (1961) considered NIC to be a subspecies of CHE, as did Gosse and Matthes (more or less).

I have not seen live individuals. Five preserved specimens originating in Monga, Bili River, Uele drainage (de Witte coll. 1935 Congo Museum) have been studied. These, unfortunately, have been preserved in formalin, and the red pigments have been destroyed completely. The throat pattern corresponds to that of CHA in the female. There is a uniform black area in front of the corners of the mouth, and the surfaces of the head behind this dark area are white. The interopercles are edged with black. The male has much less dark pigment on the posterior part of the throat than males of CHA, and the two dark crossbars of the common *Epiplatys* throat pattern are visible. The ventrals of males are longer and more pointed than those of females, and the dorsal and anal fins are higher. Some of the individuals

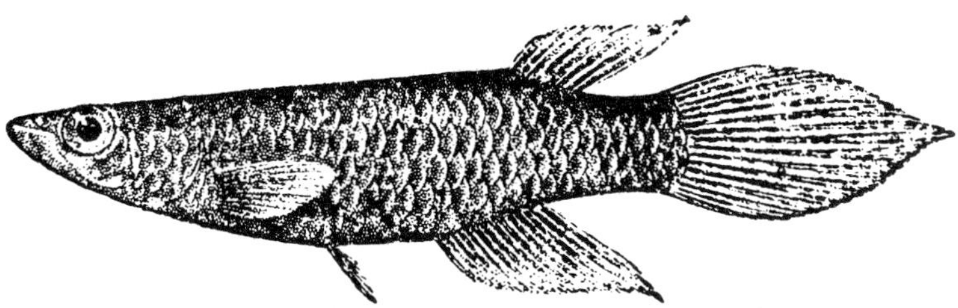

Haplochilus nigricans.
Type.

The type for NIC as figured by Boulenger 1915.

have a faint dark marbled pattern on the belly in front of the ventrals, as in CHE and MUF. The frontal scalation is normal (G-type, no H-scales). On the basis of the study of these specimens I hesitate to place NIC as a subspecies of CHE, although indeed these two forms are very similar. There are no reports of NIC in the aquarium; it has probably not been imported.

NIG: *Aphyosemion (Fundulopanchax) nigerianum* Clausen 1963 (=GAR)

Clausen (1963: 197–201) had eight specimens from a swamp near Arum, in the southern foothills of the Jos Plateau of norther Nigeria. He also had a large number of specimens from many other parts of Nigeria. He did not compare NIG to previously described species of the genus.

I have studied these specimens and about 300 preserved individuals from Nigeria and West Cameroon. As it has already been said under GAR, no constant differences in morphology have been discovered between the types of NIG or between other populations of NIG and the populations of GAR from the Cross drainage of Nigeria and West Cameroon. Thus it seems that NIG cannot be separated from GAR by details of the phenotype. It has also been mentioned that hybrids from crossing between individuals referable to NIG from Owo of West Nigeria and individuals referable to GAR from Eyomojok, west of Mamfe of West Cameroon, were fully viable and almost fully fertile. The F_2 individuals from $F_1 \times F_1$ spawnings also were normal, although in color patterns they were more variable than the quite variable Eyomojok-strain. It has, however, also been mentioned that hybrids from a crossing of males from Port Harcourt of eastern Nigeria and females from Akure of western Nigeria were rather feeble and difficult to keep and also, that these hybrids proved to be almost sterile in $F_1 \times F_1$ spawnings; many F_2-individuals died in the eggs or soon after hatching, and the few individuals that survived the hatching were just as feeble and sterile as were the F_1-individuals. Both strains were referable to NIG. One of the females and one of the males used in the crossing last mentioned are shown in color in the description of NIG, and one of the hybrid males also is shown.

It has also been mentioned that within strains referable to NIG, the haploid chromosome number is variable and that one strain (Akure) had 20 such chromosomes, whereas another strain (Owo) had only 18 chromosomes. The Eyomojok strain of GAR had 20 chromosomes.

I have studied preserved specimens referable to NIG from Riyum, Vom, Abuja, Arum, Akwanga and Lafia, all caught north of the Niger-Benue drainages: from Igbetti, Kabba, Eperin, Esa-Oke, Ilesha, Akure, Owo, Ondo, Idandre and Foriku, all individuals living southwest of the Niger River and from Ogoja, Obudu, Eyomojok, Mamfe and Port Harcourt, east of the Niger and south of the Benue drainages and from the Cross drainage. A map showing the most important of these localities has been published (Scheel 1964: 510–515) together with information concerning the Akure and the Port Harcourt aquarium strains and their hybrids (: 510–15, 519, 550, 567–68, 596, 610–12).

Eggs and live individuals of the Akure strain were distributed in 1958, and many strains were established in many countries. I distributed this strain under the CAL name. Since 1958 the Akure strain has been a rather common strain of aquarium fishes, because NIG is very easy to keep and to breed. Adult males of NIG are known to be very aggressive, and they easily kill females during spawning if the females are not willing to spawn at once. This strain has been described and figures (often in color) by many authors: Franz (1959: 234–35), Nieuwenhuizen (1960: 231–33), Franz (1960: 5–8), La Corte (1961: 106–08), Hesse (1961: 289–91), Hiller (1961: 10–11) and Hajek (1967: 48..), who all used the CAL name for this strain, and by ISIS (1963: 220), Aslanian (1966: 5–8), Cahalan (1966: 17–18) and Franz (1966: 406–07), who used the NIG name.

The reason for the easy maintenance of various strains of GAR probably is connected to the biological conditions for this species in nature. Unlike most other strains of *Aphyosemion* which have been introduced as aquarium fishes, GAR is not a species of the coastal humid forest; in Nigeria and in West Cameroon, for example, GAR seems to be restricted to the igneous rock and to be replaced on the sedimentary by other species of *Aphyosemion*. The water of the igneous complex usually contains much more dissolved mineral matter than that of the sedimentary. GAR lives in the drier parts of the tropical forest and in the derived savanna, and from these biotopes it is accustomed to larger variations of water conditions and of temperature than species living in the humid forest. See also GAR.

NII: *Aphyosemion (Fundulopanchax) nigri* Ahl 1935 **(=BIV?)**

Ahl (1935: 124) had a single specimen imported as an aquarium fish and said to come from the mouth of the Niger River. He received this fish from Siggelkow. He placed NII closest to SPU and GAR, saying that NII differed from these species by the more slender body. He also placed NII close to GUL, but separated it from GUL on the basis of the position of the anal fin, which he thought was closer to the snout in NII than in GUL.

Unfortunately, the type for NII apparently has not been inspected by other zoologists, and Radda has informed me that the type is no longer kept in the Berlin Museum and probably has been lost.

NII appear to be very well separated from GUL by the number of dorsal fin rays (11 in NII and 15–18 in GUL). In an analysis of meristic characters, NII was not separated from BIT, BRU, GAR, HOL, PAP, RIG and UNS, and the separations from BEA and SPP were weak.

The caudal fin of NII was trilobate (with three long streamers) thus separating NII from BRU and GAR. NII had a dark brown band or stripe from the gill cover to the root of the caudal and one more similar band or stripe above the first one and disappearing below the dorsal. The dorsal and the central part of the anal were reddish; the other fins were transparent. In spite of the rather low number of dorsal fin rays, the dorsal was positioned almost above the anal. On the basis of the published information, it appears to be justified to place NII very close to BIV until more material corresponding to Ahl's description has been found in Africa.

NIM: ***Epiplatys nigromarginatus*** **Schultze 1937** **(=GRA?)**

Schultze probably did not intend to describe this species, and it seems that he thought that Ahl had described and named the aquarium strain concerned. NIM first was reported by the aquarium society of Leipzig, which used the NIM name (1937: 410). The strain had been shown at an exhibition prepared by the society, and it was considered to be easy to keep and to breed. The male had a yellowish-brown basic color with a bluish metallic hue and with many brown spots. The unpaired fins were yellow, with black margins. There were three or four dark crossbars present on the sides at times.

Schultze (1937: 606) also used the NIM name for his description of this form. He said that Scholze and Poetzschke got the specimens from West Africa, together with *Neolebias ansorgei*, and that these fishes came from some locality between Cameroon and Congo. He said that Ahl had described and named this form. He published two fine photos (H. Baur photographs) and said that these fish resembled SEX by the shape of the body. The male was a variable grey, with six rows of wine-red spots along the sides, the median row extending from the root of the pectorals to the root of the caudal, other rows being shorter. The gill covers were blue, with red worm-like markings. The mouth was dark blue, and the eye was brilliant green. The pectorals were yellow, edged with white; the ventrals were similarly marked. The anal was yellow to greenish, with red spots and a dark submarginal band with a white edge. The dorsal was grey, with red spots and a dark margin. The caudal was yellow near the base, being grey with red spots and a dark margin on the remainder. The rays of the upper part of this fin were dark, with dark spots in between. The central rays of the caudal were extended. There were sometimes four or five rather indistinct dark crossbars. In the female, the dark bars usually were more marked, and the rows of spots along the sides sometimes were irregular.

Ahl (1938: 56) also described NIM and indicated that NIM came from the Niger drainage. He placed NIM closest to ORN, GRA and MUF. He said that the specimens had seven to nine dark cross-bars and that the anal and the lower part of the caudal were broadly edged with black. In Baur's photo of the male, there is a narrow dark margin on these fins.

Schultze, and not Ahl, is considered the author of the NIM name, because he described and named NIM first. Also, it seems that his description (including the photos) is more informative than that of Ahl.

NIM appears to correspond very well to GRA in all morphological details except for the (probable) absence of red lines across the throat in males. I have already mentioned under GRA that the males of some populations of GRA-like fishes lack the conspicuous red cross-lines of the throat and that in such males no trace of red pigments is visible under the microscope. Such males occur within the Niger Delta, and my three males from the Mboumboula River, about 30 km north of Douala in eastern Cameroon, belong to this phenotype of GRA. My crossing experiments with different populations of BIV, of GAR and of ARN and FIL, inhabiting different localities in southern Nigeria, strongly indicate that within this geographical area some zoogeographical border exists and thus it is likely that the swamp-living GRA of Nigeria and Cameroon will prove to be composed of two or more major groups of populations that are reproductively isolated and thus represent different bio-species or—if they are morphologically sufficiently different—different morpho-species. As the throat-patterns of African rivulins seem to play an important part in species recognition, the marked differences between the throat-patterns of males of GRA and NIM may indeed indicate reproductive isolation. Until crossing experiments between natural strains of GRA and NIM have been carried out, I prefer to incorporate NIM in GRA.

NIS: *Fundulus nisorius* Cope 1877 **(New World??)**

Cope (1871: 456) had seven specimens, labelled "Gaboon, W. Africa." D. 12, A. 10–11, Sq-l. 36. The anal originated below the middle of the dorsal. They were a uniform light-brown, yellow below. The larger individuals, reaching 10 cm, were females, and in these the oviducts were prolonged in a tube near to the extremity of the first anal fin-ray. These tubes contained eggs. Smaller, faintly cross-barred individuals were considered to represent males; they did not develop the tubes mentioned above. Boulenger (1915: 39) refers to Cope's description without comments, but in spite of this I do not think that NIS could represent a rivuline species of the Old World, or any rivuline species at all. It probably represents some mislabelled funduline species.

NOR: *Panchax normani* Ahl 1928 **(=OBS)**

Ahl (1928: 120–21) had a single specimen from Yaunde, in eastern Cameroon. This individual had been collected by Zenker. He placed NOR

The type for NOR as figured by Ahl 1928.

Abb. 6: *Panchax normani* spec. nov. — Type.

closest to JAC. Holly (1930) placed NOR as a synonym for CAM, but Myers (1933: 182) placed NOR in *Epiplatys.*

Although I have not seen the holotype, the ratio for the caudal peduncle (almost two times as long as deep) and the length of the pointed caudal fin (not quite as long as the head, which seems to measure about $28\frac{1}{2}\%$ of standard length) indicate that NOR belongs to *Aphyosemion.*

Holly corrected some of the data given by Ahl (D. 12, not D. 8, A. 17, not A. 15, etc.). The color of NOR was uniformly yellowish.

In 1966 I caught rivuline specimens in many localities around Yaunde and found numerous individuals of EXI and of OBS. In a few localities, a species of *Epiplatys* resembling SAN was present. No further species of Rivulinae appears to be present near Yaunde. Morphologically, NOR appears to be closer to OBS than to EXI, and it is here considered to represent a simple synonym for OBS.

NUC: *Paecilia nuchimaculata* Guichen 1866 **(=HOM)**

Guichen (1866: 444) had specimens from Madagascar. No further information concerning the type locality was given. Boulenger (1915: 50) said that NUC came from brackish water. Morphologically NUC cannot be separated from HOM.

NYO: *Panchax nyongensis* Ahl 1928

Ahl (1928: 119) had only a single specimen. This individual was caught by Rosenberg within the Nyong drainage. Ahl placed NYO closest to CHI. Holly (1930:214–15) inspected the holotype and changed some of the data given by Ahl (D.8, not D.7, A.16, not A.13) and placed NYO closest to CHA, EXI and DEC. A strange combination of phenotypes.

In an analysis based on meristic characters, NYO was not separated from 50 of the 120 nominal species of rivulins of Atlantic Africa. The preserved NYO was yellowish, with red pigments basally in the pectorals.

The type for NYO as figured by Ahl 1928.

Abb. 5. *Panchax nyongensis* spec. nov. — Type.

There was a broad band from the root of the pectorals to the root of the caudal, this band being made up of very large central brownish spots on each scale of the median rows. The fins were colorless, with spots at their far edges. On the basis of the description, which is not very informative, it seems to be impossible to determine the systematic position of NYO. The ratio for the caudal peduncle (about 1.8 times as long as deep) and the length of the caudal fin (a bit longer than the head, which measures about 27% of standard length) both indicate that NYO should be placed in *Aphyosemion*.

NYO cannot be properly defined on the basis of published information.

OBS: *Panchax obscurus* Ahl 1924 **(Valid sp.??)**

Ahl (1924: 55) had a single specimen from Yaunde, eastern Cameroon. He placed OBS closest to CAM. Holly (1930: 211–13), however, considered OBS to represent a simple synonym for CAM, and he changed some of the data given by Ahl for OBS. D.11, not D.12; A.15, not A.11, Sq-l. 31, not 29, etc., etc.)

The present author also considered OBS to represent a synomyn for CAM during the Cameroon expedition of 1966. This point of view also was held by Clausen (1967: 24), who said that the geographical distribution of CAM almost corresponded to that of EXI. The specimens of CAM-like

Adult male of OBS from stagnant brook, 18 miles west of Yaunde, east Cameroon.

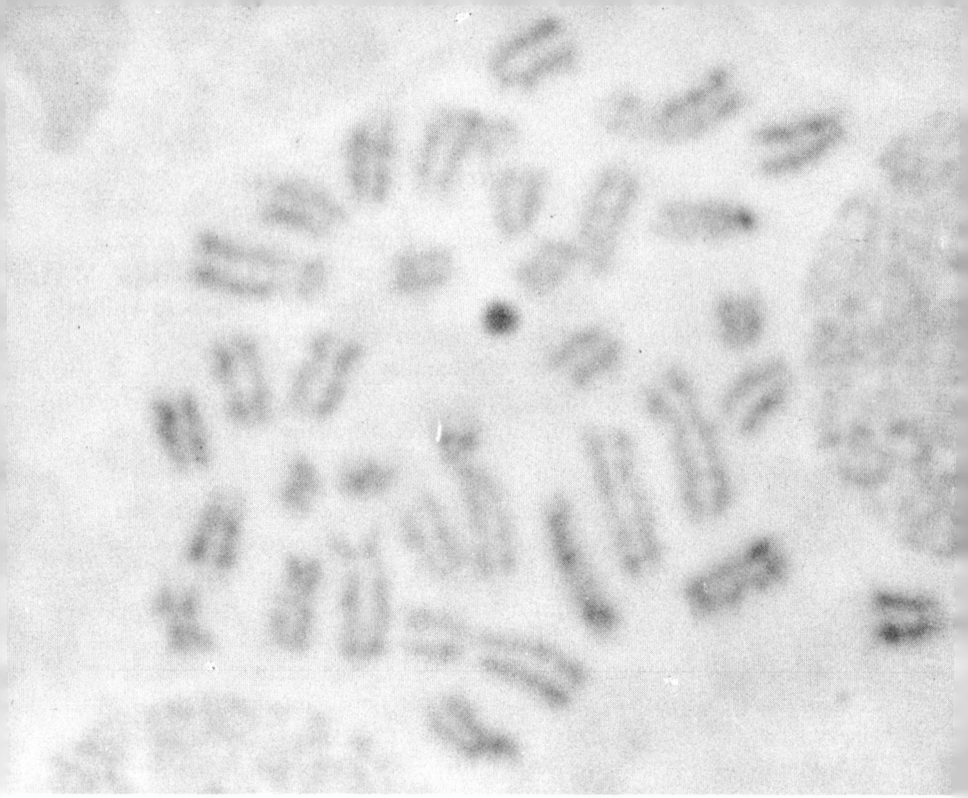

Mitosis in OBS. $2n = 34$.

forms that we caught in very many localities in eastern Cameroon were, however, so small and their color patterns so weakly developed that usually it was impossible to determine if these individuals belonged to CAM or to some similar species. A small number of live specimens from a locality 18 miles east of Yaunde, towards Edea, were brought to Copenhagen alive and an aquarium strain was established. The biotope for this population was a dirty, muddy, almost stagnant brook with many red deposits, probably from iron oxides. When the males of this sample matured in my tanks, they did not correspond to CAM, but to a rather different species. This form corresponded rather well to CAM by morphological characters, excluding color patterns, but this is also true of a number of species in *Aphyosemion* and in *Roloffia*. CAM cannot be separated from about 30 of the 120 nominal rivuline species of Atlantic Africa, on the basis of meristic characters.

I identified this strain of CAM-like fishes as OBS, considering CAR, NOR and PRE as junior synonyms. BEL might represent a synonym for a more southerly locality (Sangmelima).

OBS resembles the strain of aquarium fishes that was imported into Germany in 1913 together with the second shipment from Cap Lopez (see AHL and AUS). The OBS-like form was said to come from Gaboon, and Arnold considered it to represent CAM (see CAM), although Boulenger previously had identified the other *Aphyosemion* species of this shipment as CAM. See AHL and AUS. Arnold's point of view appears to have been the basis for the public's idea of CAM's appearance. The photo in Innes' book (1935–48: 284) labelled CAM shows the OBS-like form. This form again was imported (into the United States) in 1933, according to Innes.

At the present time it is impossible to determine whether the OBS form is restricted to the Yaunde area, being replaced by similar forms in other localities of the inland plateau or, as it seems more likely, that it has the same range as EXI and thus extends all over the wooded part of the inland plateau of eastern Cameroon and northern Gaboon. OBS-CAM-like populations were common in all localities visited in 1966 within the Cameroon part of the plateau, but the color pattern of males appeared to be much more variable than within EXI.

If OBS extends right down to northern Gaboon, it is likely that the 1913 (and 1933) strains mentioned above are OBS, but it is also likely that OBS then represents a synonym for STR.

The male OBS has a very strong cold blue sheen on the sides and on the back, especially on the back. The red spots on the sides are numerous and produce incomplete red lines along the sides. The unpaired fins are yellowish, changing into a saturated golden in the edges of the fins. No fin pattern

Primary meiotic division in OBS. n = 17.

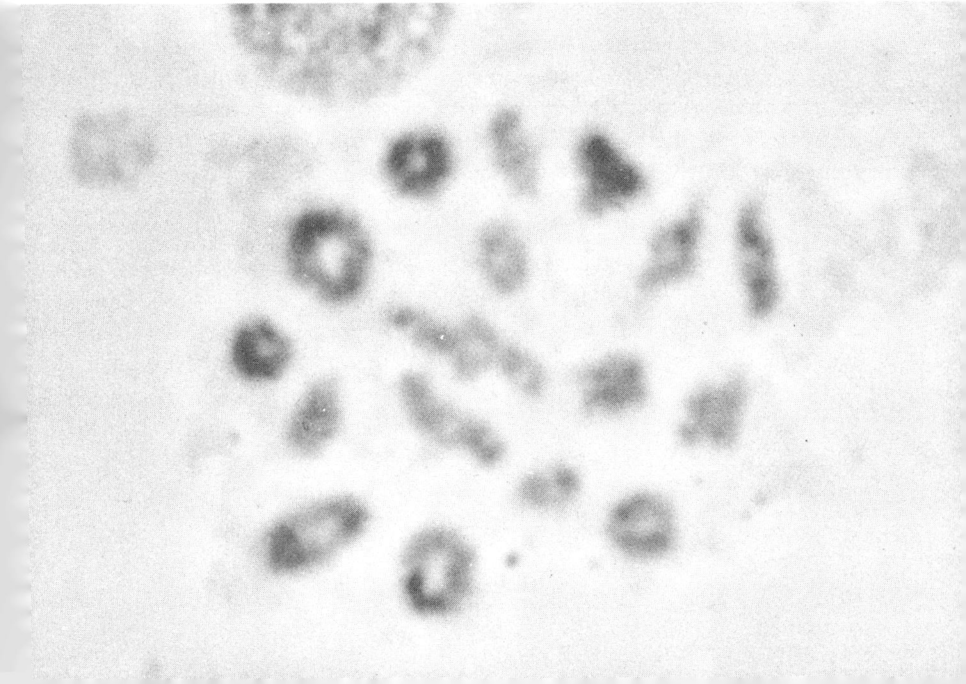

develops in these fins, although generally the many red dots and streaks tend to build up narrow submarginal or marginal red bands in the anal and in the upper and lower part of the caudal. The female OBS sometimes develops the rounded dark marking anteriorly on the throat, which seems to characterize females of the CAL superspecies and females of *Roloffia*.

OBS has 17 haploid chromosomes. The karyotype is characterized by a smoothly graded series of chromosomes, the longest measuring about 2.3 times the mean length of the basic chromosomes (for n = 25) or 3.7 times the length of the shortest chromosome. There is a metacentric (or sub-metacentric) element almost as long as the longest chromosome, which is an acrocentric element, and probably there are also two more metacentric elements slightly longer than the longest acrocentric of the basic complement.

Individuals of OBS are rather sensitive to attack from mycobacteria; although several breeding pairs have been distributed to friends in Europe, OBS has not produced stable aquarium strains, but further attempts will be made to make sure that OBS will be established within the international groups of killifish fanciers. Use clean, hard and alkaline water for keeping and slightly acid water for reproduction. Adult individuals are not feeble in my tanks.

OCC: *Aphyosemion occidentale* Clausen 1966 (Valid sp.)

On the 9th of June, 1909, the German aquarist Grote (1910: 222) of Hamburg, serving on a German ship returning from the Niger Delta area, caught some "*Fundulus*" specimens on Sherboro Island, off the coast of Sierra Leone. Grote brought some of these specimens alive to Hamburg.

Arnold (1911: 137, 604) reported that an aquarist (not named) offered him some colorful *Fundulus* to buy, but as he had had bad experiences with African *Fundulus*, he refused to buy any. He was told that these fishes had been taken on Sherboro Island. On the 26th of October, 1909, Grote's aquarium society ("Ludwigia" of Hamburg) reported that the new *Fundulus* had been shown to the members and the published description of color corresponded well to those of OCC.

In November, 1909, Arnold saw a pair of this species at the house of his friend Memmler, and in December of that year the new species was shown at the aquarium exhibition in Hamburg. In February, 1910, Memmler's pair died and was preserved and presented to Arnold. About November, 1910, Arnold sent them to Boulenger, who identified Grote's species as SJO. This identification is difficult to understand, because Grote's fish differs rather dramatically from Lönnberg's SJO.

In January, 1911, many more specimens were imported and Gellner (1911: 73) described and pictured the "Red *Fundulus*" or "*Fundulus*

Young pair OCC. Photo by Dr. Herbert R. Axelrod.

Old male OCC of the common aquarium strain.

The holotype for OCC. Male from Blama, eastern Sierra Leone. P.35.2572 of the Copenhagen Museum. Roloff collection.

Syöstedtie?" He said that his material came from the Gold Coast (Ghana). See also SJO. His description of the "Red *Fundulus*" and his figure are not very informative, but it is quite clear that his strain corresponded to Grote's strain from Sherboro. One month later, Arnold (1911: 137) described and figured Grote's fish, for which he now used the SJO name. Rachow (1911: 101) also described and illustrated (F. Mayer) this fish under the SJO name and said that the individuals pictured by Mayer came from Axim in the Gold Coast. Because the fish had become popular in Germany, professional collectors gave incorrect information concerning the collection sites. Mayer's figure shows a dark longitudinal band in the female.

Boulenger (1915: 38) redescribed Lönnberg's SJO, but the preserved material on which he (in part) based this description probably did not contain any specimens of SJO. They probably all were OCC and BAT. He figured OCC under SJO, and (1915b: 203) he also reported OCC (under the SJO name) in the collection of fishes from Matca of Sierra Leone. This collection contained the types for ANN. By this time, the name SJO had been firmly attached to the species now named OCC. Other species reported under the SJO name are mentioned under SJO.

In 1966 Grote's Sherboro *Fundulus* was named OCC by Clausen (1966: 331–41), who pointed out that Grote's species did not correspond to Lönnberg's SJO, which had been called COE by aquarists and zoologists since 1915. Grote's species was found to represent an unnamed and undescribed species.

It is strange that although OCC has frequently been sold as an aquarium fish since 1909 and probably has been exported from Sierra Leone, only two reports of this form has been referred to Sierra Leone (Grote 1910: 222 and Boulenger 1915b: 203). OCC is not so easily kept and bred that it is likely to have survived in the aquarium since 1909.

In 1962, Roloff rediscovered OCC in Sierra Leone. He found this species at Magbenta, near Rokel of central Sierra Leone and near Blama of eastern Sierra Leone. In 1963 Schiötz collected additional material at Kasewe and Makeni of central Sierra Leone. Roloff kindly offered me live individuals from Magbenta and from Blama, and as we both considered these two strains to be different by some traits of color patterns and by size of adult individuals, I prepared a crossing; 123 very active hybrids were raised, and they proved to be fully fertile. In the sperm of the males, the pairing of the 23 chromosomes from each strain was complete. No univalents or other abnormalities were discovered.

OCC has been reported from Sierra Leone only. The range of OCC probably extends from the extreme eastern part of the country to the central part. In the eastern part of Sierra Leone, OCC is replaced by the very similar TOD, which probably represents a distinct, but closely related species, OCC and TOD appear to be restricted to swamps and the drier part of the forests and the derived savanna, being replaced at higher altitudes by GUI.

The very handsome colors of the OCC male are shown in the color-photos, which show young fish. Old males are more handsome, because in

Head of preserved male of OCC. Aquarium strain of 1960/61 in Europe. Notice wrinkled scales on opercle, series of small neuromasts just below rim of eye and neuromasts round preopercle in closed canals with pores.

Male OCC from crossing of a strain from central Sierra Leone to a strain from eastern Sierra Leone. Fertile.

them many broad snow-white fringes develop on the posterior and distant part of the anal. In the dorsal fin, the fin rays extend beyond the edge of the fin, producing white spine-like filaments. Streamers do not develop in the fins.

Both sexes may develop a broad, dark, rather regular band from the gill covers to the root of the caudal. In females, a rounded dark marking usually develops anteriorly on the throat in connection with the development of the dark lateral band.

The frontal scales often are protected by a thick layer of skin, especially in males. The pattern is normal (G-type, no H-scales). The karyotypes of OCC and TOD are similar and contain 23 haploid chromosomes. Two or three of these exceed the maximum length of the basic cyprinodont complement; they probably are "double-sized" elements. In OCC, all chromosomes appear to be acrocentric, whereas in TOD, one of the "double-sized" elements seems to be metacentric. Both forms seem to have a metacentric element of mean basic length.

Ctenoid fin ray papillae frequently occur in the anal fin of males, but microscopic studies of the supporting bony element indicate that this is not a ctenoid spine, but a bud-like rounded body, rather similar to the supporting

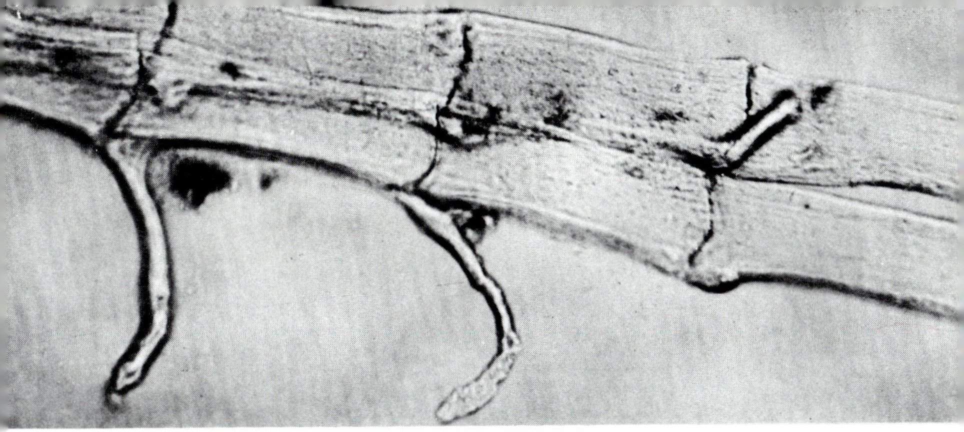

Ctenoid spine-like structures on distant part of first anal fin ray of male OCC. These structures generally develop very close to junctions of segments.

element found in fin ray papillae of the pectorals of species of *Cynopoecilus* (South American Rivulinae). No ctenoid spines develop on the scale edges; instead, another ctenoidy is present, which I elsewhere have called the "wrinkled scale" type. (Scheel 1966: 141). This type of ctenoidy also occurs in *Cynopoecilus* (*C. melanotaenia* Regan and *C. ladigesi* Foersch) and in males of most species of *Roloffia*. See also BUR. The hemoglobin pattern is normal (4 lines) for *Roloffia* and *Aphyosemion*.

Male OCC from Sierra Leone. Roloff 1962.

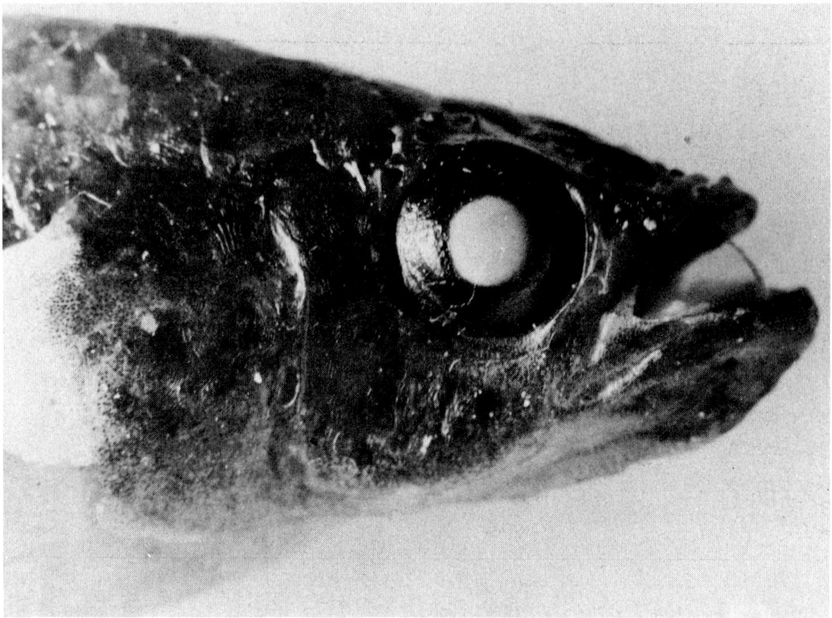

Head of holotype OCC from Blama, Sierra Leone. P.35.2572 in the Copenhagen Museum. E. Roloff collection.

OCC is an annual rivuline species, as Gellner (1911: 73) observed. He noted that the male used the dorsal and the anal fins to hold the female and that the eggs were deposited about 1 cm below the surface of the bottom layer (sand). He also noted that eggs which had been kept at 23°C. for more than eight weeks did not exhibit any indication of development. I kept eggs from OCC in slightly moist peat for 273 days and hatched fully viable fry. Four hundred and twenty-six eggs were studied in one experiment. These eggs first were spawned by several pairs in fine and dirty mud; to prevent untimely development, the eggs were kept for some weeks under anaerobic conditions. Then they were washed out from the mud and counted. They were placed in clean water (slightly acid, zero hardness) in open containers at 22–23°C. Eggs which developed a pigmented embryo were sorted out and counted. The result is given as the total number of eggs removed after a certain number of days:

10% after 58 days	20% after 62 days	30% after 67 days
40% after 70 days	50% after 71 days	60% after 73 days
70% after 75 days	80% after 78 days	90% after 83 days
100% after 100 days		

The initial period is rather short (about 50 days) compared to *Pterolebias peruensis* Myers from the upper Amazon River of Brazil, which was studied at the same time. In the species last mentioned, the initial period was 85 days. The rate of development of the eggs also is more quick in OCC (3.2%/day) than in PER (2.3%/day). Twenty-eight eggs died during the initial period and six more eggs died during development of the embryo. The eggs of OCC has a very distinct hexagonal pattern and measures 1.4–1.5 mm.

Males of the hybrid strain mentioned above developed a slightly higher number of fin rays than did females. The mean number of fin rays for 21 males was $D_{m}. = 20.2$ and $A_{m}. = 18.8$, and for 17 females: $D_{m}. = 19.1$ and $A_{m}. = 18.1$, the mean number of scales in a longitudinal series being 34.3 and 33.9, respectively. Similar conditions appear to exist in the old aquarium strain of OCC and in individuals from Magbenta. The N'Zérégoré species of *Roloffia* (see MAE) also had more fin rays in males, but in other nominal species of *Roloffia* the males and females appear to develop similar numbers of fin rays. Much more conspicuous differences are found in *Cynolebias*. See Peters (1964a: 175–184 and 1966: 337–47).

Old male OCC of aquarium strain 1960/61 in Europe. Notice the pattern of dorsal and anal fins.

Front of preserved male of OCC. Aquarium strain of 1960/61 in Europe. Notice the wrinkled D- and E-scales, the closed frontal neuromast pattern and the rather large unscaled area between these neuromasts and the G-scale.

Clausen (1966: 338) made OCC the type-species for *Roloffia* and placed all species of *Aphyosemion* (except for WAL = SPU) living west of the "Dahomey Gap" in *Roloffia*. He also said that OCC takes in a somewhat isolated position within *Roloffia*, with GUI as the closest relative. On the basis of my own studies, I agree with Clausen that OCC represents an isolated species among Rivulinae of the Old World, probably just as different from other species of nominal *Roloffia* as these species are different from *Aphyosemion*, but see also GUI.

OCC undoubtedly represents the most ancient of the species of rivulins of the Old World, that develop "double-sized" acrocentric chromosomes, and OCC probably is close to the ancient form from which *Roloffia* and *Aphyosemion* originated.

OES: *"Panchax oeseri"* **(Invalid name)**

The history of the OES name is difficult to unravel. In the description of VEX, Meinken (1929: 257) said that VEX appeared to be closely related, although not quite identical to *Panchax oeseri* E. Ahl. He also said that he had not been able to find any description of this species in the literature. He referred to a photo of OES published in the *Aquarium* (which I have not seen) and to a color description by Herbert Schmidt of Berlin. Meinken said that Schmidt's description did not correspond to VEX, and he concluded that VEX and OES were different.

Arnold and Ahl (1936: 300–01) shortly described OES as *Aphyosemion oeseri* E. Ahl, 1928, and said that OES originated from the eastern part of the Niger Delta and from the Calabar district, and that it was imported into Germany in 1928. Beyer (1931: 73) also said that OES originated from the eastern part of the Niger Delta and from Old Calabar and that this species had been discovered by Dr. Oeser during his collecting voyage to West Africa a few years previous. He also said that Oeser brought live material to Germany.

Hesse kindly produced some old printed notes which came from German aquarium publications (the name of which are not known, as the notes had been cut out). One ink drawing by Meinken (1930, signed) and labelled OES shows a phenotype similar to VEX. Another note said that Grelk of the "Platy-Tischrunde" had offered the author of that note a strain of rivulins which that author had found to be similar to the strain described by H. Schmidt in the October issue of *Aquarium*. He said that Oeser brought to Berlin two live males of this form during the summer of 1928 and that Ahl named this strain in honor of the importer (Oeser). More were imported in autumn of 1928 by the "Platy-Tischrunde," and many live individuals were imported from Port Harcourt by "Wandsbeker" during the summer of 1929. He also said that OES, VEX and CAL were rather similar.

Oeser (1928: 374) published a detailed report of his voyage to West Africa in 1928. There are no indications that he collected fishes in Nigeria or in Cameroon, but he published a detailed report of the collection of SEX and a species undoubtedly of *Aphyosemion* very near Sta. Isabel of Fernando Po; he said that one live male of this population was brought alive to Berlin.

Thus it seems that Oeser's *Aphyosemion* did not come from Nigeria or Cameroon, but from Fernando Po and also that it resembled VEX (=CAL) and CAL. CAL-like fishes are known to live in southern Nigeria and probably also at Port Harcourt, whereas in West Cameroon and maybe also on Fernando Po CAL is replaced by the very similar AHL. It also seems that OES never was properly described by Ahl and that the little informative description in Arnold and Ahl (1936: 300–01) was based not on Oeser's fish only, but also on the specimens from the importations of the "Platy-Tischrunde" and by "Wandsbeker." OES thus appears to represent a *nomen nudum* (or a synonym for CAL or AHL).

Adult male of the species of *Aphyosemion* which Dr. R. Oeser caught in a brook (Rio Consul) within Santa Isabel, Fernando Poo in 1928. This male was taken by the present author in February 1968. Sides and back are brilliant bluish-green. Pectorals, ventrals and almost the whole anal fin are a saturated golden in fine males. Upper and lower edge of caudal are white, narrowly edged with golden in some males. This species appears to be closest to OBS.

OGO: *Haplochilus lujae ogoensis* Pellegrin 1930 **(Valid sp.??)**

Pellegrin (1930: 201–02) had 24 specimens from the Leconi and the Passa Rivers of upper southern Ogowe. These individuals were collected by Baudon. In 1931 (:220–21) he reported another series of 23 OGO from the Louesse River, a tributary of the Kouilou River.

Pellegrin said that the caudal fin of the OGO male resembled that of CAS (=SCH? =CHR? etc.) from Stanleyville of the central Congo. OGO had a lyre-shaped caudal fin, with an upper and a lower streamer. The male was brown on the back and sides, the belly being yellow. There were three or four parallel orange lines along the sides, following the rows of scales. The dorsal and anal had small orange spots; the caudal had stripes of dark and light color alternating. All unpaired fins might develop black edges. Of the Louesse individuals he said that they corresponded in color patterns to those from the Upper Ogowe River, and that there were three or four longitudinal carmine lines alternating with brown lines on the sides. See MAG and STR. The few meristic data published for OGO do not permit a separation of this form from about 60 of the nominal rivuline species of Atlantic Africa.

The data published for OGO are very close to the average figures for the genus *Aphyosemion*, and more information is needed before the systematic status of OGO can be fixed. Lambert probably soon will supply these additional data.

OLB: *Epiplatys oblrechtsi* Poll 1941 **(Valid sp.?)**

Poll (1941: 139) had four specimens from a small river named the Maseni River. They were taken at the point where this river empties into the Bon-de-Boan River, an affluent of the Nuon River. This locality is situated close to the Ivory Coast/Liberian border.

It has already been mentioned under FAS, that Daget and Iltis (1965: 190–92) considered OLB to represent a subspecies of FAS.

Clausen (1967: 31) said that the frontal lobe on the dorsal part of the snout of OLB occasionally has a shallow furrow. The present author also studied many preserved specimens referable to OLB from Herold Olsen's collections from N'Zérégoré of High Guinée and found no individual with a furrow on the central fleshy lobe in front of the two neuromasts. But in some of the specimens referable to LAM (see LAM) such a furrow was indeed present. I had specimens of the form without furrow on the fleshy lobe identified as OLB by Poll in 1965, so probably Clausen thought that the other N'Zérégoré form (also?) represented OLB.

Poll described the color patterns of the types for OLB. There were seven to eight (dark) crossbars on the hindmost part of the side, the bars being slightly oblique. Each scale had a small red spot, and these spots produced parallel rows along the sides. The posterior part of the dorsal and the upper part of the anal had black spots. The anal also had red spots. There was a dark oblique (submarginal) line in the posterior part of the anal.

In the large collection of preserved rivuline specimens from Herold Olsen's collection from N'Zérégoré there are very many of OLB caught during 1963 and 1964. These resemble CHA, in particular the SHE sub-

Preserved male (above) and female (below) OLB from N'Zérégoré, Guinée.

species in morphology and in color patterns, and probably they are not-referable to any known population of FAS. The N'Zérégoré OLB is a large fish. Adult males measure up to 97 mm total length, 81 mm standard length. The preserved male is very dark all over. The sides and the unpaired fins are a brilliant cold blue, changing into violet at certain angles of light. There are numerous red spots on the sides; almost every scale has such a spot, even the scales on the back and on the belly, but the spots last mentioned are considerably smaller than those of median scale rows. The spots usually are vertically oblong or crescent shaped in upper scale rows. In some males, an almost perfect red reticulation develops on the back.

In spite of the considerable number of large red spots, the spots do not tend to produce red lines along the sides. In some males, there are many thin dark vertical bars or lines posteriorly on the side; in other males, these lines are completely absent.

No distinct throat pattern is present in most males, except for a rather uniform over-all black pigmentation. No red pigments appear to be present on the throat.

In the anal fin, a row of large, very dark-red spots is present just below the root of anal. This pattern corresponds well to that of SHE. There are also a number of small red spots posteriorly in that fin, and the distant edge often has a narrow red marginal band. Other unpaired fins have a broad uncolored marginal band. The dorsal is marbled with red spots or streaks on the dark bluish background. The caudal usually has some large, very dark-red spots near fin base, the lower edge being like the anal. The ventrals are short; they usually do not reach the anal, but in a few males the ventrals are pointed and reach the 5th anal fin ray.

Females are more lightly colored, and the belly and the lower part of the side is almost pure white. There are up to ten narrow vertical dark lines or bars reaching the back and the keel posteriorly on the side. The throat pattern corresponds to that of CHA. The red pattern on the sides corresponds to that of males, but the red spots are much smaller.

The Nuon River is an affluent of the Ceston River of eastern Liberia. This river rises on the southern slopes of Mount Nimba. The fresh waters round N'Zérégoré are not connected to the Ceston River, but it is not unlikely that a rivuline species which occurs in the Ceston River also occurs around N'Zérégoré. See also MAE.

ORN: *Panchax ornatus* Ahl 1928 **(Valid sp.?)**

Ahl (1928: 118) had a single specimen from the aquarist Gerlack of Hamburg. He placed ORN closest to GRA, MUF and NIC. The origin in nature of this strain is unknown. It is likely that it originated in Gaboon or in what used to be the French Congo.

According to Velke (1930: 829), ORN was imported prior to 1928 by the "Platy Tischrunde." He said that the eyes were brilliant golden, and when the fish was in the dark parts of the aquarium, only the eye was visible (see GRA). There were six or seven rows of dark-red spots along the sides, the upper row being very short and not reaching the root of the dorsal. The lower row was not parallel to other rows (see also GRA and MAC), but was just above the root of the anal and low on the keel. The sides of the male were brilliant golden, bronze or greenish-golden. The fins of males were a faint yellow, with a greenish hue and had many dark spots. The anal was faintly edged with red; the caudal also had a red margin. The eye was green above, golden below, more brilliant in the female than in the male. Apparently, adult individuals did not develop dark crossbars, but Velke said that in the juveniles of ORN there were six or seven oblique dark bars, which disappeared with age.

Thus ORN is intermediate to GRA and MAC by color patterns. ORN cannot be separated from these and other species by meristic characters. For the present time, we probably should consider ORN to represent a distinct species within the GRA superspecies. This superspecies seems to have more forms than GRA, ORN and MAC.

ORT: *Cyprinodon orthonotus* Peters 1844 **(Valid sp.)**

Peters (1844: 35) had several specimens from Quellimane in Mozambique. This town is situated on the coast, north of the Zambeze River.

Peters described the colors of the (live?) male. The male was dark green, with a metallic brilliance on the upper part of the sides and golden on the lower part. Each scale had a blackish-brown spot. The unpaired fins were yellowish, with similar spots. According to Boulenger (1915: 33), who studied two of the types and an additional specimen also from Quellimane, this species reach 70 mm. The pattern of the teeth have been mentioned under GUE.

As has already been said, the systematics of the species of *Nothobranchius* is highly unsatisfactory, and none of the species of this genus can be identified to satisfaction at the present time. First of all, ORT must be defined morphologically and biologically on the basis of live individuals from Quellimane. When this has been done, other species, similar to ORT, can be defined on the basis of biological studies, including crossings. It seems that the species of *Nothobranchius* are very variable by meristic characters and very similar in color patterns. Further studies of the pattern of the teeth and number of rows of scales on the interopercle below the center of the eye may prove to be useful, because these characters appear to be distinguishing features.

Under GUE it has been said that one of the aquarium strains of *Nothobranchius* which is kept at the present time seems to correspond rather well to ORT, although the males of this species develop an almost perfect red reticulation on the sides, whereas the types for ORT and other Mozambique populations near Beira appear to develop square or crescent-shaped red spots on the sides and on the lower part of the head.

The largest of these "ORT" in my own collection measures 42 mm standard length and 51 mm total length. The strain had: D.15–16, A.16–19, Sq-l. 26–28 in ten individuals. Ctenoidy was well developed in fins and long fleshy filamentous ctenoid fin papillae occur in the dorsal (few) and anal (many) fins. The ctenoid spines on the body and the head were rather short and rudimentary. The anal fin papillae were very dark. There are several rows of rather short teeth in the jaws and there are no conspicuous differences in size between some of the teeth of each row. There is one, rarely two, rows of scales on the interopercle. This form has 18 haploid chromosomes and one of these, at least, is a metacentric element. The color pattern corresponds to that of GUE and is close to the pattern of the GUE-like strain of PAL.

The aquarium "ORT" mentioned here seems to correspond to the strain described by Goldstein (1966: (9): 5–7). The photo (Lugenbeel photo) corresponds well to my strain. KUH and TRO probably are synonyms for ORT.

PAL: *Fundulus palmquisti* Lönnberg 1907 **(Distinct sp.?)**

Lönnberg (1907: 7) had 28 rather small specimens from Tanga of

Preserved male (above) and front of that male (below) of the 1957 strain known under the PAL name by aquarists. The frontal scales are very small, the anteriormost of these scales is situated far behind the frontal neuromasts which develop the "open" pattern.

Pair of PAL of the 1956 strain usually called GUE by Aquarists. W. Foersch photo.

Pair of PAL of the 1957 strain. W. Foersch photo.

Adult male PAL of the 1957 strain. Franz Werner photo.

Tanganyika. The largest individual, a male, measured 42 mm total length, and PAL appears to belong to the group of small forms in *Nothobranchius*. The males had a perfect red reticulation on the sides. The dorsal and anal had rows of oblong red spots, producing irregular transverse lines. The caudal fin was bright red. The females were without dark markings.

Lönnberg placed PAL closest to GUE, but he separated PAL from GUE by the lower number of scales along the sides (27–28 scales versus 30–32 scales), the less deep body and the juxtaposition of the dorsal and anal fins, which are on the same vertical line in GUE, whereas in PAL the dorsal is slightly in advance of the anal. He also separated PAL from GUE by the length of anal fin rays, which are almost the depth of the body in PAL but much shorter in GUE.

An aquarium strain referable to PAL was imported from East Africa in 1957 (Tropicarium of Frankfurt, Germany). I received live individuals from the importer and raised an aquarium strain which was distributed to very many countries by egg-sendings. Full grown males of this strain rarely exceeded 50 mm total length and very few adult females reached 40 mm total length. The eggs were small, measuring about 1.0 mm. The males of this handsome strain had a perfect red reticulation on the sides. There were no red markings on the lower part of the head or on the belly. The red pattern of the dorsal and anal fin corresponded to the description of PAL. The caudal fin was bright red all over, without any trace of a dark distant margin. In some males, the dorsal, anal and even the pectorals developed a bright lemon color, whereas in other males of this strain no lemon color was seen.

Unfortunately, individuals of this handsome strain produced fully viable and fully fertile offspring in crossings with individuals of a rather similar strain, which I have already mentioned under GUE. This strain also belongs to the small species of *Nothobranchius*, and adult individuals correspond to the 1957 strain of PAL. The males of the strain develop a color pattern rather similar to that of GUE, and they differ from the 1957-males by the broken red pattern of the hindmost part of the sides, where thin vertical red lines are present, and by the broad black distant margin in the caudal. Also, there are more red spots and streaks in the dorsal and anal, and the lemon

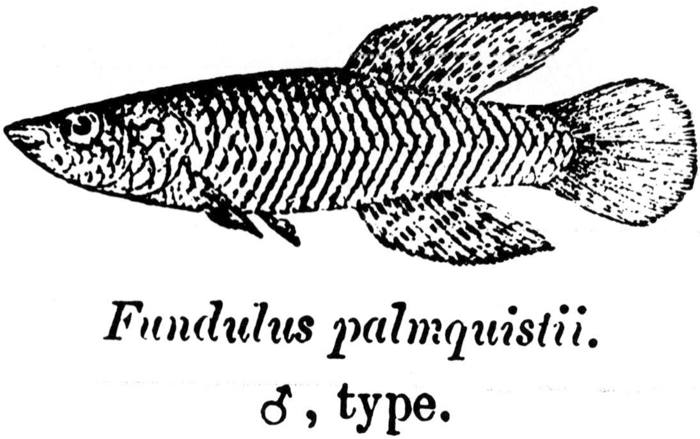

One of the "paratypes" for PAL as figured by Boulenger 1913.

fin color has not been observed in this strain. The strain probably was imported by Aquarium Hamburg in 1956. Eggs measured 1.0 mm.

The 1956 strain usually produced a large surplus of females in broods, whereas in the 1957 strain, an even larger surplus of males occurred; as the females were almost similar, it is highly likely that most of the breeding pairs sold to other aquarists were mixed. My male hybrids all developed the black margin in the caudal fin and were very similar to males of the 1956 strain.

The 1956 strain is still to be had. This strain has 18 haploid chromosomes. The largest chromosome is a metacentric element, measuring about 2.8 times the mean length of the basic chromosomes (for n =25). There are also two (to four) clearly acrocentric elements, larger than the longest chromosomes of the basic complement, both (the largest) measuring more than two times the mean basic length.

These two strains have a single row of scales below the eyes. There are about four rows of teeth, the anterior row containing rather large and curved teeth, whereas in other rows and in particular in the posterior row, some teeth are rather large.

Ctenoid spines are well developed on the scales of the sides and of the opercle and preopercle. Prominent anal papillae are present. Dorsal papillae are present in the males of the 1957 strain, but these papillae are smaller and less numerous than those of the anal fin. On the dorsal part of the head, the lateral line pattern differs somewhat from that of the larger species of the genus. The supraorbital groove often (but not always) is interrupted between pore 2a and pore 2b of Gosline's system (1949: (519)), and the neuromasts appear to be larger than usual in *Nothobranchius*; they resemble those of *Epiplatys* and *Aphyosemion*. The differences in the development of the supraorbital groove appear not to be so large and constant that they can be useful in the systematics of the genus. The few preserved specimens of the 1956 strain are quite similar, but the pattern of teeth is more like that of GUE, without enlarged teeth in posterior rows.

PAN: *Esox panchax* Hamilton-Buchanan 1822 **(Valid sp.)**

Hamilton-Buchanan (1822: 211, 380) based PAN on individuals from the Ganges drainage of India. As opposed to other Asian rivuline species, PAN extends over huge geographical areas. According to Job (1940: 53–67) PAN has a wide range in India and has been found in the Bengal, Bihar, Orissa, Assam, Punjab, United Provinces, Sind, Cutch and Central Provinces. Thus PAN occurs in the Indus, Ganges and Bramaputra drainages. PAN extends eastward from India, reaching Burma and Thailand, where it is common in the lowlands near the coast. It also occurs on the large islands of the Gulf of Siam. From Thailand PAN extends southward into Malaya and Indonesia. It has also been taken on the Andaman and the Nicobar Islands of the Indian Ocean.

If PAN has inhabited this huge geographical area for a long time, it is likely that different populations have become reproductively isolated. The haploid chromosome number for one strain from (?) Thailand indicates this, because in this eastern strain there were 18 haploid chromosomes only. See SIA. Other Asian rivuline species have much higher chromosome numbers, close to the ancient chromosome number for the family. If PAN originated in the India-Ceylon area, where the other four species of *Aplocheilus* live at the present time, it is likely that Indian populations of PAN have more than 18 chromosomes, because as it appears in Africa, the chromosome number decreases along the path of extension of a certain species. See BIV.

For that reason, some importance is attached to the rather many subspecies of PAN, which have been described, because some of these may indeed represent bio-species within PAN, which, if sufficiently different from other bio-species of PAN in preserved material, might become distinct morpho-species, when once in the future the structure of the populations of PAN is biologically studied. Also, a number of names created for PAN-like populations, which at the present time are considered to represent synonyms, might become important in the future.

In 1839 McClelland (: 301, 324) described *Aplocheilus chrysostigmus*, in 1846 Cuvier and Valenciennes (: 383) described *Panchax buchanani* and *P. kuhlii* (: 384), in 1850 Bleeker (: 22) described *Panchax melanotopterus* and in 1906 Köhler (: 190) described *Haplochilus andamanicus*. All these names are considered to represent synonyms for PAN at the present time.

Five different subspecies of PAN have been created by various authors. See MAI, LUT, MAN, DOM and SIA. All these subspecies have been based on color patterns only. The described differences in color patterns between the different nominal subspecies are not large, if compared to the individual variation found within most African rivuline species. These differences in colors refer to the development of yellow or orange pigments in unpaired fins and of black pigments on fin margins generally, but differences in the pattern on the sides also are used. These differences may or may not reflect important genetic differences between differently colored populations and strains.

The color pattern of the Ganges PAN of India (type-area) was described by Mackessack (1958: 171–73), who lived in Bihar and studied the local population of PAN. The sides of the male are olive-green above, changing into mauve in the middle of the sides, with small patches of iridescent light-green on the gill covers. The unpaired fins are tinged with pastel shades of light-blue, light-orange and black. The dorsal is gleaming white at the base, with a black bar at the center and a light blue color at the top. The anal fin is of creamy color on the inner edge, shading off to delicate orange on the outer. The caudal fin has a bluish-black outer rim enclosing an area of light-blue, from the center of which radiate several bright orange streaks towards the blue-black border.

PAN was imported into Germany in 1899. The first description of this strain was given by Schmidt (1901: 213, 235), who said that the basic color was blue to sky-blue. There was a dark stripe along the back. The dorsal and anal were spotted and had an orange margin, the former also with a dark marking. The caudal had a dark margin, the central part being bluish.

There were some brilliant reddish spots on the gill covers. The female was without dark margins in the caudal. This strain was said to originate in Calcutta.

Pair of PAN from Djakarta, Indonesia. Thung Kim Tek's strain.

Köhler (1906: 190, 383) reported two strains, which he said had been imported from Bombay and New Goa, and which he thought differed sufficiently from PAN to be named as a subspecies. See MAI and LUT. As Köhler probably did not know the color patterns of Hamilton-Buchanan's PAN, he probably compared the new strains to the so-called Calcutta strain of 1899.

Köhler (1907: 239) also created a subspecific name for an aquarium strain of PAN, which was called "*Haplochilus panchax var dayi*" by some German aquarists. See MAN.

Arnold (1906: 635) said that the data concerning the origin in India of Köhler's MAI and LUT were wrong. Unfortunately, he did not inform the reader as to what evidence he based this idea. He published very detailed data concerning two strains of PAN, which he said originated from a locality near Colombo in Ceylon. He said that a German sailor first brought him two males. One of these had yellow ventrals, a greenish-yellow anal with orange spots and a caudal with a broad milky-white submarginal band and a black edge, the center of the fin being spotted with brown. The dorsal was white and had a crescent-shaped dark spot at the base and a narrow dark edge. The pectorals were colorless. The body was yellowish brown, with a strong golden metallic hue. Each scale had a violet-red spot. There was a dark line through the eye.

The second male was darker; its ventrals were dark yellow. The anal also was dark yellow, with brown spots and streaks and a red margin. The caudal had a broad orange-yellow submarginal band and a dark margin, the central part of the fin being spotted with brownish-red. The dorsal was orange-yellow, with a crescent-shaped dark spot at the base. The body was dark brownish-olive, with a metallic hue that changes from yellowish-green to bluish-green.

The next year, the same sailor brought Arnold three females, which he said had been collected in the same pool as the males. These females had an orange caudal margin; in one of them, the anal was orange and the color of the caudal was more saturated than in the other females, in which the anal fin was colorless.

He paired the female with the orange anal to the male with much orange and yellow color, and the other females to the second male. From the combination first mentioned, he had very variable offspring, and the colors of the caudal submarginal band varied from sulphorous-yellow to orange-red. From the second combination he had variable offspring also, and the color of the caudal submarginal band varied from milky-white to bluish. Arnold and his friends were not able to establish any aquarium strain that bred true to color.

Stansch (1908: 172) had a strain from the Tjilatjap Island near Java, which was brought to him by a naval officer. Within this strain three color varieties developed. One of these resembled Köhler's MAN from Ceylon. Another resembled his LUT, and the third phenotype Stansch considered to represent a new color variety, which was closest to MAN, but with the anal fin edged with black in both sexes. He spawned a pair belonging to the "new" phenotype and found the offspring to be very variable, and he was not able to establish a strain of that phenotype, which bred true to color pattern.

Arnold (1911: 669) reviewed the discussion concerning the different color varieties of PAN and said that he had never been able to establish any strain of this species that bred true to color. He referred to a shipment of PAN from Calcutta in 1911. He was sure that this shipment originated in pools near Calcutta. In the males, the caudal fin color varied between "white" and "orange" in the females, the orange color was always present in the dorsal, anal and caudal. The colors of the females were stable. The offspring from one pair of this shipment was very variable in colors.

Gerlack (1912: 239) said that Köhler and Arnold went too far in their conclusions. He himself had bred the 1899 strain and found that it bred true to color. He thought that Arnold's 1906 strain probably did not come from Ceylon and that the individuals used for the reproduction came from different populations.

Smith (1945: 422–24) described the color patterns of two populations from Thailand. Males from the Tachin River, at Tachalom of Central Thailand had a yellow dorsal fin with a jet-black basal blotch. The anal was white at the base and with a bright red color in the distant third of the fin, which was pale red in females. The caudal fin was dusky yellowish-red, with a narrow blackish edge in males. The ventrals were rich lemon-yellow in males and plain white in females.

Males from a tidal pool at Lem Sing in southeast Thailand had a yellow dorsal with a black spot, bright orange anal, and rosy-orange caudal with a narrow blackish edge. The ventrals were yellow. Females had a pale orange dorsal with a black spot. The caudal was pale yellow, without a dark edge.

The Thai names for PAN are Pla-hua-ngern or Pla-hua-ngon.

Roloff (1957: 302–03) caught many specimens of PAN in Thailand, but he considered them poorly colored and uninteresting. In the southern part of the country, near Patalung, he discovered a more handsome and smaller phenotype, which he brought back to Germany. This variety reached 6 cm in total length. The strain from Patalung was described as a new subspecies of PAN by Klausewitz (1957: 196–99). See DOM.

Meinken (1964: 146–48) described a new subspecies of PAN on the basis of aquarium-kept individuals which were said to come from Thailand. Individuals of this strain had rows of red spots along the sides. He used the name *Aplocheilus panchax rubropunctatus* Meinken for this strain, but the name "*rubropunctatus*" appears to be preoccupied. See SIA.

I have myself kept various strains of PAN, including SIA. These strains bred true to their color patterns. In a strain from Djakarta, Indonesia, which I got from a shipment of eggs from Thung Kim Tek of Djakarta, the colors of the male were very faint and very constant. The SIA strain and other aquarium strains bred true to colors.

The SIA strain of PAN had 18 haploid chromosomes; about six of these were metacentric elements, which were longer than the longest acrocentrics of the basic complement. The hemoglobin pattern of this strain was simple and rather similar to that of LIN. The pattern of lateral line organs of the dorsal part of the snout resembled that of LIN by the "open" structure and by the presence of more than one pair of neuromasts. These were housed in a mutual depression in PAN, whereas in LIN the large neuromasts are isolated in their depressions. The pattern of frontal scales is normal (G-type, no H-scales) but in some individuals the two large E-scales fuse and produce one very large central scale.

Pair of PAN from Djakarta, Indonesia. Thung Kim Tek's strain of 1962.

A bright lemon color develops anteriorly on the throat of males which have been preserved in alcohol. This color soon is lost. In females a faint lemon color sometimes develops on the throat under similar conditions.

The throat pattern is simple. There is a single dark crossbar just behind the lower lip. The intensity of this band does not seem to change according to the mood of the individual. Ventrals of both sexes are short.

PAP: ***Fundulus pappenheimi*** **Ahl 1924** **(=BIV)**

Ahl (1924b: 142) had a single specimen from Bipindihof of Cameroon. The fish had been caught by Zenker. Ahl placed PAP closest to BIV, BIT and RUS. Holly (1930: 208–09) inspected the holotype and changed minor details of the description. The color of the individual was described as a uniform yellowish-white.

Bipindi is situated on the Lokundje River of southern Cameroon, in the lowlands (about 63 meters above the sea) of the coastal plains. I caught numerous individuals of BIV in the geographical area in question, and there were no indications of two or more different phenotypes of BIV-like fishes. Also, PAP cannot be separated from BIV by meristic characters; PAP should be considered to represent a simple synonym for BIV. See also LOE and BIV.

PAR: ***Panchax parvus*** **Sundara Raj 1916** **(=BLO)**

Sundara Raj's (1916: 268–69) material came from Madras of India. Myers (1933: 181) considered PAR to represent a synonym for BLO, and this point of view has been generally accepted.

Beyer (1931: 38) said that two different strains of PAR were kept by German aquarists of that time. The two strains differed by the size and color of the brilliant spots on the sides. In the "old strain" (=BLO) the spots on the sides were red and greenish-golden, whereas in the "new strain" or "bluish-green variety" from Madras, the spots were larger and of a bluish-green color. Also the fin colors were different in males, but Beyer did not describe the differences. The new strain was imported in 1929.

If the type-locality for BLO was correctly given by Capt. Block (Cochin of Kerala, situated on the southwestern part of the Indian peninsula), Sundaja Raj's PAR from Madras (situated on the eastern coast of the peninsula) might be reproductively isolated from BLO, because of the widely separated geographical localities from which the types from these two forms came.

PAS: *Panchax (Aphyosemion) pascheni* Ahl 1928 **(=AHL?)**

Ahl (1928b: 122–23) had four specimens from Longji of Cameroon. He placed PAS closest to BEL. Holly (1930: 212–13), who examined the types, changed some of the data of the definition (A.14, not A.12).

Longji is a small town situated close to the coast in southern Cameroon, some miles north of Kribi. In 1966, I collected rivulins in several localities near and in Longji and found many individuals of BIV, SEX and AHL and apparently one more species, belonging to *Aphyosemion* and being rather CAM-like. The specimens were, however, too small for a proper identification.

PAS is very ill-defined, and in an analysis of meristic characters PAS was not separated from 72 of the 120 nominal species of rivulins from Atlantic Africa.

The individuals of PAS were yellowish; each scale of the upper part of the fish's side was dark edged. The unpaired fins were colorless or with a faint red color, with rows of red spots in one individual and with a narrow red submarginal stripe and a broad white edge.

The type for PAS as figured by Ahl 1928.

Panchax (Aphyosemion) pascheni spez. nov.

PAS cannot be separated from AHL by meristic characters or by color patterns, and because AHL occurs in the type-locality for PAS we may, for the time being, incorporate PAS in AHL. If, however, future studies of the Longji rivuline fauna disclose that PAS and AHL are the same species, AHL becomes a synonym of PAS.

PAT: *Fundulus patrizii* Vinciguerra 1927 **(Valid sp.?)**

Vinciguerra (1927: 254–57) had many specimens from temporary pools within the Harenaga swamp area, to the left of the Giuba (Juba) River in southern Somali (formerly Italian Somaliland). The pools desiccate completely during the dry season. March Patrizi also told him that this species had also been taken in small temporary pools along the caravan tracks at higher altitudes in sandy areas where during dry years no pools are produced. In such pools at the foot of the Cut (hill) Gelindi between Fakia and Ilescid he found small (17 mm) specimens probably referable to PAT.

The individuals of PAT appear to be small; the largest specimen measured 34 mm standard length. Vinciguerra placed his PAT very close to PAL, ORT, NEU and GUE and closest to PAL. Apparently PAT cannot be separated from PAL by any morphological character, but the geographical distance between the type-localities is too large to permit any incorporation of PAT in PAL until both forms have been biologically studied. The colors described for PAT correspond to those of PAL.

PET: *Haplochilus petersii* Sauvage 1882 **(Valid sp.)**

Sauvage (1882: 324–25) had five specimens from the southeast Ivory coast. They had been taken in the same locality as the types for CHA. See CHA.

Arnold (1908: 314), in this first account of CAL (see CAL), suggested that the SEX-like individuals present in the first shipment of CAL represented PET. If this shipment originated in the Niger Delta, these SEX-like individuals might have been LON, because SEX and GRA were well-known to Arnold at that time.

Keilhack (1910: 119) reported PET from Ossidinge of Cameroon and from Misahöhe of western Togo (Togo Hills), and he thought that PET represented a subspecies of SEX. He had one of Sauvage's types at hand, and he found that it was not in agreement with Sauvage's description of PET, so he suggested that an accidental exchange of individuals had taken place in Paris.

Boulenger (1913: 82–83) wrote the editor of *Blätter*, Dr. Wolterstorff, that the preserved individuals from the Gold Coast, which erroneously had been called PET by Träber (1912: 33), resembled SEN (now SPI). Boulenger is right in this. Thumm's fine ink-drawing (in Träber 1912) is a very characteristic figure of a pair of SPI, and the description of the

"PET" strain corresponds well to SPI. SPI is known to occur near the coast, just east of Accra of Ghana.

The type for PET as figured by Sauvage 1882 (above) and in the present state in the Paris Museum (below).

Boulenger (1915: 59) reported PET from Dunkwa of Ghana.

Myers (1924: 2–4) placed PET in *Aphyosemion*, but in 1933 (: 182), he referred PET to *Epiplatys*. Clausen (1967: 23) removed PET from *Aphyosemion* and placed it in his *Roloffia*.

Meinken (1952: 310–13) identified an aquarium strain from the southeast Ivory Coast as PET. This strain is still kept in the aquarium, although it is not common. Gras (1961: 580) tentatively referred some rivuline individuals from the Djigbe forest of Dahomey to PET.

Daget and Iltis (1965: 198–99) reported 35 specimens of PET from the coastal areas of eastern Ivory Coast, PET being replaced in western Ivory Coast by another species which they identified as LIB (see MAE).

I have seen Sauvage's types for PET. They belong to the collections of the Paris Museum. They are not in good condition, and only one specimen is not a juvenile. The largest specimen measures 37 mm standard length, 47 mm total length. Sauvage said: D.7, A.14, but there are at least D.9–10, A.16 in the types. In spite of the poor state of preservation, the types for PET undoubtedly correspond to the strain identified as PET by Meinken in 1952. I have also seen one individual of PET from the Bokitsa mine of Ghana. This individual, which is in the British Museum, was erroneously identified as SEX. In the British Museum there are also the eight specimens from Dunkwa, reported by Boulenger (1915: 59), I found D.9–10, A.15–16, Sq–l. 31–32 in these.

Pair of PET. G. Wolfsheimer photo.

Three small specimens collected by Roberts in swamps near Elubo, Nzima, southwestern Ghana in 1963 were also studied, together with five specimens of PET caught by Schiötz in 1961 at the Kakum Forest Reserve near Mansu, Ghana and six specimens of the 1952 strain. These individuals had: D.8–11, A.15–16, Sq–1. 29–32.

The rounded dark marking anteriorly on the throat, which characterize females of *Roloffia* (and some species of *Aphyosemion* and *Pachypanchax*) is present in both sexes of PET, but in this species, the intensity of the over-all dark pigmentation of the lower surfaces of the head of males is faint and does not seem to differ from the corresponding pigmentation of the female. The frontal scalation is normal (G–type, no H-scales). No ctenoidy seems to develop in the male. The central rays of the caudal fin extend somewhat in some individuals, and in these individuals the shape of the caudal fin resembles that of most species of *Epiplatys*.

The red pattern of the sides of males is very faint, except for the "wound-like" marking behind the root of the pectorals. In males of PET, this marking is very conspicuous indeed. Other parts of the sides have a red reticulation or crescent-shaped red spots.

Bruce Turner of the United States told me some years ago that he had received some PET-like individuals from Monrovia, Liberia. See MEL.

The pattern of lateral line organs of the dorsal part of the snout is "closed," thus placing PET in *Roloffia* rather than in *Aphyosemion*. The color pattern and the general appearance of PET support its reference to *Roloffia*. PET differs markedly from all other known species of this genus by the pattern of the egg-membrane, which does not develop any pattern at all (having an hexagonal reticulation in all other species). Also in crossings to individuals of *Aphyosemion*, PET appears to be closer to that genus than any other species of *Roloffia* which has been used in crossings to *Aphyosemion*. Although PET resembles the species of the LIB group in some traits, it probably should not be incorporated in this group and should be placed as a rather isolated species of *Roloffia*.

PIC: *Panchax (Aphyosemion) pictus* Meinken 1932 **(=MEI)**

This name was created for a strain of aquarium fishes in Germany by Meinken (1932: 56–57). Myers (1933: 181) found the name "*pictus*" to be preoccupied and created a new name for Meinken's species, see MEI.

PLA: *Haplochilus playfairi* Günther 1866 **(Valid sp.)**

Günther (1866: 314) based PLA on individuals from the Seychelles Islands in the Indian Ocean. There were many individuals collected by Lieutenant Colonel Playfair. Günther described the colors: brownish with three or four indistinct, serrated, dark longitudinal bands. Sometimes a black line across the base of the middle dorsal rays.

Boulenger (1915: 51) had 51 specimens from different localities in the Seychelles.

The first live PLA were imported into Germany in 1924. These individuals were taken in a park in Zanzibar, according to Rachow (1925: 17). Since that time strains of PLA have often been kept in the aquarium, and young individuals are rather popular, whereas full grown individuals, because of their extreme aggressiveness, are not so popular.

Adult male of PLA. Franz Werner photo.

Morphologically PLA is close to HOM, but PLA generally has a shorter head and more fin rays in the dorsal and anal.

The pattern of lateral line organs of the head appears to be very stable (see HOM), and the neuromasts of the dorsal part of the snout develop the "closed" or "*Roloffia*-pattern." The fleshy lobes are large and tend to cover the grooves, in which the neuromasts of the dorsal part of the head are situated. This situation is quite unique in Rivulinae, but a rather similar condition also occurs in DAY, although not to the extent seen in PLA. The neuromasts of the sides of the head are housed in closed canals with pores, as in most other rivuline species of the Old World.

The postorbital neuromasts, however, are situated in open grooves. (See also HOM.)

The frontal scalation resembles that of HOM, and all individuals studied developed the E-pattern, which is very rare in rivuline species of the Old World. See HOM, BLO and GUI. The scalation of PLA differs, however, from that of HOM by the way that the G-scale is larger and the H-scales, when present, are smaller than in HOM. Not more than two H-scales have been observed in PLA, whereas in HOM there are usually three such scales, the third scale being in central position. The scale pattern above the eyes is similar to that of HOM in some individuals of PLA. The large frontal scales always are rather wavy and resemble the "wrinkled type" found in males of most species of *Roloffia*. In *Epiplatys* and in *Aplocheilus* rather wavy scale edges are sometimes seen within the frontal pattern.

Male PLA. n = 24, 2n = 48.

The hemoglobin pattern is complicated; thus far, it appears to be unique among rivuline species of the Old World. The pattern has six lines and thus probably is closest to the different *Aplocheilus-Epiplatys* patterns.

PLA has 24 haploid chromosomes. No large metacentric elements or double-sized acrocentric elements appear to be present in the karyotype. The karyotype of PLA probably corresponds to the ancient cyprinodont karyotype.

The color pattern of the throat resembles that of GUI, but it seems that the dark crossbar behind the line of neuromasts is generally longer than in GUI and more "*Epiplatys*-like". After preservation in alcohol, a strong lemon color develops anteriorly on the throat, as in DAG and some more rivuline species. As usual, this color soon is lost.

The male PLA has numerous red spots on the sides, the spots tending to build up red lines along the sides. There are also many similar spots, usually irregularly arranged, in the unpaired fins. The caudal and anal usually are edged with dark red. In females, a certain dark marking develops in the basal part of the dorsal fin, resembling the dark dorsal marking that characterizes all species of *Aplocheilus*. This marking is absent in the HOM female, however.

PLA develops traits in morphology and in color patterns which place it close to HOM, GUI and *Aplocheilus*.

POL: ***Panchax polychromus*** **Ahl 1924** **(=AUS)**

Ahl (1924: 308–09) had 27 specimens said to come from Cap Lopez (of Gaboon), but actually they undoubtedly were aquarium-raised individuals of the strain which aquarists now call AUS. See AUS.

Rachow (1924: 753) considered POL to represent a synonym for AUS. Ahl placed POL closest to CAL and said that POL had been known under the name "*Haplochilus* from Cap Lopez" by German aquarists since 1913. It has already been mentioned under AHL and AUS that correct definitions of these names cannot be prepared at the present time, but it was also recommended that the names now in general use for very many years should be maintained. Thus I consider POL to represent a synonym for AUS.

PRE: ***Panchax preussi*** **Ahl 1924** **(=OBS)**

Ahl (1924: 53) had a single specimen from Yaunde, East Cameroon. He placed PRE closest to CAM. Holly (1930: 211–13) redescribed PRE on the basis of the type and changed some of the data published by Ahl (D. 11, not D. 10; A. 16, not A. 14). He made PRE a synonym for CAM.

PRE is an ill-defined form. In an analysis of meristic characters, PRE was not separated from 62 of the 120 nominal species of *Rivulinae* of Atlantic Africa. It has already been said under OBS that in the Yaunde area there seem to be only two species of *Aphyosemion* (EXI and OBS). PRE appears to be closest to OBS and is here considered to represent a simple synonym for OBS. The type was almost colorless (yellowish-white), with a red spot in the dorsal fin.

RAC: ***Nothobranchius rachovii* Ahl 1926** **(Valid sp.)**

Ahl (1926: 346–47) had three specimens from an aquarium strain kept by Rachow and one additional specimen from the Berlin Aquarium. All were said to come from Beira in Mozambique. RAC has frequently been imported from Beira in recent time. See Hansen (1957: 129–32) and Roloff (1959: (6): 12–19). These modern strains appear to correspond well to Ahl's RAC.

The color pattern and the meristic characters published for RAC place this form close to TAP and BRI, but in RAC the orange colors are much more conspicuous than in BRI and probably also in TAP. The basic color of the RAC male is a deep orange, especially anteriorly. The red pattern of the sides is an irregular red reticulation, which breaks down into a number of fine vertical red lines posteriorly. The fins are blue, with dark red spots, which often are confluent, producing transverse bars in fins. The caudal fin has a broad black distant marginal band and a broad orange submarginal band. The female resembles the female PAL and has no dark markings on sides.

One of the present strains of RAC has nine haploid chromosomes. (Post 1965: 47–93, found eight haploid chromosomes in another strain). The six largest chromosomes are metacentric. The largest metacentric element measures a bit more than four times the calculated mean length of the basic

Adult male of RAC. Franz Werner photo.

Adult male of RAC (?). Kremser photo.

chromosomes (for n=25), and thus this metacentric element might have been produced by a centric fusion of two double-sized acrocentrics. The hemoglobin pattern corresponds to the common rivuline pattern (4-lines) and thus differs markedly from the pattern found in GUE. See GUE. The eggs are small, usually less than 1.0 mm. There is a single row of large scales below the center of the eye. The teeth are arranged in several rows, which all seem to contain enlarged curved teeth, as in ORT and rather different from the pattern of the 1957 strain of PAL. Ctenoid characters are well developed in the RAC male and correspond to those found in PAL. The dorsal fin has well developed sensory papillae.

RIG: *Fundulus riggenbachi* Ahl 1924 **(=BIV)**

Ahl (1924b: 142–43) had four specimens from a source near Yabassi in East Cameroon. He placed RIG closest to BIV, BIT, RUS and PAP. Holly (1930: 209) examined the types and changed some of the data published by Ahl. (D. 13–15, not D. 11–13. A. 13–14, not A. 11–13, Sq-1. 28–29, not 26–28 etc., etc.). The types were a uniform yellow, with rows of faint spots in some of the fins.

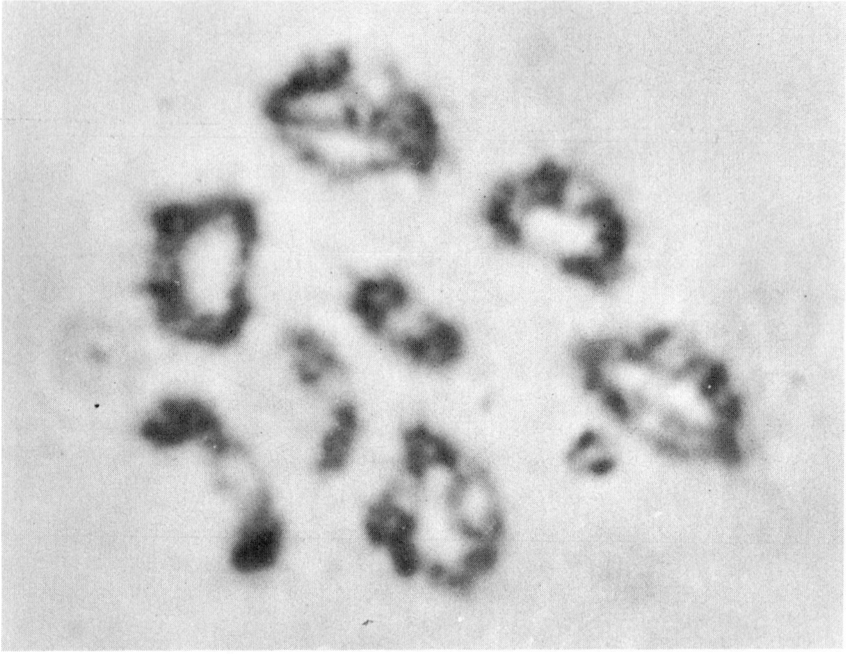

Primary meiotic division of RAC. n = 9.

Secondary meiotic division of RAC. n = 9.

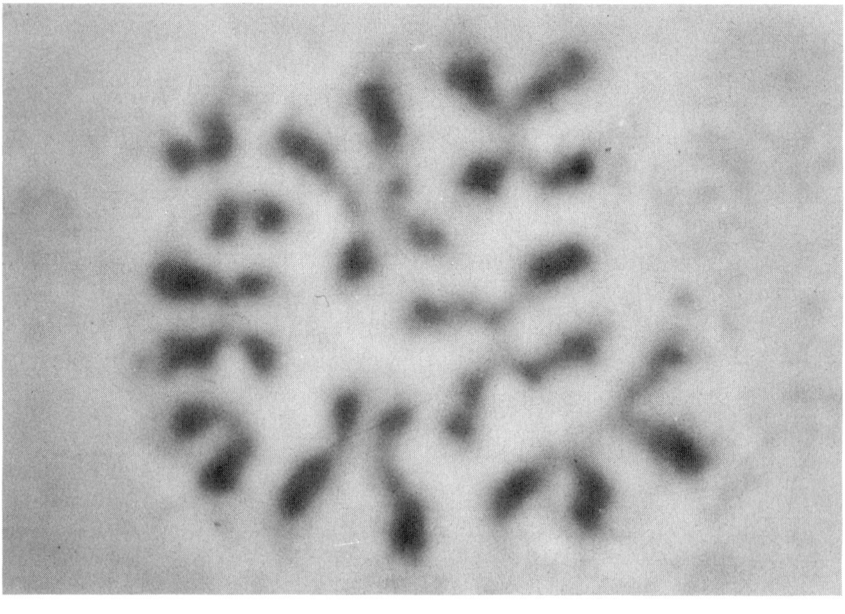

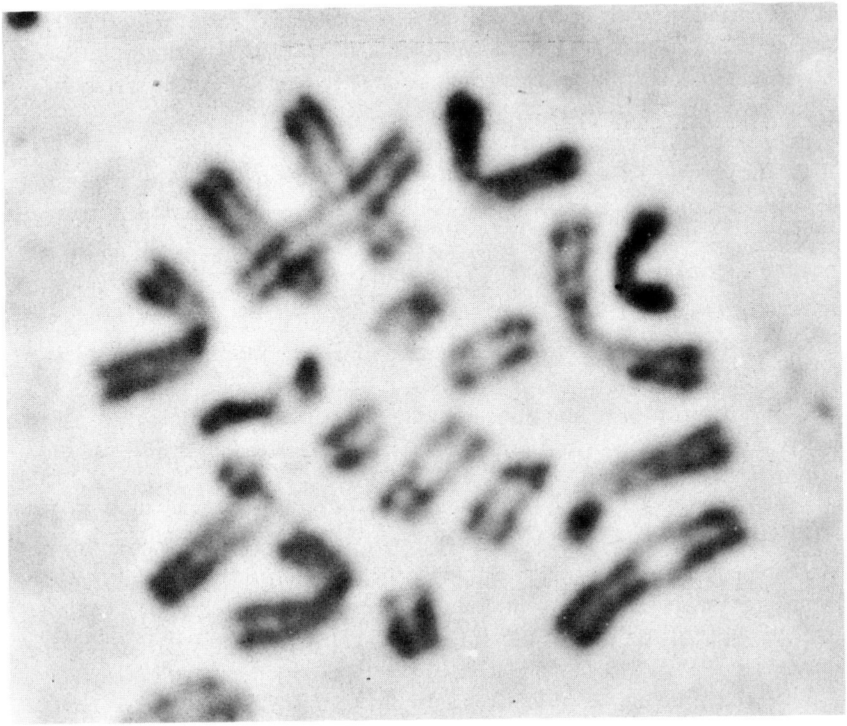

Mitosis of RAC. 2n = 18.

The modern Yabassi is situated north of the Wouri River, which is very wide at this place. I have not visited the northern bank, but I have caught many rivulins near the southern bank, opposite Yabassi. Among them were many BIV, and these did not differ from the usual lowland phenotype represented by the Mboumboula strain, taken 35 km north of Douala, within the western part of the Wouri drainage.

In other localities southeast of Yabassi, near the road towards the Yabassi-Douala-Edea junction, a quite different BIV-like phenotype occurred. In this population, no dark lateral bands were visible, not even in females and juveniles. Also, this phenotype appeared to be more robust and somewhat larger than the usual BIV of this area. At the junction above mentioned, the usual BIV phenotype was present. The two phenotypes of BIV did not occur together in any locality. The aberrant phenotype can be taken 17 and 26 km north of the road junction towards Yabassi, whereas the normal BIV can be taken just north of the junction and 51 km north of the junction. Populations of SEX were present in these localities and appeared to be unchanged, whereas the populations of *Procatopus cf similis* Ahl were very variable in fin colors of the males, although in each locality the fin color was constant. Within the geographical area in question, there seems to be no geographical barriers. The landscape is rather flat and covered by a very luxuriant rain-forest.

The observed aberrant populations of BIV-like fishes in the apparently very limited geographical area southeast of Yabassi may indeed represent RIG, but until more is known I prefer to consider RIG to represent a simple synonym for BIV.

ROB: *Nothobranchius robustus* Ahl 1935

Ahl (1935c: 128–29) had a single individual, collected by Stuhlmann (in 1892) at Tschangarra of Nord-Usinja, probably in modern Tanganyika. I have not been able to find these localities on my maps. Stuhlmann described the colors. The individual was light blue and had dark-edged scales and brownish-red fins.

Ahl separated ROB from SEY, GUE and ORT by the position of the dorsal fin, which in ROB is more anteriorly inserted than in the other forms. This point of view is difficult to understand, because he also said that the first ray of the dorsal was inserted midway between the eye and the root of the caudal, and this position of that ray corresponds exactly to the position given by Pfeffer (1895: 47) for the type (not the figure!!) of GUE, and which Boulenger (1915: 34) published for Günther and Playfair's individuals (SEY according to Ahl) from the Seychelles Islands. So apparently ROB is not separated from GUE and SEY.

Adult male ROL from unknown locality of Sierra Leone.

Adult male ROL from the Freetown area of Sierra Leone.

Adult male ROL from Giema of eastern Sierra Leone.

ROL: *Aphyosemion roloffi* Roloff 1936 **(Valid sp.)**

The first report on this species probably was published by Grote (1910: 220), who during his collecting voyage to West Africa caught some "ELE" specimens in pools, four to six hours by foot from Freetown, Sierra Leone. These were taken on the 7th of December, 1908. There are no indications that they reached Germany alive.

The second report was published by Roloff (1936a: 324–26), who caught live individuals near Freetown. He hired a motorboat and sailed for more than five hours up the Sierra Leone River towards Porto Loko. In a swampy area near the river he found a population of ROL and brought 40 live individuals back to Germany. He said that Ahl had already named this form by the ROL name, but he was not aware that the description had not yet been published. He also (1936b: 387) published a short description of ROL and of the reproduction of this species. He also (in Henzelmann 1937: 65) published a fine photo showing the ROL male and female. On the basis of all these publications ROL appears to be sufficiently defined and described, and thus I place Roloff and not Ahl as the author of the ROL name.

Mayer (1937: 313) also described and pictured ROL and compared ROL to CAB (=LIB), finding these two forms to be different. Ahl (1938: 240–41) also described ROL and published the meristic characters.

Just after the introduction of ROL as an aquarium fish in Germany, some discussion concerning the validity of ROL arose among German aquarists. These people thought that ROL and CAB were identical, and Mayer's article mentioned above is part of that discussion. See also "Aq. Soc. Wiesbaden-Mainz" (1936: 751) and "Aq. Soc. Plauen" (1936: 783).

Daget (1964: 142–44) reported ROL from various localities of Lower Guinée and Fouata Dialon (Mamou, Friguiadi, Kolenti River, Konkure River, Boffa, Thia and Koba). He said that he had compared the paratypes for GER with his own material of ROL and thought that these two forms were conspecific, although the types for GER were more slender than his ROL specimens. Although the present author has not seen Daget's material of ROL from Guinée, he thinks that this material probably belongs to GER, not to ROL.

During three expeditions to Sierra Leone between 1962 and 1965, Roloff collected rivulins in many localities all over the southern part of the country. He kindly sent me live and preserved material of most of the forms. Roloff distinguished between seven different forms of Sierra Leone *Aphyosemion* (now *Roloffia*): OCC, TOD, GUI, ROL, BER and two more forms, one of these being GER. I agree with Roloff in that six of these forms undoubtedly represent valid biospecies and probably also valid morpho-species, which deserve names. As for the seventh form, I have not seen sufficient material to develop any exact idea on the systematic position of this form, which may represent a valid species indeed.

When the individuals (live and preserved) referable to ROL were studied, this assemblage of strains and samples was not at all homogeneous, and it was difficult to imagine that all these individuals belonged to a single morpho-species. Even the strains from the rather limited Freetown area were quite different, not only by color pattern but also by general appearance. My breeding capacity, however, did not permit any crossings between these Freetown strains, and all efforts were concentrated towards a biological isolation of BER from ROL. The crossing experiments strongly indicated that the type-strain of BER was reproductively very distinctly isolated from the sympatric strains of ROL (?) and from ROL from the Freetown area. See BER.

The true range of ROL is not known in detail. From the Freetown area, ROL probably extends north and west. The ROL-like populations which in the Barmoi district are sympatric with GER, and the ROL-like populations which in the Kenema district are sympatric with BER and GER, cannot be incorporated in ROL without much hesitation. These problems probably can be solved only on the basis of extensive biological studies, including crossings and cytology.

I separate ROL from BER by the pattern of the anal fin and of the sides, and from GER by the color pattern of the throat of both sexes, etc. See BER and GER.

The chromosomes of two strains of ROL have been studied. The karyotypes were similar and corresponded to that of BER. See BER. These two strains of ROL are allopatric, and one of them (Giema, near Kenema of east Sierra Leone) might represent a different species.

Also the throat pattern corresponds to that of BER. There are no red pigments on the throat of the male, which becomes very dark during fighting and pre-mating display. In the female, the rounded dark marking anteriorly on the throat often is seen. Just after preservation in alcohol, the lower surfaces of the throat of the individuals of both sexes become an intense lemon color, but this color soon disappears. All individuals studied had a normal pattern of frontal scales (G-type, no H-scales). The two E-scales are very large, and the scales just above the eyes (F-series) are very small or even absent. The character last mentioned is a character of the species of the LIB-group in *Roloffia*.

RUF: *Aphyosemion rubrifascium* Clausen 1963 **(=BUA)**

Clausen (1963: 203) had eight specimens from a swamp between Ndop and Bamunko in the Bamenda Highlands (altitude 1200 meters). He considered RUF to be very common in the grasslands of the Cameroon Highlands, and in the adjoining savannas between the rivers Vuri, Sanaga,

The holotype for RUF in the Copenhagen Museum.

Mbam and Nun. He did not compare RUF to other species of the genus.

It has already been said under CAR, that Pellegrin (1929: 360) identified some fishes from Banyo, Foumban, Bamkin and Yoko (western part of the Sanaga drainage) as CAR, but these individuals undoubtedly belong to BUA. It has also been said (under BUA) that on the basis of external morphology, including color patterns, RUF cannot be separated from BUA. It was also said that individuals of RUF from the type-locality were reproductively isolated from individuals of BUA from Bamkin.

Clausen and the author visited the type-locality for RUF in March, 1966. The swamp or marsh is situated in an isolated valley, and it probably measures less than ten square miles. The area available for fish life seems to be only a small fraction of this area. It was indeed surprising to find a rivuline species in such a highland biotope, but it seemed that even at the end of the dry season there were very many individuals, although almost all were juveniles or semiadults. These individuals were very slender and badly fed. The water was turbid and measured 3.0 German degrees of hardness and pH 6.4.

An inspection of the types and paratypes disclosed a rather irregular pattern of frontal scales, but in most individuals these scales have been lost to such extent that a reconstruction of the pattern was impossible. In some of the few individuals of the aquarium strain so far preserved, there are prominent H-scales, situated below the G-scale as in GUL and SJO. The pectoral sensory (ctenoid) papillae mentioned by Clausen in the description are prominent in large males, but in these no ctenoid spines on scales or within the papillae have been observed.

RUF has 19 haploid chromosomes. The largest of these is an acrocentric element measuring about 2.5 times the mean length of the basic chromosomes (for n=25). This element probably is not present in EXI and BUA-Bamkin. The next chromosomes from the top seem to correspond to the largest (acrocentric) chromosome of EXI and BUA. All the chromosomes develop a smoothly graded series, and only one chromosome appears to be a metacentric element. This element corresponds to the largest acrocentric of the basic complement by size, and it also seems to correspond to the similarly sized metacentric element in the EXI and BUA karyotypes. See also EXI and BUA.

The hemoglobin pattern corresponds to the common rivuline pattern (4 lines).

RUM: *Aplocheilus rubrostigma* Jerdon 1849 (=LIN)

Jerdon (1849: 331) based RUM on specimens from the Malabar Coast of India. RUM is generally considered to represent a synonym for LIN. The first aquarium strains of LIN were identified as RUM. See LIN.

RUO: *Fundulus rubrofasciatus* Brünning 1930 (Invalid name?)

Brünning (1930: 78) created this name for a very ill-defined species apparently belonging to *Aphyosemion.* I have not seen the original description, but Meinken (1930: 598) mentioned Brünning's new species and said that this newly imported species was sea-green, without bands in the anal and caudal fins, but with numerous crossbars on the caudal peduncle. Meinken thought that RUO might represent SPU, which had not been imported at that time. Maybe he was right.

RUR: *Nothobranchius rubroreticulatus* Blanche & Miton 1960 (Valid?)

Blanche and Miton (1960: 215) had six specimens, three from Koundoul, near Fort Lamy of the Chad Republic, and three from Bahr-Marako, also near Fort Lamy, but situated in Cameroon. RUR is the sole "true" species of *Nothobranchius* in this part of Africa. The authors placed RUR very close to TAP from the Lake Victoria area.

The presence of a species of *Nothobranchius* within the Lake Chad drainage is not too surprising, because it is believed that this drainage once was connected to the Nile drainage and that a few more cyprinodont species occur in both drainages (see BIF and SPI), *Aplocheilichthys normani* Ahl representing the procatopodines in both drainages.

Although the number of individuals studied by Blanche and Miton is rather limited, and from a limited geographical area, the individual variation of meristic characters is surprising, even within Rivulinae. They found: D. 16–20, A. 16–20, Sq-1. 29–34 and Sq-t. 24–30. The variation of dimensions is similar.

The side of the male is green, with a mother-of-pearl hue and a purple reticulation, each scale being edged with red. The head is purple, with large areas with mother-of-pearl brilliance on the gill covers. The dorsal and anal are densely spotted with purple, the spots often merging, producing a median band in these fins. The margin of these fins is black, and there is also a broad white submarginal band. The basal two-thirds of the caudal is red; the distant edge is black. The females do not develop any color pattern.

RUS: *Fundulus rubrostictus* Ahl 1924 **(=BIV)**

Ahl (1924b: 140) had three specimens from the aquarist Matte, who said that these individuals came from the Niger drainage. Matte was not, however, a collector of freshwater fishes in Africa. He lived in Berlin (professional breeder), and his data on the type-locality for RUS were undoubtedly not first-hand.

Meinken (1930: 221) said that this aquarium strain probably was imported together with the strain, which Ahl used for his description of BIT and that no one knew when these fishes were imported. He said that RUS resembled BIT but appeared to be a bit more flattened. Ahl placed RUS closest to BIV.

Meinken apparently changed this idea, because in *Die Aquarien Fische* (page 1075) he said that RUS resembled CAM, MEI and OES, and that RUS was imported as an aquarium fish (once more?) in 1932 by the "Platy Tischrunde."

Also Ahl apparently changed his mind, because in 1936 (Arnold and Ahl 1936: 301) he said that this form from the Niger was first imported in 1932. It is not known which authority is responsible for the identification of the 1932-strain as RUS, and this strain appears to represent a species different from the one which Ahl had in 1924. RUS cannot be separated from BIV and 21 other nominal species from Atlantic Africa.

SAK: *Panchax sakaramyi* Holly 1928 **(=HOM?)**

Holly (1928: 313) had 20 specimens from Diego Suarez, in the extreme northern part of Madagascar. These individuals had been taken in mountain brooks at the 500 meter level and 30 kilometers from Diego Suarez. Holly placed SAK closest to PLA from the Seychelles and Zanzibar, but he thought that SAK differed from PLA by the position of the dorsal and the larger number of scales along the sides and around the body. The present author does not agree with Holly in this, because the differences published are too small to be sufficient to separate two African rivuline species. Also, the differences in meristic characters between PLA and HOM are not conspicuous, and probably these two forms grade into each other in meristic characters. The very short description of the color pattern places SAK closer to HOM than to PLA, because no dark marking was described for the basal part of the dorsal and no dark (or red) spots for the sides. The HOM individuals have no spots on their sides and no dark marking in the lower part of the dorsal. PLA has many spots on the sides, and in the female there is a black marking in the lower part of the dorsal. Little is known concerning the biotopes for HOM, but the data available indicate that HOM is a species

The type for SAK as figured by Holly 1928.

of the coastal lowlands and that it also occurs in brackish water. PLA is known to occur at the 330 meter level in Seychelles.

SAN: *Panchax sangmelinensis* Ahl 1928 **(Valid sp.?)**

Ahl (1928: 118) had a single specimen collected by Schreiner at the upper reaches of the Lobo River, near Sangmelima, East Cameroon. He placed SAN closest to CHE and SUP. Holly (1930: 213–14) studied the type, and (as usual) he had to change some of the data given by Ahl in the description. (D. 10, not D. 9. A. 17, not A. 15. Sq–l. 31, not 28, etc.)

Myers (1933: 182) thought that SAN belonged to *Epiplatys.*

Some information concerning the identity of the Lobo River is given under BEA.

In 1966, I found individuals of three Rivuline species at the upper reaches of the Lobo River, near Sangmelima, and in many other localities round Sangmelima. Two of these species were species of *Aphyosemion* (EXI and "OBS"); the third species was an *Epiplatys.* This species of *Epiplatys* was found in many localities all over the wooded part of the inland plateau, but it was much less common than EXI and OBS-like individuals. It was found within the Sanaga, Nyong, Ntem and Congo drainages. In the north it occurs round Yaunde, in the south we found it very close to the Gaboon border, and in the east it was taken right up to Doumé (between Bertoua and Abong Mbang). But I was much surprised to find an adult male of this phenotype in some *Tradescantia* dipping into the water of the Mambaka River, an affluent of the Mbéré, flowing into the Logone River of the Chad drainage. This locality is situated more than 300 kilometers north-northeast of the forest and in the true savanna.

The description of SAN is so vague that it is difficult to determine to which genus this nominal species belongs. The position of the anteriormost dorsal fin ray above the last third of the anal indicates that it belongs to *Epiplatys*, whereas the ratio for the caudal peduncle (1.75 times as long as deep) indicates that SAN should be placed in *Aphyosemion*; the length of the caudal fin (shorter than the head) also supports this idea. The snout is broad and somewhat longer than the diameter of the eye, thus again indicating that SAN is an *Epiplatys*. Ahl described the colors of SAN as yellowish, with red markings on the head, each scale with rows (bands) of red spots. Pectorals and ventrals dark reddish. This description is in agreement with the Cameroon species mentioned above.

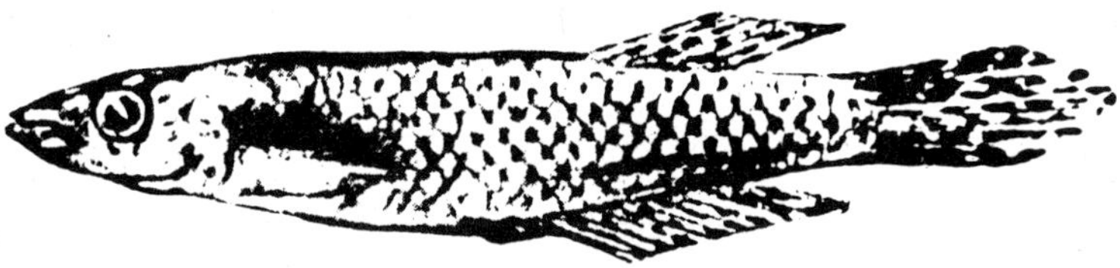

The type for SAN as figured by Ahl 1928.

Live individuals from a population living 30 km north of Sangmelima, towards Mbalmayo, were taken to Copenhagen alive. These individuals were extremely shy and difficult to keep alive.

No trace of dark crossbars were seen in these half-grown individuals, but in the females a broad, very dark band frequently was visible between the root of the pectorals and that of the caudal.

In the male, the fins were a faint yellow, the sides having a very faint bluish-green hue (stronger in some of the males from other localities, especially within the Nyong drainage) and with regular rows of crescent-shaped red spots; almost every scale of the median rows had this red marking. The spots were edged with black. The basal part of the unpaired fins had many small dark spots. The throat pattern was simple. No distinct dark or red bands occurred. The lower surface of the head of the male was rather dark. There were some red pigments near the corners of the mouth and just behind the lower lip (probably attached to the minute neuromasts). Juveniles had a fine red line just behind the lower lip.

The eggs were large (1.7 mm) and were characterized by a hexagonal membrane pattern, a pattern very rarely seen among eastern species of the

Adult male of SAN from brook about 30 km north of Sangmelima, towards Mbalmayo, east Cameroon. $n = 24$, $2n = 48$.

genus, but rather common among western species, and in *Aplocheilus.*

Nine fry were hatched, but the juveniles were difficult to keep, and they died one by one. In the juvenile phase, these individuals developed a distinct dark color pattern resembling that of juveniles of BIF, but unlike the development of patterns in BIF, the dark median blotches extended vertically and produced dark crossbars of the GRA type. The dark pattern characterizing the GRA-ORN-MAC group developed just above the root of the anal. Still later, and before the individuals were half-grown, these dark markings disappeared completely. Similar dark patterns were observed in half-grown individuals from a small stream just north of the Ntem River, south of Ambam of the extreme southern Cameroon. This strain has become extinct.

The male had 24 haploid chromosomes, which all seem to be acrocentric; the lengths of the longest and the shortest chromosomes appeared to correspond to the basic cyprinodont karyotype. No ring bivalents occurred in primary meiotic divisions.

There are four large specimens in the British Museum of an *Epiplatys* species caught by Bates in 1903. These individuals had been identified as SEX, which they are not. They correspond well to the individuals here reported under SAN. They resemble MAC from Portuguese Congo by the general appearance and color patterns. As in the Sangmelima strain, the red spots are not rounded, but crescent-shaped (perfectly rounded in MAC and its allies), and the row of red spots just above the root of the anal and on the keel is not present. This row of red spots characterizes males of the GRA-ORN-MAC superspecies. Bates' individuals came from the Nyong River.

Although the juveniles, not the adults, of SAN resemble the species of the MAC-GRA group, I doubt that SAN is closely related to this generally coastal group in *Epiplatys*. The eggs are much too different in size and in membrane pattern. SAN is more likely to be a northern member of a species-group which is little known at the present time. I have seen one male caught by Lambert at Nzingmeyong of northern Gaboon. This individual seems to correspond well to SAN, but in this adult male the dark pattern which I observed in the juvenile SAN is present. Probably Lambert's Gaboon individual represents another form of the SAN group. The two Gaboon males caught by Higgins in 1872, which I have mentioned under MAC, also may belong here. Maybe ANS is one more member of this group in *Epiplatys*.

SCH: *Haplochilus schoutedeni* Boulenger 1920 **(=CHR?)**

Boulenger (1920: 36) based his description of SCH on individuals from the Medje River, an affluent of the Ituri River of the northeast Congo. Poll (1952a: 46) said that SCH was known from numerous individuals from the Central Congo drainage and that individuals from the Leopoldville area had very few red spots on the sides, compared to other local populations in the Congo. He mentioned that this species had been imported as an aquarium fish into Germany and had erroneously been identified as SIN. He also said (1952b: 166) that CAS represented a synonym for SCH (see also CAS and CHR). He reported SCH from Ibembo, Lisala, Buta, Bambesa, Yangambi, Stanleyville, Monga, Kunugu and Mongende, the localities representing, in particular, the northern-northeastern part of the Central Congo drainage. Poll (1959: 140) reported individuals from Leopoldville. Lambert (1961: 33–36) considered SCH to represent a synonym for CHR and said that by an odd coincidence the types for the latter belong to the phenotype having very many red spots in the male, whereas those for the former had very few such spots.

The Leopoldville population of nominal SCH was imported as an aquarium fish in 1949; a stable aquarium strain soon was established, under the SIN name. The present author kept this strain for several years and found it to breed true to color patterns, except for the yellow color of

The type for SCH as figured by Boulenger 1920.

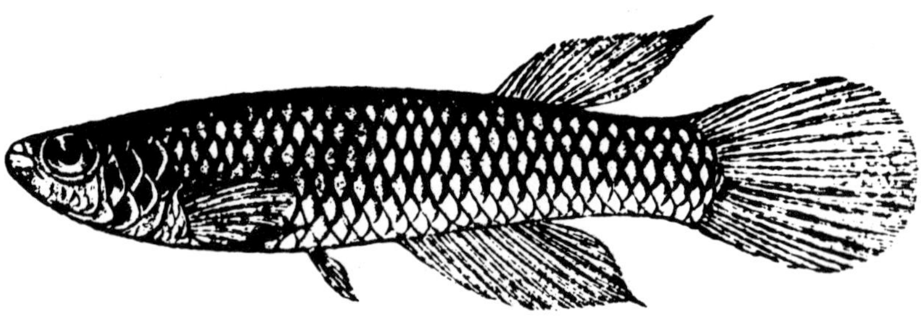

Haplochilus Schoutedeni

the unpaired fins (anal in particular); this color was absent in some males. As it had been pointed out by Poll, this strain is characterized by the very few red spots on the sides and in the fins of males. It seems that this strain has become extinct in the aquarium and that it has been replaced by strains in which the males develop many red spots on the sides. See also CHR.

SRE: *Aphyosemion (Fundulopanchax) schreineri* Ahl 1935 (=BEA)

Ahl (1935c: 113–23) had six specimens collected by Schreiner at the upper reaches of the Lobo River, near Sangmelima of southern Cameroon. It has already been said, under BEA and GUS, that SRE should be considered a synonym of BEA. The types for these three names, a total of nine specimens, were all collected by Schreiner, probably in one sample. It remains, however, an open question whether BEA (GUS, SRE) represent poorly described individuals of BAT, or another *Fundulopanchax* species, perhaps from a locality different from the one indicated for these three nominal species. Ahl placed SRE closest to COE and WAL.

SEN: *Haplochilus senegalensis* Steindachner 1870 (=SPI)

Steindachner (1870: 559) had three specimens from Dagana. This locality is situated near the mouth of the Senegal River. The largest of them had many narrow dark oblique bars and a longitudinal dark band from the posterior part of the eye to the root of the caudal. This band was situated below the median line. The two smaller individuals had no dark crossbars, and the lateral band was more distinct; in these individuals there was also a much more narrow dark line on the lower edge of the body reaching just to the lower edge of the caudal. These two small specimens undoubtedly are BIF; see Svensson (1934: 84–86) under STE.

Garman (1895: 129–30) considered SEN to represent a valid species with MAR and BIF as subspecies (varieties) and CHA as a probable synonym. He undoubtedly is right as far as MAR is concerned. Pellegrin (1911: 183) reported SEN from the Bafing River of Senegal, but in 1962 Daget (see LIB) considered these individuals to represent LIB. In 1913 Pellegrin described ACU as a subspecies of SEN. See ACU. Boulenger (1915: 71) had individuals from Dabola, Guinée, from the northern territories of Ghana, from Ghana and from Old Calabar (these were referred to LON by Boulenger in 1911, see LON). He said that SEN was characterized by a carmine or dark streak around the lower jaw. Myers (1924: 2–4) considered SEN to belong to *Aphyosemion*, but in 1933 (: 182) he said that SEN might represent an *Epiplatys* species.

Pellegrin (1925: 98) reported SEN from Madjingo, on the Djoua River, an affluent of the Ivindo River and part of the Ogowe drainage of Gaboon. It is highly unlikely that SEN occurs in the locality indicated by Pellegrin. Holly (1927: 132) reported SEN from Lake Debo, south of Timbucto in

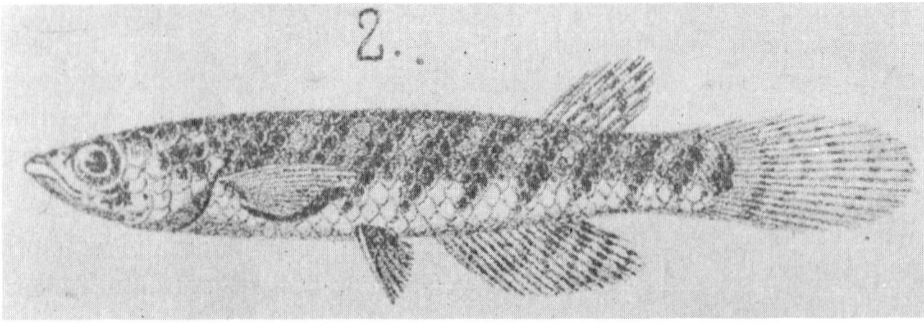

The type for SEN as figured by Steindachner 1870.

Mali. Pfaff (1935: 294) reported SEN from Jebba on the Niger River in Nigeria. I have seen this individual, which corresponds well to SEN (= SPI), which is common in this area. Svensson (1934: 84–86) reported SEN from the Gambia River. He redescribed and figured SEN on the basis of 40 specimens. He said that the two lines across the throat were carmine. This information might be important, because in all individuals of SEN (=SPI) which I have seen these bands are black, with no visible red pigments. Males from some populations of the closely related FAS, which inhabits the savanna biotopes in Sierra Leone, may develop such red lines on the throat. As I have already pointed out under FAS, the SEN (=SPI) form probably represents a derivate of FAS which might have originated from some of the savanna populations of FAS in Sierra Leone or in Guinée. The Gambia population of SEN might represent a form which has maintained the red lines of the throat.

Poll (1939: 35–36) extended the range of SEN to the central Congo and reported individuals from the Stanley Pool area. I have seen these individuals (coll. Vincke 1932) and found them to be in full agreement with SEN, except for the throat pattern. Although they probably first were in formalin (which renders full development of black pigments) there are no dark bars across the throat in the two largest individuals, whereas in the nine smaller individuals there are traces of such bars. Other dark pigments are fully developed in these 11 specimens.

Monard (1940: 98) reported SEN from Ponte Robalo and Pitche of Portuguese Guinea. He mentioned the dark crossbars in the caudal fin. Daget (1948: 42–43) reported SEN from several localities of Sudan (Mopti, Macina, Diafarape) and Senegal (Velingara) and in 1951 (: 1147) he had more Senegal specimens (Zinguinchow and Sangalkam). Poll (1952a: 159 and 1952b: 296) reported more specimens from the central Congo drainage (Gankama River, Kwamouth). Daget (1954: 318) reported other SEN individuals from the Upper Niger drainage (Auana, Siguiri, Bissikrima,

Mopti, Ke Machina, Dio Bozo) and said that this species was frequently found near the Niger River, among plants and usually near the surface, where the individuals rested immobile, sometimes with the posterior part of the body curved. They were not difficult to catch by hand, if the hand were moved very slowly, but when frightened, they escaped very quickly in zig-zag course, sometimes passing obstacles by jumping out of the water. He said that these Niger specimens did not differ from those from Gambia (Svensson's individuals). He also (1957: 159) reported SEN from the Upper Volta drainage and said that SEN did not occur in the Upper Comoé River. He reported individuals from Samandeni, Black Volta and said that he considered DEF to represent SEN. See DEF. Lambert (1958: 51) reported SEN from coastal rivers of Lower Guinée, but Daget (1962: 127) thought that these individuals were FAS. Daget (1960: 326) reported other specimens of SEN, not only from the Black Volta but also from the Upper Comoé. He said that in the Black Volta River, SEN was found above the junction of this river and the Bougouri-Ba River and in the Upper Comoé it was found above Leraba.

The Copenhagen Museum has many individuals from different localities in Nigeria (Maiduguri, Kontagora, Agenenode, Abugi, Pategi and Jebba). These individuals have been studied together with samples from the British Museum. [Ghana (coll. Wolterstorff 1911–12) and the Kaduna River of Nigeria.]

Live specimens from three major river drainages have been studied (see also Scheel (1962: 397–400)). These came from Maiduguri (Chad drainage) and Kontagora (Niger drainage), both in northern Nigeria, and a locality 15 miles northwest of Ketta in southeast Ghana (Volta drainage). The individuals from these three populations were very similar. The Chad males had almost no yellow-orange color in their fins, whereas the Volta males had much orange and yellow pigment in the tips of the fins. The Niger males were intermediate. No important morphological differences were observed. Individuals from the Volta and the Chad strains were crossed to individuals from the Niger strain, and the hybrids were fully viable and fully fertile. These two hybrid strains were crossed and the individuals from this crossing also were fully viable and fully fertile. These crossings indicate that these strains of SEN are rather similar genetically.

Individuals of SEN are not frequently imported as aquarium fish, in part because populations do not live near the coast or near areas from which aquarium fishes are frequently exported. Also, the individuals of SEN are not as handsome as most other species of African rivulins.

The first live individuals were imported into Germany in 1912. Träber (1912: 33) said that he bought specimens from Härtel under the FAS name, but the editor (Wolterstorff) noted that the figure (Thumm's) and some preserved specimens had been sent to Boulenger, who identified the strain as PET, although he had not seen the types for PET. Thumm's fine figure shows a pair of SEN indeed, and as I said just before, I have seen Wolterstorff's individuals and they are SEN, not PET. Arnold (1912: 61) reported SEN in Germany. This strain was said to come from Boma of the lower Congo area, and the identification had been prepared by Boulenger. In a letter to the *Blätter*, Boulenger (1913: 82–83) corrected these identifications and said that the Gold Coast (Ghana) individuals (first identified as PET) probably represented SEN, and that the Boma individuals (first identified as SEN) were MAC. See MAC and PET.

There were no further reliable reports of SEN in the aquarium literature until after the second World War. Individuals, probably from Ghana, were imported during the 1960's; I received photographs of some. My own strains have not been distributed.

SEN (=SPI) is very closely related to FAS of Liberia, Sierra Leone and Guinée. Unlike FAS, which generally is restricted to the forests, SEN has not been taken in the forest and seems to be a pure savanna form, which in most biotopes is sympatric with BIF. If the synonyms for SPI are included (ACU, DEF, MAR) this species extends over huge geographical areas, including all major and most minor river drainages of West Africa, the Chad, the Congo and the Nile drainages. Thus SPI is the most widespread rivuline species in Africa. In the savannas of Sierra Leone, SPI is replaced by some populations of FAS and in the savannas of East Cameroon (south of N'Gaundéré) SPI is replaced by BUA.

The Kontagora strain from the Niger drainage had 17 haploid chromosomes. Seven of these are metacentric elements, longer than the longest acrocentric of the (remaining) basic complement. Calculations on the length of the chromosomes of SEN indicates that it originated from a form which had 25 haploid chromosomes. Among the chromosomes of basic length there is also a metacentric element, slightly shorter than the largest acrocentrics. It appears that the five largest chromosomes (metacentrics) of SPI are similar to the five metacentric elements in the FAS karyotype. Unfortunately the sterile but fully viable hybrids of the FAS/SPI crossing were not cytologically studied.

The most conspicuous character of SPI is the dark crossbars in the caudal fin. This bar-pattern might be absent or hardly noticeable in a few individuals, but generally it is present. No other rivuline species of the Old World develops *dark* crossbars in this fin.

SEN is not an ideal aquarium fish. First of all, it is not a handsome species, although the Ghana individuals are rather colorful. Usually, individuals are very shy and dive deep into the mud of the bottom, where they may stay for hours. See also the slender and colorful form under FAS, which exhibits similar behavior.

SEF: ***Poecilia sexfasciata* Peters 1863** **(=DAG)**

Peters (1863: 396) had a single specimen from Liberia. This individual had been given to him by the President of Liberia, Mr. Benson.

Peters apparently was not aware that the name "*sexfasciata*" was preoccupied, as it had been used for a quite similar species from Gaboon by Gill in 1862. Although Peter's SEF and Gill's SEX are rather similar, they are distinct species, and for that reason Peters' species deserves a new name, which it accidentally received in 1953.

In 1868 (: 146) Peters received an additional specimen (from Gaboon!!) which probably did not belong to his own species, but to that of Gill, and on the basis of these two individuals he created the new genus *Lycocyprinus* Peters. From the text of the description of this new genus it appears that Peters had the Liberian individual and not the Gaboon individual in mind. Thus it seems that the species (DAG) to which the Liberian individual belongs is the type-species for *Lycocyprinus*.

Garman (1895: 157) considered SEF to represent a synonym for *Aplocheilichthys spilauchen* and also in his opinion *Lycocyprinus* represented a synonym for *Aplocheilichthys*. Keilhack (1910: 119) discussed at length the names SEF, SEX, INF, PET, SPI and *Aplocheilichthys spilauchen*. He corrected the description of SEF on the basis of the types. He also had 26 individuals from 12 different localities for his revision. The two specimens from Peters' description have been mentioned. These are probably DAG and SEX. He had one of the paratypes for PET, but he found that it did not agree with Sauvage's description (see PET), and he placed PET as a subspecies of SEF. The "paratype" for PET probably was one of Sauvage's individuals of CHA. He also had 25 specimens from Cameroon. The specimens from Victoria (coll. Preuss), Logobaga (Monke, coll.), Lokundje (Zenker, coll.), Longji (Paschen, 28 specimens) and Efulen (Bates) probably all were SEX. The individual from the Lobe River (Bates) might be either SAN or SEX, depending on where this individual has been caught. The individual from Fernando-Po (Conradt) is SEX, and the Ossidinge (Mansfelt) individuals probably also were SEX. He also had three specimens from Misahöhe in Togo. These had seven to nine dark crossbars, D. 9?–12, A. 15–16, Sq-1. 25–28. They are difficult to place, but they might be SEX. He redescribed SEF on the basis of this "mixed" lot, which seems to contain a single individual of SEF, only the holotype.

Myers (1933: 182) considered SEF to represent a synonym for CHA and placed *Lycocyprinus* as a synonym for *Epiplatys*.

Poll (1953: 262) described and named a SEF-like form from the Ivory Coast as DAG, and Daget and Arnoult (1964: 596) considered the population from which the type for SEF probably originated to represent a subspecies of DAG, which they named MON. Thus Peters' 1863 SEF probably is *Epiplatys dageti monroviae* Daget and Arnoult 1964.

Male of SEX from southern Nigeria.

Adult male SEX from Lagos of west Nigeria.

Adult male SEX from Umudike, near Umuahia of east Nigeria.

Adult male of SEX from southern Nigeria.

Front of preserved individual of SEX from Ijebu-Ode, west Nigeria. The pattern of frontal neuromasts is open. G-type.

Clausen (1967: 27–28) divided the genus *Epiplatys* Gill into two subgenera, with SEX as the type-species for the subgenus *Epiplatus* and GRA as the type-species for the subgenus *Parepiplatys* Clausen. By this division DAG and MON become part of *Parepiplatys*, because in this subgenus the pattern of lateral line organs of the dorsal part of the snout is "closed." But as SEF=DAG still is the type-species for *Lycocyprinus* Peters 1868, this name and not *Parepiplatys* probably is the name of the subgenus, if *Epiplatys* has to be divided at the present time.

SEX: *Epiplatys sexfasciatus* Gill 1862 **(Valid sp.)**

Gill (1862: 136 footnote) had numerous specimens from Gaboon, but still his description of SEX is not very informative: as a consequence, rather many mistakes have been made concerning the identity of SEX-like

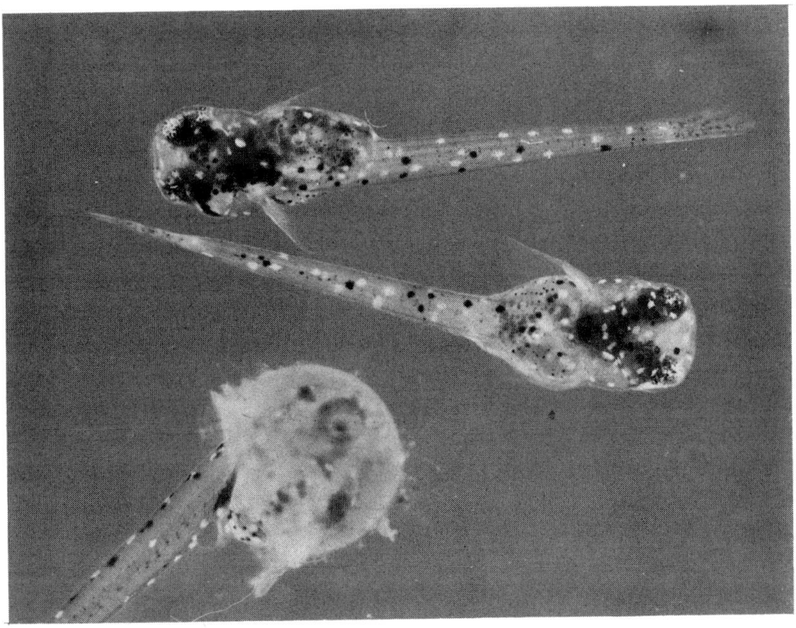

Fry of SEX just after hatching. One fry still with a rather large yolk sac has head in egg.

species. The redescription prepared by Boulenger (1915: 54–55) was based on many specimens of SEX, but he also used many individuals of CHA, MUF, MAC and probably two more species for that redescription. The redescription prepared by Keilhack (1910: 119, 121) for SEF was based on an even more heterogeneous material, although the majority of the individuals used were SEX and not SEF.

Adult male of hybrid strain: SEX-Ibadan/Barombi. First generation.

Male of hybrid strain: SEX-Ibadan/Barombi. First generation.

Adult male of SEX from southern Nigeria.

Adult male of hybrid strain: SEX-Ibadan and SEX-Barombi.

Two of the syntypes for SEX, No. 7129, 1–5 in the Philadelphia Museum. Duchaillu collection. The largest (above) measures 63 mm standard length.

Fowler (1930: 78) prepared a full redescription on the basis of 30 of Gill's cotypes. I have seen five of these cotypes from the Philadelphia Museum and found these individuals to correspond to the phenotype, which I believed to represent Gill's species (Scheel 1962: 365–75). These five had D. 10–11, A. 16–17, Sq-1. 29–30; the pattern of lateral line organs of the dorsal part of the head was fully "open" in all individuals. The ventrals were short in males, and the six dark crossbars were present and arranged as in Nigerian SEX.

It has already been mentioned under SEF that Peters (1863: 396) also used the name "*sexfasciata*" for a rather similar but different species from Liberia (see SEF and DAG).

The geographical range of SEX is rather well known at the present time. The western limit of expansion is situated in the Togo Hills of eastern Ghana (Clausen 1967: 27–28). West of these mountains and also in some localities situated within the Togo Hills, SEX is replaced by its close relative, the CHA species. Keilhack's report of SEF from Misahöhe (see SEF) might indeed concern SEX. SEX also has been reported from central and eastern Dahomey (Gras 1961: 580) and from very many localities in southern Nigeria. In Nigeria, SEX is very common in different biotopes within the forest. In southwestern Nigeria, SEX also occurs in the derived savanna (Meko, Aiyetoro, Imala, Igboro and Abeokuta), reaching almost the true savanna at Meko. Within the Mamfe basin, SEX reaches Mamfe but apparently is not represented east of this city. This phenotype of SEX seems to differ somewhat from the usual phenotype in Nigeria and Cameroon by color patterns and by a reduction of the fleshy lobes of the lateral line system on the dorsal part of the head. This form was also mentioned by Clausen (1967: 27–28), who considered it to represent a distinct species. Both sexes of the "Mamfe-Ikom" form usually have more than six dark crossbars on the body. The ventrals are long in males, D. 10–13, A. 15–18, Sq-1. 28–29 in 12 specimens from the Eyomojok area, near the Nigerian-Cameroon border. On the basis of these specimens and observations of live specimens in the field, it is impossible to determine whether the observed phenotypical differences reflect important genetic differences. South of the Oban-Roumpi-Manengouba mountain range in West Cameroon the usual phenotype of SEX is very common in all small streams of the forest, right up to the steep slope that separates the Mamfe plains from the Kumba plains. The ventrals of males from all these populations are short. In the isolated Lake Barombi-ma-Mbu (Trewavas 1961: 146 and Scheel 1962: 365–75), a population of SEX occurs in bays. Trewavas pointed out that the color pattern of the single specimen from this lake in the British Museum differed somewhat from the usual dark pattern of SEX. The present author received a single specimen also from this lake from Birket-Smith (living in

Mitosis of SEX. Male of Cameroon/Ghana crossing. 2n = 48.

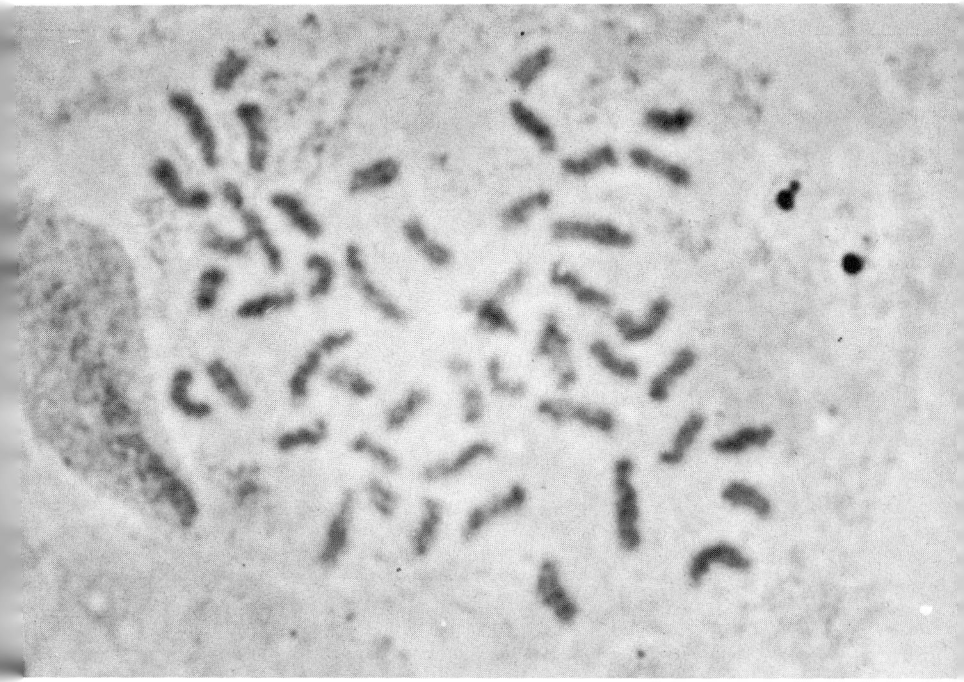

Adult, sterile hybrid male: SEX/DAG MON. The figures show different color patterns and it is seen that the dark crossbars are almost rudimentary and the dark longitudinal band is incomplete. Both photos by Dr. W. Foersch.

Sex reversal in female of the hybrid strain: SEX-Ibadan. Barombi/Barombi. Second generation.

Adult hybrid male: SEX/LON.

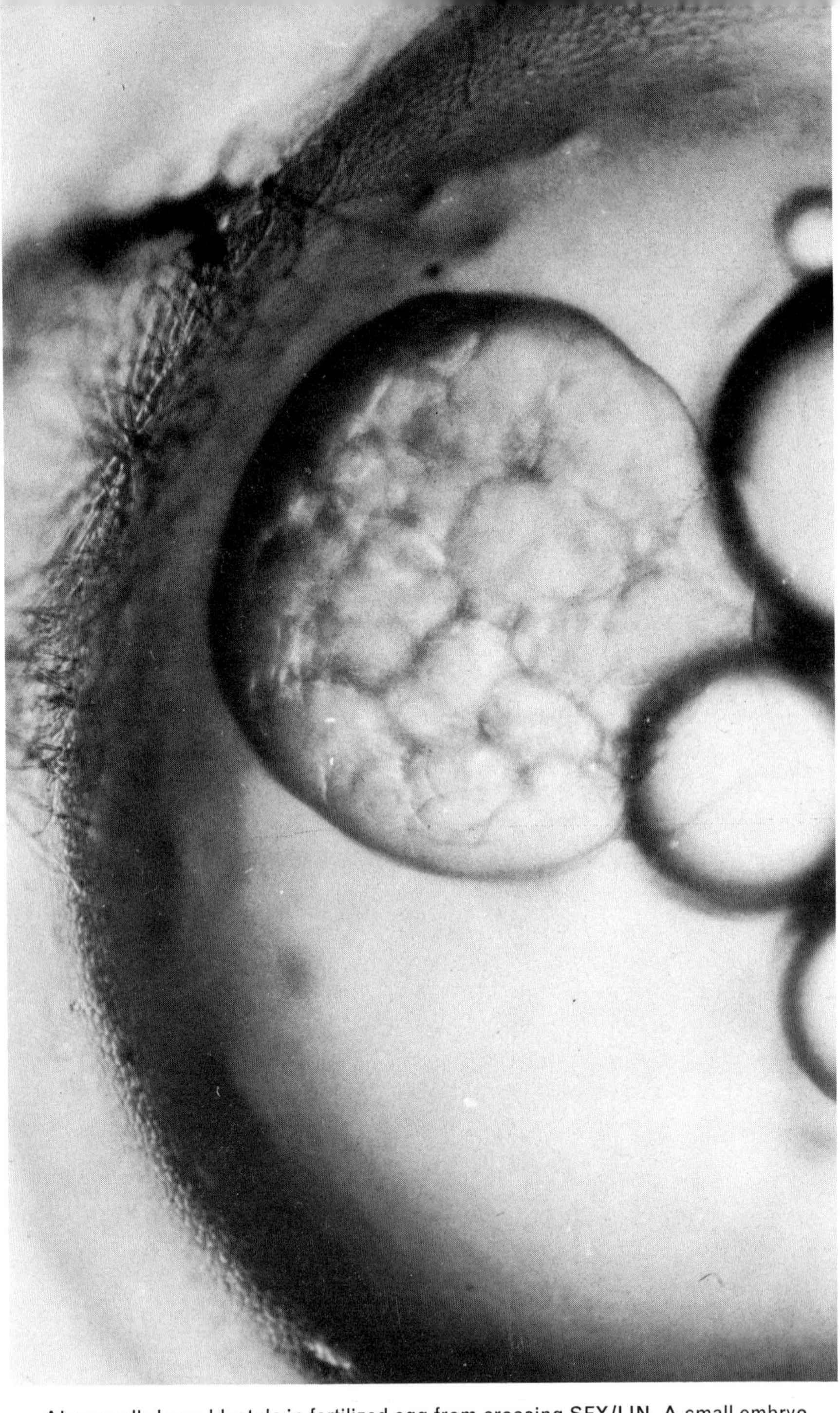

Abnormally large blastula in fertilized egg from crossing SEX/LIN. A small embryo developed in this egg.

Nigeria) in 1959. This individual, a female, was crossed to a SEX male from Ibadan of western Nigeria, and a hybrid strain was raised. The offspring from this crossing were very viable. One male of the hybrid strain was back-crossed to the Barombi female. The individuals produced from this back-cross were very feeble and difficult to raise and, once matured, to keep alive. Sex reversal occurred in a female which developed the colors and behavior of a male. The strongest male of this generation was crossed to a female of SEX from the Ologbo-River, south of Benin City in the western part of the Niger Delta. The viability of the next generation was much improved. Their fertility was subnormal, however, and many individuals had to be raised in order to preserve this mixed strain for further study. In 1965, (Scheel 1966: 132–33) some males of this strain were cytologically studied and 25 haploid chromosomes (instead of 24) were found, with no indications of univalents. The fertility of this strain was very low at that time. From 133 fertilized eggs, only 31 fry were had, and most of these died before they reached maturity. By 1966, only one male and two females had survived; 98 eggs were had from these individuals. Of these, 24 developed a blastula, 20 had a multi-disc, 18 developed an embryo, 6 fry hatched out and all died shortly after. The male was cytologically studied, and the usual 24 haploid chromosomes were present and the meiotic divisions appeared to be normal. No univalents were observed. The two females were tested with a male from the Abeokuta area of southwestern Nigeria. More than half the number of fertilized eggs did not develop normally; the fry which hatched were weak, and the mixed strain became extinct seven years after the creation.

In 1966 I visited Lake Barombi-ma-Mbu and found many individuals near the outlet. Live specimens were brought to Copenhagen for further study. Inidviduals of this population resemble the "Mamfe-Ikom" phenotype mentioned above by color pattern and by meristic characters, but in the males the ventrals were short. Orange pigments were well developed in the tips of the vertical fins. The number of dark crossbars, often irregularly distributed, was highly variable, ranging from the usual six to ten or even more. The karyotype (n = 24) corresponds to that of Ghana, Nigeria and East Cameroon males.

A male from this new Barombi strain was crossed to a female from Ghana. The hybrids were rather viable, but they grew slowly. All but one of the 25 hybrids were males. The number of crossbars was highly variable, ranging from six to eleven. Ventrals were short. The sole female developed the LON pre-mating dark pattern (see LON). No trace of the dark crossbars remained on the sides when the lateral band was visible. Using two of the males, 117 eggs were had. From these, 46 more or less viable juveniles resulted. Thirteen of these were still alive after six months, but of these only five appeared to be normal. Other individuals were very feeble and did not grow. The pattern of dark crossbars was just as variable as in the first generation.

For the control of these results, one male of the usual phenotype of SEX from the Mboumboula River, 35 miles north of Douala in East Cameroon, was crossed to the Ghana female. Twenty-five very viable and quick-growing hybrids were raised. All were males with short ventrals and six dark crossbars on the sides. As there were no females in this brood, some of the males were crossed to the single female of the Barombi/Ghana strain. The fertility appeared to correspond to that of the Barombi/Ghana males, but the offspring were even less viable. Of 35 eggs, 27 of which had an embryo, 20 hatched and two individuals survived after six months. One of these was very feeble.

Hybrids were also raised from a crossing of the Ibadan strain and a strain from Benin City. These individuals were fully viable and fully fertile. Males of the first generation developed extremely long ventrals, exceeding the size found in any male of SEX known to me.

These initial crossings indicate that widely separated populations of SEX are more or less reproductively isolated when in direct contact in the aquarium. In nature, these distant populations are connected by a chain of populations, and genes probably are transmitted through the chain. On the basis of the crossings above mentioned, SEX cannot be divided into a number of biologically isolated subunits, and also the Barombi populations probably cannot be separated from the usual SEX bio-species.

SEX is very common in both Cameroons. In East Cameroon (formerly French Cameroun) it occurs in the coastal plains only, because it has not overcome the steep slope that separates these lowlands from the inland plateau, but see also ESE. SEX has been reported from Fernando Po (Keilhack 1910: 119, 121 and Boulenger 1915: 65–66) and from Spanish Guinea (Boulenger 1915: Only two of the nine specimens are SEX; the other specimens are mentioned under MAC) and from Gaboon (Gill 1862: 136, Peters 1868: 146 and Boulenger 1915: Abanga River). The southern limits of the range of SEX are unknown, but SEX seems to be absent in southern Gaboon, being replaced by MUF and other forms. SEX has not been taken within the Congo drainage.

I have studied about a thousand preserved specimens of SEX from all major parts of its geographical extension. Apart from the Mamfe-Ikom and the Barombi populations, the phenotype of SEX is rather constant within its huge range. The number of dorsal fin rays and of scales in a longitudinal series undergo some clinal variation.

Within Gaboon and both Cameroons, both characters (means) develop maximum figures (D.11.0–11.3 and Sq-l. 30.2), and in these populations the ventrals of males are short. The color patterns are somewhat variable and in the males, there is usually much orange-yellow color distantly in unpaired

Two hybrid males: SEX/CHA SHE. The red pattern of the anal fin and the shape of the body and the head are slightly different. The male figured above usually was very SHE-like, whereas the other male resembled SEX more than SHE. Both were sterile.

fins. The red spots on the sides are rather large. In the Niger Delta area and in the extreme southern part of western Nigeria the two characters reach a minimum (D.10.5 and Sq-l. 28.8–29.1) whereas in the northern, less humid part of the forest of this area, the two characters remain rather unchanged (D.11.0–11.5 and Sq-l. 28.8–29.2) although the means for various populations are rather variable. Whereas males from Umuahia, near the watershed between the Cross and the Niger drainages, have rather short ventrals, much orange in distant parts of fins and an over-all bronze sheen in males, the populations of the Niger Delta develop quite different colors. In these males, no orange color occurs in fins, which are lemon almost all over. The metallic sheen on the sides is a grassy green, changing into golden anteriorly; the red spots are small and inconspicuous, and these males are very similar to those of LON. All these males have long and pointed ventrals. West of the delta, all males have long and pointed ventrals, but the color of the sides changes. All yellow and orange fin colors disappear; the red spots on the sides are large and often edged with black. The metallic sheen on the sides is a cold blue. The Ghana population is quite similar. The Meko population differs somewhat from other western populations. In the fins of males there is rather much yellow, and the red spots are small. Further information is given under LON.

The throat pattern does not reach the distinctness seen in CHA and many more species in *Epiplatys*, but in SEX this pattern is of the usual *Epiplatys* type, with two dark crossbars anteriorly on the throat. The posterior band often is absent in males and females from the Niger Delta and other western populations. Males from some populations develop rather much red pigmentation within the dark crossbars of the throat, and these become almost red. In other males, the whole anterior part of the throat develops a faint red pigmentation. In the males there are almost always six dark crossbars on the sides, whereas in females additional bars may develop, but often these bars are temporary. During the pre-mating display females of SEX usually do not develop the dark longitudinal band, also mentioned under LON, in western females, whereas in those from Cameroon, this pattern is more often seen. In the 1959-Barombi female this band was conspicuous; below the band, the dark crossbars were very distinct, whereas above the band, they had disappeared.

I have studied about 18 different strains, all originating from known localities in Ghana, Nigeria and Cameroon. The hemoglobin patterns for many of these strains have been tested by Dr. Sick, who found them to be similar to those of CHA, LON, ESE, CHE and DUB. In all these individuals six distinct bands developed. Two (females) from the Togo-Hill population from Ghana developed a different pattern with four lines only, but this pattern did not correspond to the common rivuline pattern. The foremost band was displaced and in advance of the usual position. The Barombi/Ghana and the Mboumboula/Ghana hybrids, however, developed the usual six-line pattern.

Except for the aberrant hybrid males mentioned, all males from Ghana, Nigeria and Cameroon had 24 haploid chromosomes. No super-sized chromosomes have been found, and no ring-bivalents develop in primary meiotic divisions. The chromosomes of SEX are rather "woolly" and the karyotype cannot be correctly reconstructed. Clausen (1967: 27–28) separated SEX, LON and two undescribed species (one of these being ESE) from other species of the genus and placed these four forms in a separate subgenus (*Epiplatys*). Other *Epiplatys* species, except for ANN and DUB, were placed in a new subgenus named *Parepiplatys*. This theme is discussed under SEF.

In the aquarium, SEX is very easily kept and bred, but the full colors of the adult male do not develop at once. After 18 months most males have developed all colors. Males from western populations are less handsome than those from the Niger Delta and from eastern and southern populations.

Live specimens of SEX were present in the very first shipment of aquarium fishes from tropical Atlantic Africa, according to Arnold (1906: 319) and contemporary descriptions and illustrations show that Arnold identified these individuals correctly. This shipment originated from Warri in the western part of the Niger Delta. Köhler (1906: 398) described the long and pointed ventrals of the males of this strain of SEX. Zimmermann (1908: 201) crossed SEX and GRA from the Niger Delta and proved that these two rather similar forms were reproductively isolated. His results correspond to my results of a similar crossing in 1961/62. Stansch (1908: 77) separated the aquarium strains of SEX and GRA by the length and shape of the ventrals in males, by the number of fin rays and by color patterns. His excellent diagnosis of SEX and GRA apparently has been overlooked by zoologists and aquarists, or the frequent mistakes concerning the identification of these two forms would not have occurred. Gerlack (1912: 741) described a color variety of SEX that was different from previously described strains. His strain probably originated in some locality east of the Niger Delta area. Seleuthner (1936: 71) described another strain, which he called SEX, but his strain seems to be closer to the red color varieties of CHA. INF represent a synonym for SEX.

SHE: *Epiplatys sheljuzhkoi* Poll 1953 (Valid ssp.)

Poll (1953: 262–65) had one male and one female, presented by Roloff· These specimens came from a collection of aquarium fishes, which Sheljuzhko (1954: 312) caught near Abidjan of the Ivory Coast. Poll placed SHE closest to MAC and OLB.

It has already been said (under CHA), that SHE appears to represent a distinct subspecies of CHA, from which it differs markedly by color patterns, but not by meristic characters. The differences are well preserved in formalin and in alcohol.

SHE and CHA are not, however, separated by post-mating isolating mechanisms. Hybrids are fully viable and fully fertile and in subsequent generations no serious abnormalities were discovered. Also the spermatogenesis of some of these hybrid males appears to be normal. These two forms also appear to be rather strictly allopatric, probably being restricted to certain biotope-types and/or geographical areas.

The aquarium strain of SHE was identified as MAC by some German aquarists and Meder (1953: 143) and Heger (1953: 253) in Germany, and Roberts (1954: 217–18) in the United States used the MAC name for this strain. One of Roberts' photos (two males and one female) was published in Axelrod and Schultz (1955: 428) under the DAG name.

Sheljuzhko (1954: 312) published notes concerning the geographical distribution and ecology of SHE in the Ivory Coast. He found SHE to be very common in many biotope-types near Abidjan and up to 85 kms northwards, towards Agboville. The individuals were not difficult to export.

In 1962 Clausen sent me live specimens which were placed in my tanks. These individuals came from Ayenasi and Tikawbo of southwest Ghana.

Throat pattern of female SHE from southwest Ghana. The two dark bands converging.

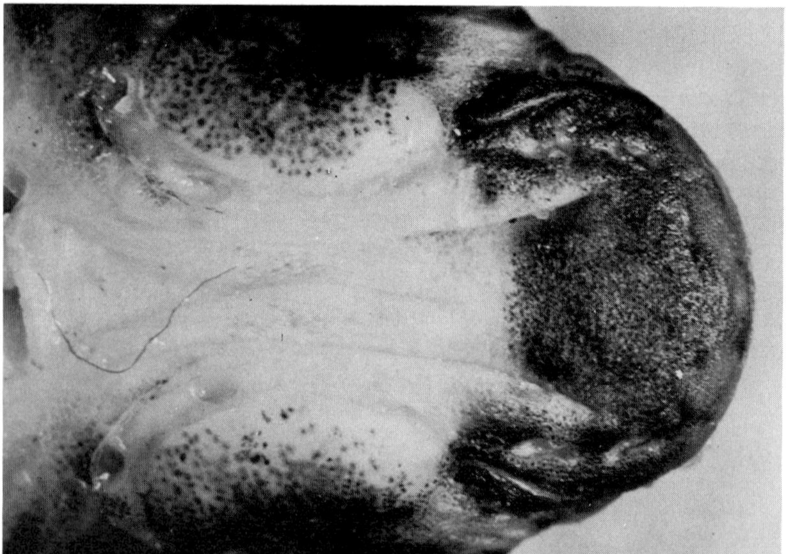

Adult male of CHA SHE from Tikawbo of southwest Ghana.

Clausen also sent specimens referable to CHA; these came from Abra and Angona, also of southwest Ghana. Four aquarium strains were established on the basis of these four lots. In 1963 Bruce Turner of New York sent me photos showing individuals from Akwatia in central Ghana (Pra River system), which he received from a friend in Ghana. These individuals probably belong to the CHA phenotype. Foersch sent me color slides showing the male and the female of a strain imported into Germany about 1962. These individuals also are CHA-like, but not in agreement with southwest Ghana populations of CHA.

The Ayenasi and Tikawbo strains were very similar and not different from Sheljuzhko's strain of 1953. In the male, the sides and the unpaired fins develop a strong cold bluish metallic brilliance, rather violet at times. The red pattern of the sides is a broken red reticulation, which on the median part of the side breakdown into vertically oblong or crescent-shaped large red spots which often are edged with black. The throat pattern is simple and of the usual *Epiplatys* type, well marked in this form. There is almost no red pattern on the throat of males, but in the female the dark lines contain rather much red pigmentation. Most males develop a row of very large dark-red spots just below the root of the anal, and these spots or markings often are edged with black. This particular pattern does not develop in other subspecies of CHA within Ghana, but it is present in the OLB-like population from N'Zérégoré of High Guinée. The anal fin and the lower part of the caudal have thin red lines in between the fin rays, whereas in the upper part of the caudal and in the dorsal, these lines decompose into a number of small red spots. The pectorals are colorless, with dark margins.

The anal fin and the lower part of the caudal fin usually are rather broadly edged with black. Yellow pigments, so well developed in the CHA subspecies, are absent in most individuals of SHE and if present these pigments are not visible to the naked eye. The red pigments of the SHE male are well concentrated and produce distinct patterns, whereas in CHA males these pigments are more diffused. As it has been said under CHA, the SHE individuals almost always develop a distinct black crossbar just behind the root of the pectorals, whereas in CHA this bar is absent. In SHE, there usually is no dark bar above the root of the ventrals. This bar also is absent in CHA, whereas in the Kumasi subspecies (unnamed), the bar above the ventrals usually is present.

I have studied a large number of preserved specimens from Ghana and the Ivory Coast. The most westerly distribution of the SHE phenotype seems to be near Lamto of the Bandama River in the central Ivory Coast and at the western limits of the savanna wedge of Bouaké (see also SPM). In females from the Abidjan area, a dark crossbar above the ventrals often is present. There were also individuals referable to SHE (or to the Kumasi phenotype) from Wenchi and Ejura near the northern limits of the forest in Ghana and from the Nkasupir River, about 20 miles south of Dormaa-Ahenkro of Ghana.

The morphology of SHE corresponds to that of CHA. Hemoglobin patterns, chromosomes and eggs are similar.

SEY: *Nothobranchius seychellensis* Ahl 1935 **(Valid sp.?)**

Ahl's (1935c: 128) description of SEY corresponds to Boulenger's (1915: 34) description of Günther and Playfair's three specimens of ORT from the Seychelles Islands, which Boulenger considered to represent (in part) the types for MEP. Ahl probably simply translated Boulenger's text into German. Ahl placed SEY closest to GUE, but he thought that it differed from GUE by the reduced number of dorsal fin rays (D.14–15 versus D.15–17) and other characteristics. He also said that the types were in the British Museum and that the males of SEY were unknown.

SIA: *Aplocheilus panchax siamensis* nom nov. **(Valid ssp.?)**

Meinken (1964: 146–48) described *Aplocheilus panchax rubropunctatus* on the basis of some aquarium fishes said to originate from Thailand. The name "*rubropunctatus*" appears to be preoccupied by Steindachner's *Haplochilus rubropunctatus* (1867: 345), which, according to Myers (1933: 181) might represent a mislabelled African *Epiplatys*. The name SIA is proposed to replace Meinken's name for this strain. SIA seems to differ from other known populations of PAN by the many red spots on the sides. See also LUT, which seems to develop red spots along the sides, and also Arnold's notes on his Ceylon strain of PAN (1906: 635) under PAN. The anal fin of the SIA male has a very broad orange-red marginal band. More information concerning biological characters for SIA is given under PAN.

SIN: *Haplochilus singa* Boulenger 1899 **(Valid sp.)**

Boulenger (1899: 113) had a single individual from Boma, Lower Congo. In 1902 (: 269) and in 1915 (: 78) he reported two more specimens of SIN from the Lindi River, near Stanleyville in the northern part of the central Congo area.

Nichols and Griscom (1917: 725) reported about a dozen specimens of SIN from Stanleyville, from pools in a clear running river, an affluent of the Tschimbo River. Myers (1933: 182) placed SIN in *Epiplatys* after having seen the type, but in 1924 (: 2–4) he said that SIN might belong to *Aphyosemion*. Poll (1952: 160) had the holotype, one individual from Leopoldville and one more from "Congo." But the same year (: 293) he removed the Leopoldville specimen from SIN and placed it in SCH.

The first strains of SCH (=CHR) in Europe were identified as SIN about 1949. See SCH and CHR. Myers (1954: 89) said that any stock of *Aphyosemion* which may come in under the SIN name (there are some in Europe) must be considered as misidentified. . . SIN is an *Epiplatys*.

The type for SIN as figured by Boulenger 1915.

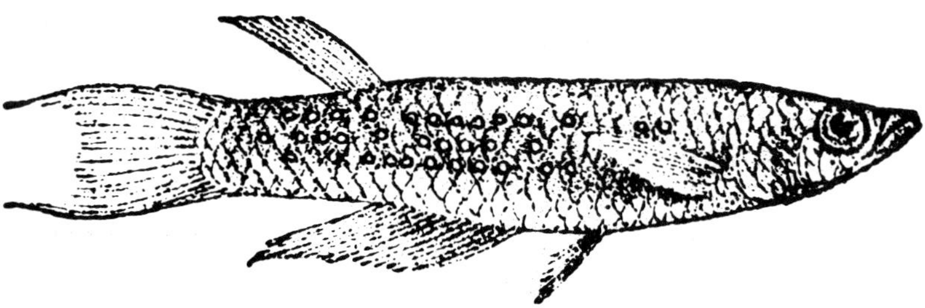

Haplochilus singa.

♂, type (A. M. C.).

Poll and Gosse (1963: 96) said that SIN is an *Epiplatys* and not an *Aphyosemion*, and that SIN is restricted to the Lower Congo area.

Boulenger (1915: 78) figured the holotype with a bilobate caudal fin, but the shape of that fin probably is the result of damage. The type for TAE (=BIF), also an *Epiplatys* species, was figured with a caudal fin quite similar to that of the SIN type. Poll (1952) said that the holotype for SIN was badly damaged. The position of the dorsal fin is that of an *Epiplatys* and color patterns and meristic characters place SIN very close to the Boma population of MAC, which has been mentioned under MAC. These Boma individuals differ, however, somewhat from the types for MAC from Portuguese Congo. Until live material of these populations has been studied, I hesitate to place MAC as a simple synonym for SIN, although undoubtedly these two forms are very close.

SJO: *Fundulus sjoestedti* Lönnberg 1895 **(Valid sp.)**

Lönnberg (1895: 191–92) had two specimens. A large male came from a small brook near the waterfall of the Ndian River. The paratype, a female,

Adult male of SJO. Photo by Dr. Herbert R. Axelrod.

badly damaged, came from Bonge, also of West Cameroon. According to Sjöstedt (1904) he reached these two localities by boat in 1891–92, and it appears from his book, that the types for SJO and BIV were taken below the fall, where the Ndian factory then was situated.

In 1966, Clausen and the author tried, in vain, to catch individuals of SJO around the present town and plantation named M'Bonge, which is situated on the Meme River. It might be that the locality, which Sjöstedt called Bonge was situated somewhat closer to the coast than the present M'Bonge, but both localities were on the Meme River.

Young male SJO, probably from southwest Nigeria.

The SJO name has not until very recently been used for Lönnberg's fish, which since 1915 has been named COE by zoologists and aquarists. Boulenger is responsible for this misidentification, and before 1915 he was sure that the aquarium-kept individuals of SJO were simple varieties of GUL. See also COE, GUL and OCC.

I have seen both types for SJO and found these individuals to be in almost full agreement with the aquarium strains, which have been named COE since 1915. The types for SJO, however, originate from localities situated near the eastern limits of the range of this phenotype of *Fundulopanchax*. The western limit of range appears to be near Lagos of southwestern Nigeria.

Ctenoid spines on scales above the root of anal in holotype of SJO (male) of the Stockholm Museum.

Holotype for SJO in the Stockholm Museum.

It is not unlikely that the morphologically defined SJO, which indeed is highly variable in meristic characters and more or less in colors, is composed of a number of reproductively isolated populations inhabiting different geographical areas. For that reason, future research on populations of SJO may disclose that this morpho-species can be divided into a number of biologically and morphologically definable units, and thus the COE name might prove to be useful for one of these. For the biological structure of South Nigerian rivuline species, see BIV, ARN-FIL, GAR and GRA.

Head of holotype SJO of the Stockholm Museum.

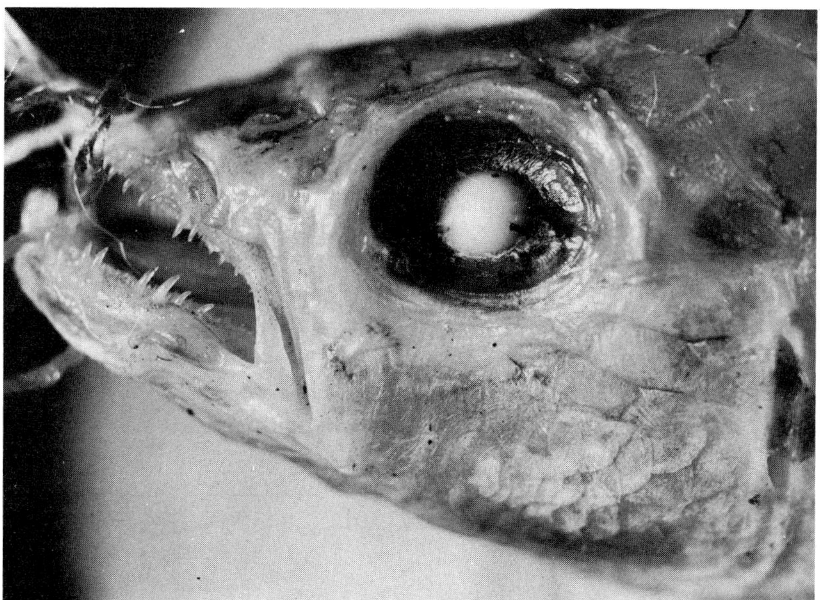

Front of preserved individual of SJO. The two prominent H-scales are visible. Frontal neuromast pattern is open. F scales above eyes have not reversed.

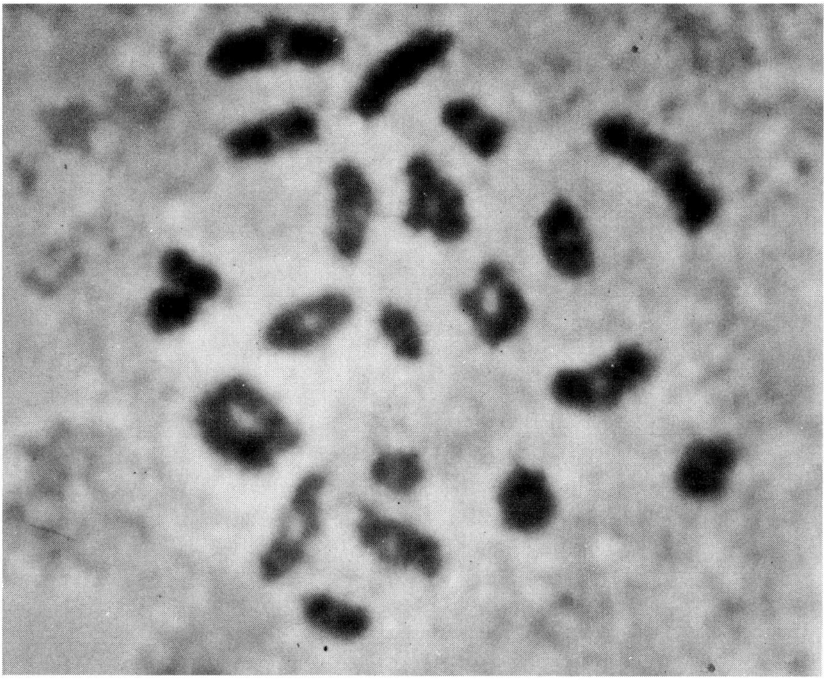

Primary meiotic division of SJO. n = 20.

Mitosis of SJO. 2n = 40.

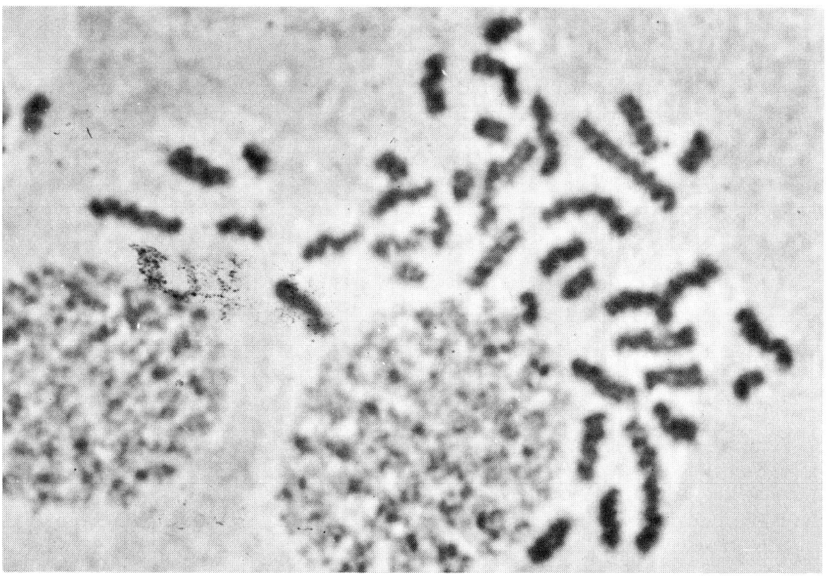

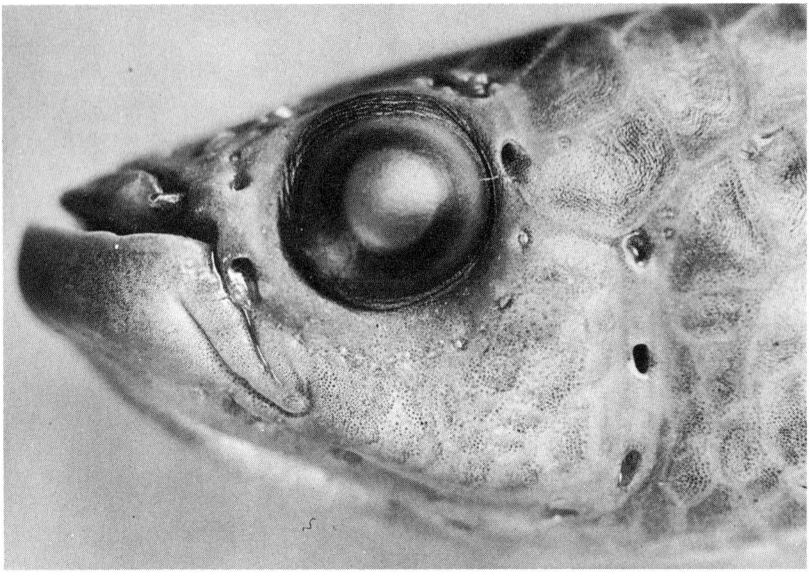

Head of preserved individual of SJO. Compare this photo to head of preserved male OCC.

The SJO name has been used for populations and strains of OCC since 1910, when Boulenger identified a quite different phenotype from the Sherboro Island off the coast of Sierra Leone as SJO. See OCC. Boulenger (1915: 65–66) listed specimens of BAT from Akonolinga of East Cameroon under SJO. I have seen these specimens and they are BAT indeed.

A strange "information" concerning SJO was given by Krüger (1913: 754), who said that Professor Y. Sjöstedt told him in 1912, that the fish named after him was caught in the Njong (Nyong?) River, above the falls

Adult male SJO.

and about 200 kms from the sea and near a station named Atiga-Samboso. According to his book (1904), Sjöstedt did not visit the Nyong drainage in 1890–92, but he might have been there later on, and he might have caught individuals of BAT. When 200 kms are measured from the coast into the Nyong drainage, a locality between Mbalmayo and Akonolinga is found. Here BAT is likely to occur.

Myers (1933: 180–81) made COE (= SJO) the type-species for the subgenus *Fundulopanchax*, and he also made SJO (probably OCC) the type-species for *Callopanchax*, another subgenus of *Aphyosemion*. Thus *Callopanchax* appears to represent a synonym of *Fundulopanchax* or an invalid name. See Clausen (1966: 324), who also said that all individuals known under the SJO name from localities west of the "Dahomey-Gap" including all SJO strains owned by aquarists, belong to OCC, whereas SJO is confined to the eastern area and is known from Cameroon and Eastern Nigeria only.

SPI: *Poecilia spilargyreia* Dumeril 1861 **(Valid sp.)**

Dumeril (1861: 258) had two (small) specimens presented by M. Schill, captain of a Russian merchantman, who took them in freshwater within the Mandingue Coast area of West Africa. The Mandingue Coast corresponds to the Senegambia area. According to Pellegrin (1925: 231), Rochebrunne (1883: 139) said that the type locality is situated either on the Falema River or near the falls of the Gouma River, both localities being situated rather far inland in Senegambia. It is not known where Rochebrunne got that information.

One of the types for SPI in the Paris Museum.

Pair of SPI from Kontagora, northern Nigeria.

Adult male SPI from Kontagora of northern Nigeria.

Young male SPI from the Lower Volta drainage in southeastern Ghana.

Young male SPI from Kontagora of northern Nigeria.

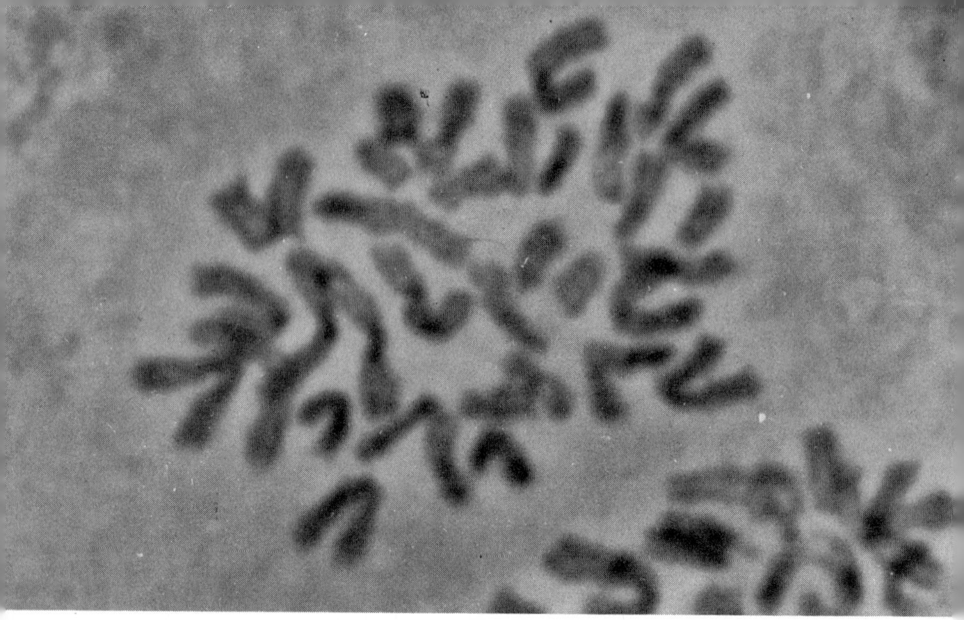

Mitosis SPI from Kontagora, northern Nigeria. 2n = 34.

Gill (1862: 136) placed SPI in *Epiplatys* together with SEX, HOM and *Aplocheilichthys spilauchen*. Garman (1895) placed SEX and INF as synonyms for SPI. Boulenger (1915: 65–66) said that SPI probably represented a distinct species (sic!) but that the types in the Paris Museum were too badly preserved to draw up a complete description. Pellegrin (1924: 231) also said that the types were very badly preserved, and he placed SPI between SEX and FAS.

I have seen Dumeril's two small individuals and they are, indeed, not in fine condition, but certainly they are not so damaged that they cannot be used for identification. The pattern of lateral line organs on the dorsal part of the snout is fully "closed" and separates SPI from the SEX group in *Epiplatys* and from *Aphyosemion.* The caudal peduncle is 1.0 to 1.1 as long as deep, thus separating the individuals of SPI from *Roloffia.* Also, the anteriormost dorsal fin ray is too far back on the back for a species of *Roloffia*, except for PET. The rather dark pigmentation of the throat and on the sides does not produce any distinct pattern. There are some irregularly distributed red pigment cells on the sides, but the pattern cannot be disclosed. The head of both individuals resembles *Epiplatys* and differs from *Roloffia.* The details of the head are rather well preserved and are useful for further identification. The ratio: interorbital width/length of the premaxillary (plate) is useful to separate SEN and BIF, which both occur in Senegambia. This ratio is 5.8 and 5.8 for SPI and 4.5–4.6 for BIF, whereas in SEN I found 5.4–5.8. Thus it seems that SPI is separated from BIF, and this idea is supported by the general appearance of SPI and BIF. This ratio is variable in FAS (4.4–6.2) in various Sierra Leone material, but SEN and FAS can easily be separated by the predorsal length or by the number of dorsal fin rays. The character last mentioned cannot be found for SPI, because the fins are rather damaged. Pre-dorsal-length (in % of standard length) is 67–74 for FAS (mean 71.5); 75–80 for SEN (mean 77.5) and 78 for both individuals of SPI. Thus SPI is much closer to SEN than to FAS.

Although no real proof can be worked out on the basis of the two types for SPI, it seems that SPI can be separated distinctly from *Roloffia* and *Aphyosemion* and from all subsequently described species of *Epiplatys* inhabiting the western part of West Africa, except for SEN and its synonyms. Also, SEN appears to be the most common rivuline species in Senegambia, and for these reasons I do not hesitate to place SEN as a synonym for SPI, but see also Gambia individuals of SEN reported by Svensson (1934: 84–86) which developed a red throat pattern.

The SPI name has been used for some aquarium strains of GRA from the Niger Delta area. See GRA. Synonyms for SPI are ACU, DEF, MAR and SEN.

SPL: *Fundulus splendidus* Pellegrin 1930 **(Valid sp.)**

Pellegrin (1930: 208) had 56 specimens which had been taken by A. Baudon within the Sangha drainage and 23 specimens from a locality on the road between Suanké and Gara-Binzam. The exact type-locality for the types (Sangha individuals) unfortunately is not known. The locality for the additional material seems to be situated on the upper reaches of the Ivindo-Ogowe drainage. Pellegrin placed SPL closest to COE and GUS. By an analysis of meristic characters I was not able to separate SPL from BEA, GUS and SRE, which undoubtedly represent a single species; see Radda (1963: 70–73). SPL was found to be close to CIN, COE and GUL.

I have seen all the types and paratypes above mentioned, and I am in full agreement with Pellegrin, who said that SPL would be a wonderful aquarium fish (if introduced as such). Even after all these years of preservation in alcohol (types), these badly deformed individuals are extremely colorful and the long filaments in all fins are still present in some of the males.

One of the syntypes for SPL. No. 29.248 in the Paris Museum.

SPL. Photo by Dr. Gery.

Pellegrin said that in the male, all fins were filamentous, and so they are. The caudal has two long streamers.

Red color is abundant, but generally no distinct pattern develops. The throat, however, has a broad red band just behind the lower lip reaching almost to the eyes. There sometimes are traces of a second red bar between the corners of the mouth. Just below the eye, a large red blotch is seen. There are also three prominent red bands across the gill cover, but on the sides of the body the red pigmentation is very irregular, sometimes with a number of very irregular, rather narrow red crossbars or lines. There are intermediate dark lines crossing the posterior part of the sides. The pectorals have long filaments and are colorless, except for some red lines between the fin rays. The ventrals have many red filaments, which are rather short. The anal has many irregularly distributed red spots and streaks, which do not form any distinct pattern; the posterior rays are very filamentous. The dorsal is similar. The upper and the central part of the caudal has a similar red pattern, whereas the lower part of that fin has a broad white marginal band, probably yellow or golden in life. The total amount of red pigmentation exceeds the amount seen in all other African rivuline species, but some males of BAT have much red pigmentation also.

Females may develop a similar, but much reduced red throat pattern or no red pattern at all on the throat. The red spots on the sides are crescent shaped or rounded, and there are a number of thin brown longitudinal lines.

The pattern of lateral line organs of the head corresponds to *Aphyosemion*. The frontal scales unfortunately have been lost in most individuals, but in a few, they remain and develop a regular G-pattern, but the frontal scalation seems to be somewhat reduced, the distance between the anterior edge of the G-scale and the upper jaw being larger than in other species in *Aphyosemion*, but not so large as in *Nothobranchius*. The ctenoid spines on scales are few and short.

The Suanké/Gara-Binzam individuals seems to have been in formalin. Red pigments are totally absent, and the individuals are much darker than the Sangha individuals. But the pattern of red pigments appears to have been similar to that of the types. The "wound-like" marking just behind the gill covers is very marked in the males of this lot. Thin dark crossbars or lines are visible in some males, but these markings are very irregular. Fin-filaments are absent in most males, except for the anal and caudal fin, but all individuals of this lot also are smaller than the types. The frontal scales are present in most of these 23 specimens. G-type, no H-scales, but as in the types, the frontal scalation has restrained from the anterior part of the front, and the G-scale is rather small for the size of the individuals and other frontal scales.

It has already been said, under BAT, that BAT and SPL appear to be very closely related. They are separated by their meristic character, not by color patterns. These two forms probably make up a superspecies restricted to the wooded part of the inland plateau of East Cameroon and northern Gaboon.

The relationships between the BAT-SPL group and other groups of *Aphyosemion* are obscure, and probably the BAT-SPL group is not closely related to any known species or group of species.

SPM: *Epiplatys spillmanni* Arnoult 1960 **(=SHE)**

Arnoult (1960: 219) had 32 specimens from a locality near Bouaké of the Ivory Coast. He caught these himself. The city of Bouaké is situated in the middle of the broad savanna wedge which almost cuts the Ivory Coast forest into two parts, and which certainly did so during more dry epochs in the past. He placed SPM close to CHA, SEX, OLB and DAG, but he did not compare SPM to SHE.

I have seen the holotype (male), the allotype (female) and twelve of the paratypes for SPM. They correspond to SHE in all visible details. All have a dark crossbar behind the root of pectorals, at least on one side of the body. There are four or more dark crossbars above and behind the anal; the additional bars probably are temporary in life. The throat-pattern corresponds to that of CHA and SHE. On the basis of this material, it seems that SPM represents a simple synonym for SHE, but as I said under CHA and SHE, the geographical distribution of the different phenotypes of CHA (including SHE) is not easily understood and there might be two or more sets of population groups corresponding to the SHE phenotype.

Arnoult found SPM in very brown and acid water in swamps. He brought live specimens to Paris and established an aquarium strain of SPM. I have

The holotype, male (above) and the allotype, female (below) for SPM. No. 60.394 and 60.394-bis in the Paris Museum.

kept live specimens probably of this strain and found them to be similar to the SHE strains from Southwest Ghana. Hybrids from crossings of the supposed Arnoult-strain and one of my strains of CHA were fully viable and fully fertile. The karyotype of the supposed Arnoult-strain corresponded to that of CHA and SHE and had 25 haploid chromosomes.

SPP: *Fundulopanchax splendopleuris* Meinken 1930 **(=BIV)**

Meinken (1930: 17) had one male and one female for the description of SPP. These individuals came from Griem's collections of 1929. See also MUC. They were said to come from Tiko in West Cameroon. He placed SPP closest to BIT and LOE and probably representing an intermediate link between these two forms.

Under MUC it has been said that for some unknown reasons German aquarists soon used the MUC name for the SPP strains. The old SPP strain or a similar strain is still kept in the aquarium under the MUC name. See Sterba (1959: 288, color photo by Foersch, showing SPP, but labelled MUC). This strain also was kept in the United States during the 1950's (I received color slides from Franz Werner of Detroit).

I caught many individuals of BIV around Tiko in 1966 and found these individuals to correspond well to Meinken's description of SPP. Further information is given under BIV.

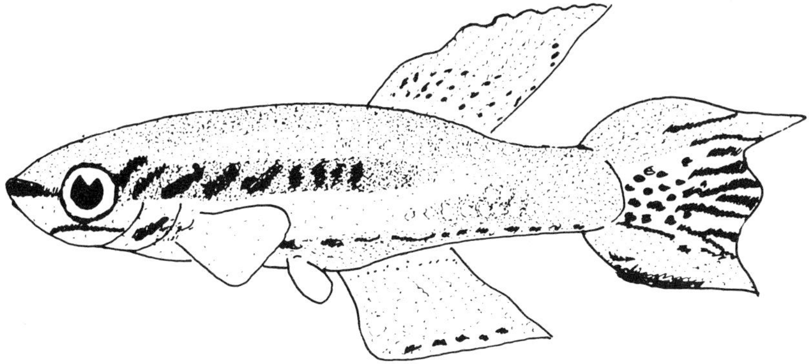

The type for SPP redrawn after Meinken's drawing of 1929.

Arnold (1933: 649) reported SPP, together with MAC, in a shipment of aquarium fishes from Equata in Spanish Guinea. The description and the figure refer this strain to BIV.

SPU: *Fundulus spurrelli* Boulenger 1913 **(=WAL)**

Boulenger (1913: 52) had three specimens from the Tano River of western Ghana. They were taken by Dr. H. G. F. Spurrell. He placed SPU closest to GAR.

Daget (1948: 39) reported SPU from Kaoue, Velingara in Senegal and from Boffa and Thia (near Boffa) of Lower Guinée. These specimens undoubtedly are not SPU, which is a forest species, but might represent THI or some other form of *Fundulosoma*.

The types for SPU as figured by Boulenger 1915.

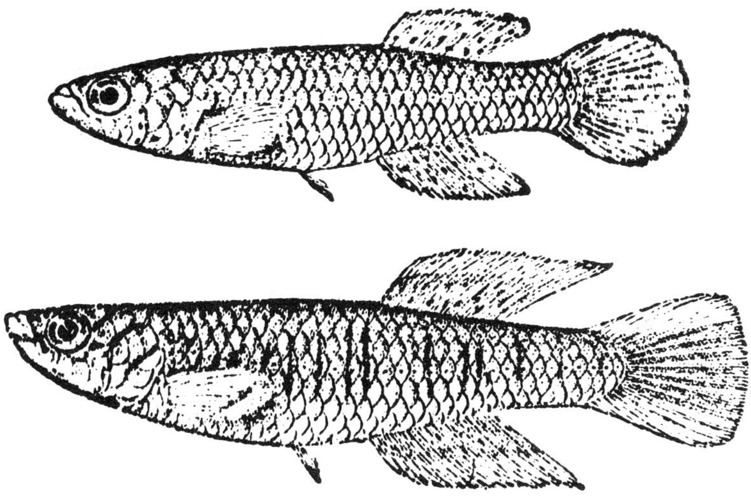

SPU corresponds well to WAL by meristic characters. WAL was based on two females from the Bokitsa Mine, Wasa, Ghana. This locality also has CHA and PET, indicating that the mine is situated in the rather humid forest of southern Ghana, and it seems that SPU should be placed as a synonym for WAL. See also RUO.

It has already been said, under FAL, that some specimens from an aquarium strain in Germany were identified as SPU (and GUL) prior to Ahl's description of this aquarium strain as FAL. FAL appears to be well separated from SPU by meristic characters and by color patterns. In 1953 Sheljuzhko collected live individuals of SPU near Abidjan. This strain was identified as GAR by Meinken. See WAL.

STE: *Panchax steindachneri* Svensson 1934 **(=BIF)**

Svensson (1934: 84–86) had about 50 specimens from swamps of the McCarthy Island area in Gambia. It has already been mentioned under SEN that Svensson was aware that two of the types for SEN probably were not this species. From his own material of SEN (=SPI) and BIF from Gambia he realized that the same differences in color pattern that Steindachner observed were present in the Gambia material, and on this basis he described STE. He said, however, that STE resembled BIF from the Nile, Sierra Leone and Cameroon, but in his opinion STE differed from BIF by the proportions of the head and the snout, by the number of rays of the dorsal, the number of scales around the body and by color patterns. His figure showing STE does not, however, differ from BIF, and appears to be different from BAR, in particular by the red crossbars in the caudal fin of the STE male. STE also is not separated from BIF by meristic characters.

Daget (1948: 43) reported STE from Thia and Boffa of coastal Lower Guinée and also from Baccoro of Guinée and from Diafarabe of Sudan. In 1954 (: 320) he placed STE as a synonym for BIF.

The type for STE as figured by Svensson 1934.

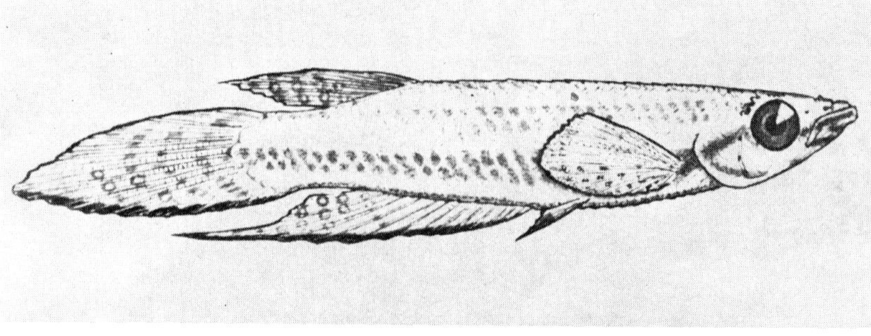

Haplochilus striatus.

♂, type.

The type for STR as figured by Boulenger 1915.

STR: ***Haplochilus striatus*** **Boulenger 1911** **(Valid sp.)**

Boulenger (1911: 266) had four specimens from the Abanga River, a northern affluent of the Ogowe drainage in northern Gaboon. These individuals had been taken by Ansorge between the first and the second rapid. In 1912 (: 22) he reported a single specimen from Lundo on the Luali River of Portuguese Congo. Pellegrin (1930: 210) reported STR from the Sangha River (Congo drainage). Poll (1952: 161) reported 17 specimens from the Mayumbe District of the Chiloango drainage of the Lower Congo area. He placed STR in *Epiplatys* and said that STR perhaps represented a subspecies of ANS.

La Corte (1961: 64–65) described and figured (in color) an aquarium strain of *Aphyosemion* from the Congo drainage under the LUJ name, but he said that the collector of this strain, Pierre Brichard, had identified this strain as STR. Further information concerning this particular strain, which still is kept, is given under LUJ.

The males of the Abanga lot mentioned above had rounded caudal fins, with one of the upper rays extended into a short filament. They were pale olive above and yellowish-white below, with carmine spots forming three or four regular longitudinal series on each side, separated by olive streaks (see also MAG and OGO) or merging into many longitudinal bands. The fins were greyish; the dorsal and anal were edged with carmine and with a carmine and a white streak above and below.

By the meristic characters STR cannot be separated from 72 of the nominal 120 species of Atlantic Africa, because by meristic characters STR is very close to the average species of the subfamily.

It has already been said, under OBS, that this species resembles STR, and as STR was taken between the first and the second rapid of the Abanga River, this species probably belongs to the fauna of the inland plateau in northern Gaboon. Thus OBS and STR may represent the same species.

SUP: *Panchax superbus* Ahl 1924

Ahl (1924: 306) had a single specimen from "West Africa." This specimen was presented by Stieler. He placed SUP closest to GRA. The description of this individual of unknown origin (probably from an aquarist) is highly unsatisfactory, and by the meristic characters SUP cannot be separated from 35 of the nominal species of rivulins from Atlantic Africa. Ahl said: "Each scale with a large silvery spot, which posteriorly is replaced by a red spot. Fins colorless. Dorsal and anal with a narrow red margin." Unfortunately, the holotype has not been re-described by other zoologists, and we do not know whether the figures given by Ahl are correct or not (see BEA, ELB, NOR, RIG, TES and other species described incorrectly by Ahl). The "silvery" spots on scales probably are the result of improper preservation or decomposition of pigments prior to preservation in a dead specimen; these spots probably were red in life. SUP appears to represent an *Epiplatys*, and among these it cannot be separated from BIF, CHE, GRA, LON, MAC, MUF, NIC, SEN and SEX. It is unfortunate that according to the rules, descriptions of a quality as that for SUP should be seriously considered. I am not able to attach the SUP name as a synonym to any particular species.

TAE: *Panchax taeniatus* Pfaff 1933 **(=BIF)**

Pfaff (1933: 294) had four specimens from Jebba on the Niger River in Northern Nigeria. He placed TAE closest to BIF, but he thought that it differed from BIF by the larger number of dorsal fin rays and by the larger number of scales around the body.

TAE cannot be separated from BIF from the Nile drainage by meristic characters nor by color patterns.

Daget (1954: 320) considered TAE to represent a subspecies of BIF, representing West African populations of this species.

I have seen the holotype and the three paratypes; they were probably preserved in formalin, as all red pigments have been lost. They correspond well to BIF from the Nile. See BIF. I found D.8, A.15–16, Sq-l. 26–27. The caudal fin is damaged and corresponds in shape to that of SIN.

Daget (1954: 320) reported TAE from many localities in the Upper Niger drainage, and in 1957 (: 159) he reported TAE from the Volta and the Comoé drainages. Gras (1961: 580) reported TAE from the Lower Oume River of Dahomey.

My first strain of BIF originated from Share, near Jebba.

The holotype for TAE in the Copenhagen Museum.

TAP: ***Fundulus taeniopygus*** **Hilgendorf 1888** **(Valid sp.)**

Hilgendorf (1888: 75–78) had seven individuals from Lake Tschaia of the Victoria-Nyanza area. This locality is situated south of Lake Victoria in Tanganyika. Pfeffer (1895: 48) said that the types were in rather bad condition and did not permit a proper description. Males of TAP had (ctenoid) spines in dorsal and anal fin rays. Preserved males had a brown reticulation on the sides. The dorsal was very dark and had a dense pattern of darker spots. The anal had three bands, one violet-brown band in basal part of fin and a marginal band of similar color. Between these dark bands there was a lightly colored band. The caudal fin was similarly colored. Females had no dark markings on sides.

Hilgendorf (1905: 416) reported other specimens of TAP from the Bubu River, above Irangi (also south of Lake Victoria).

Boulenger (1915: 37) also reported individuals of TAP from the locality last mentioned. He described fin colors:

Dorsal dark purplish brown, anal and caudal purplish brown with a white band across the middle. Females uniformly colored.

Vanderplanck (1940: 431) studied TAP within the Shinyanga District of Tanganyika (also south of Lake Victoria, not far from the type-locality). He said that in this area, TAP could be found in temporary pools only. In nature, they need six to eight weeks to become fully grown. The female deposits from 20 to 100 eggs per day. Eggs need 60 to 70 days to hatch, unless they become desiccated. In a pool measuring 40 × 50 feet and being one foot deep he estimated that there were 1100 individuals of TAP. He said that TAP stands a range from 50° to 105°F. in experimental tests.

The fin pattern and the meristic characters indicate that TAP is closely related to BRI and RAC.

TES: *Panchax tessmanni* Ahl 1924 (=BUA)

Ahl (1924: 310–11) had two specimens. The holotype came from Bosum (Bosoum) of the Ueam River area, and the cotype came from the Momo River, an affluent of the Ja (or Dja) River of the Sangha-Congo drainage. Both localities were in German Kamerun of 1911 (see ESC), but the type-locality is in what is now the Central African Republic. Ahl placed TES closest to ELB. Holly (1930: 210) redescribed the TES species on the basis of the types and changed some of the data given by Ahl (D.11, not D.6–7; A.13, not A.10–11, etc.). He placed TES closest to LUJ.

Pappenheim (1924: 134) said that the material collected by Tessmann, including the types for TES, was badly preserved. Tessmann collected specimens from 17 different species in all.

The two localities mentioned by Ahl for TES differ markedly by ecology, because the type-locality is situated in the highland savanna, whereas the other locality is situated in the humid forest of the Dja drainage, about 26 km east of Djoum in southern Cameroon, where I collected numerous specimens of EXI in 1966. I have not been in Bosoum, situated at the 1100-meter level north of Bouar, but in the major drainages of this area BUA is common. TES cannot be separated from BUA by meristic characters or by color patterns. The type for TES thus appears to belong to BUA, whereas the cotype probably should be placed in EXI.

Ahl said that on the body of TES there were ten to twelve red crossbars, and that all fins were dark, the unpaired fins having red spots and bands.

An aquarium strain identified as TES was brought to Germany from Cameroon by a friend of Mitsch of Berlin some years before the Second World War. I have not seen the original text, but I have seen a translation of Mitsch in a Danish aquarium magazine.

Mitsch's strain was said to come from a small town called Moloundou, situated at the junction of the Dja and the Boumba Rivers. In this geographical area, EXI, not BUA, is likely to occur, but Mitsch's description of his strain certainly places it closer to BUA than to EXI. The male had eight to twelve rather broad and rather regular red crossbars on the sides, the anterior-most bar being just in front of ventrals. The fins of the activated male were very dark, with red spots. The collector had told him that the fish was rare and that it lived in water that did not contain plants. Mitsch's figure shows that the dorsal fin is just behind the origin of the anal fin, as described by Ahl for TES. This juxtaposition of the dorsal and the anal does not correspond to that of EXI, but to BUA. The sides of the male were greyish olive above and yellowish below. This information Mitsch probably copied from Ahl's description and not from a study of his live material, if he had any.

THI: *Fundulosoma thierryi* Ahl 1924 **(Valid sp.)**

The identification of this species has been rather difficult because another name (WAL) has been used for it.

Ahl (1924: 54) had seven specimens from the Mangu District of northern Togo (between Kadjamba and Bogo-Moba in the south and Panpamba and

Adult male THI from Ghana. Kluge's strain.

Nacjaba in the north). This locality probably is within the Kara-Oti-Volta drainage in the rather dry savanna, with many temporary small river systems, in northern Togo. The corresponding district in northern Ghana has 40–45 inches of rain.

Ahl based the genus *Fundulosoma* (for which THI is the type-species) on the structure of the teeth, but this character probably is not suitable for the maintenance of *Fundulosoma*, which for other reasons appears to be valid.

Ahl probably had males only. They were rather small (25–30 mm) and the first dorsal fin ray was above the corresponding anal fin ray, but as in the BIV group, the number of fin rays was rather low, if we can trust Ahl's figures: D.11–12, A.14–16. Also the number of scales was low (Sq-l. 26–27). The anal and dorsal were pointed posteriorly. The caudal was truncated, slightly bilobate. Yellowish with a few red spots on sides, and similar spots and bands in fins. He did not place his THI close to other species.

Adult male Clausen's strain of THI from Sogakofe of southeastern Ghana.

Pellegrin (1934: 116) reported WAL (not THI) from Fada-N'Gourma of Upper Volta, not too far from the type locality for THI and in a similar terrain. If the descriptions for THI and WAL are compared, the meristic characters are very similar, but because WAL was based on females and THI on males, color patterns cannot be compared.

Daget (1954: 313) redescribed WAL on the basis of numerous specimens from Diafarabé on the Niger River, Southwest of Mopti. He found: D.12–16, A.14–19, Sq-l. 27–29. He figured the male and the female and referred Pellegrin's individuals to WAL. The position of the dorsal and anal fins in Daget's material corresponds to THI. The caudal fin was truncated in the male. There were red spots (not very many) on the sides of males, which were bluish to reddish. The dorsal and anal were yellow with red spots. In the figure of the male there are no bands in fins, and there are spots in the caudal. It seems that Pellegrin's and Daget's specimens should be referred to THI and not to WAL, because they correspond rather well to THI and came from similar biotopes, whereas WAL is a species of the humid Ghana forest.

Between 1959 and 1960 a very similar form was introduced as an aquarium fish into Germany. Meinken (1960: 33) figured this form and named it the "Small *Aphyosemion* from Ghana." The collector of these specimens, Kluge (1961: 234–35), published information on the population from which he took them. He said that he discovered this population near the road between Takoradi and Alenda (on the Tano River) about 30 kms from the western end of this road. The water of the biotope measured 23–27°C., pH 5.7 and 1.5 German degrees of hardness. The geographical area, pointed out by Kluge, belongs to the humid zone of Southwest Ghana, where THI is very unlikely to occur and where WAL is present. These two forms are not very similar in color patterns, but it might be that Kluge mixed up two

Adult male (above) and ripe eggs (below) of THI from Kluge's strain.

different rivuline species in his report. I have seen live and preserved material of THI from just the opposite part of southern Ghana. Clausen found a population of THI in an affluent from a swamp near Sogakofe, within the Lower Volta drainage. I kept a male from this population for some time in my tanks. It corresponded in most details to Kluge's strain. Indeed it might be that populations of THI also are found more westerly in the extreme southern Ghana, because the dry climate extends along the coast right up to the area between Cap Coast and Takoradi.

Kluge's Ghana strain became a popular aquarium fish among killifish fanciers, and it is still kept under the WAL name. The present author (1963: 162) identified Kluge's strain as WAL on the basis of Daget's redescription of the Diafarabé population. I thought that Kluge's fish should be removed from *Aphyosemion*, to which genus it had been referred by Daget, and that it should be placed in *Nothobranchius* on the basis of the egg-type, which is very similar to that of *Nothobranchius* and rather different from that of *Aphyosemion*.

Kluge's and Clausen's strains probably belong to the same species as the population from Diafarabé, and all these appear to be THI, which like these is a savanna form.

The eggs of THI correspond to those of *Nothobranchius* in most details, but they differ slightly by having a few long, non-adhesive filaments in one pole. Such elongated filaments do not occur in eggs of *Nothobranchius*. The THI eggs also resemble the eggs of ARN and FIL by their non-sticky surface, but they differ markedly by the absence of a hexagonal reticulated pattern on the membrane.

The frontal scalation resembles *Aphyosemion* much more than *Nothobranchius*, but above the eyes, there are usually no scales; if present, the scales are small. The pattern of lateral line organs of the head is difficult to study, because the neuromasts and the fleshy lobes are small and vestigial, but it seems that the pattern on the dorsal part of the snout is of the *Aphyosemion-Nothobranchius* type in individuals where the snout neuromasts are visible. The lateral line organs on the sides of the head appear to be situated in grooves, not in closed canals with pores, resembling the condition in *Nothobranchius*. The color pattern of THI differs markedly from that of (most) species of *Nothobranchius* by the absence of a red reticulation and the presence of a very limited number of red spots on the sides. Also the pattern of the caudal fin and the shape of this fin separate THI from *Nothobranchius*.

Thus THI appears to represent a form which is intermediate to *Nothobranchius* and *Aphyosemion*. If these two genera should not be united, which I think they should not, it is best to maintain the genus *Fundulosoma* for THI and similar forms in West African savanna biotopes. See GAM.

La Corte (1961: 263–65) and de Looze (1961: 265–66) both reported THI from the aquarium under the name "Ghana-Killifish."

TOD: ***Aphyosemion occidentale toddi*** **Clausen 1965** **(Valid ssp.?)**

Clausen (1965: 331–41) had two specimens, one male and one female, caught by Roloff in a pool near a spring in the forest near Barmoi of western Sierra Leone. In this locality TOD is sympatric with GER.

Clausen placed TOD as a subspecies of OCC, because of the observed differences in color patterns and because, as it seems, TOD replaces OCC in western Sierra Leone, the two forms being roughly geographically isolated. As crossing experiments have not yet been carried out, we do not know whether these two forms are reproductively isolated. On the basis of present

Adult male TOD from type locality.

evidences, I think that TOD probably is a distinct species, reproductively isolated from OCC. Otherwise, TOD probably could not maintain its phenotype in the absence of any geographical or other physical barriers between the territories of the two forms.

Superficially the male TOD differs markedly from the male OCC, but the color patterns are basically similar. As in CHA (red males) and SHE (blue males) the conspicuous differences in colors are produced by the distribution of red and yellow pigments. In both forms, the basic color pattern of the sides is a thin red reticulation, which on the back of the TOD male change into a pattern of crescent shaped spots. On the belly of both forms,

Holotype, male and head of holotype for OCC TOD. No. P.35.2573 in the Copenhagen Museum. Roloff collection.

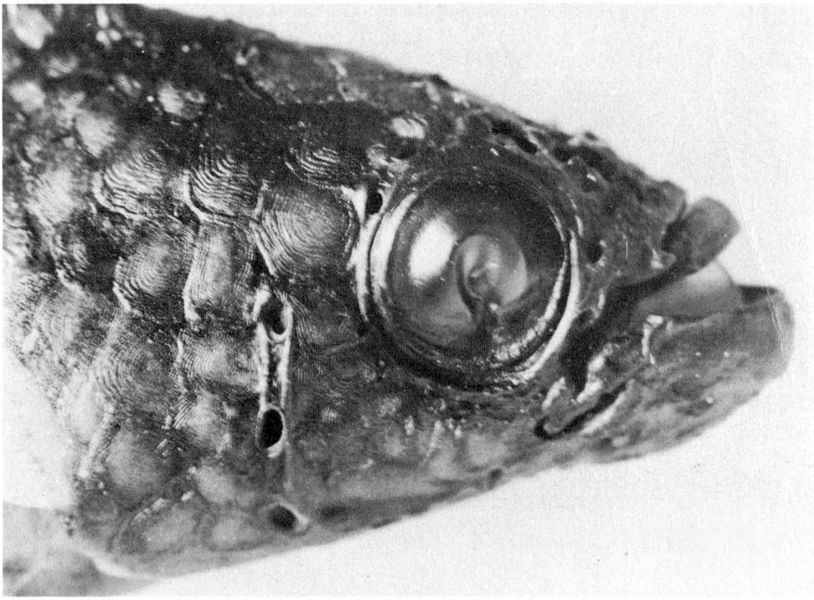

the red reticulation usually is visible to the naked eye. In the OCC male, the red pigments have an almost complete distribution, whereas in TOD these pigments do not occur outside the red pattern. In the OCC male, the metallic brilliance of the sides and basal part of the fins is much reduced by the red pigments, whereas in the TOD male the cold blue to violet-blue metallic sheen is strongly developed on the sides and in the fins. The fin patterns are rather similar, although the submarginal red band in the caudal fin continues all around the margin of the fin in TOD, whereas in the OCC male a lyre-pattern usually develops, although in some males, the sub-

Adult male TOD from the type locality.

marginal red bands tend to join posteriorly. Black pigments are well developed in the TOD male, which becomes very dark, almost jet-black (as the GUI male) when activated. Also the throat is very dark in the activated male.

The female TOD develops the broad dark longitudinal band and the rounded dark throat-marking described for the OCC female. At other times, it may develop rather narrow oblique dark crossbars on the upper part of the caudal peduncle, as in some of the females of the ROL-LIB group. This particular marking is absent or much reduced and irregular in the OCC female. Compared to the OCC female, the TOD female is darker and more greyish, less reddish. The male TOD appears to be shorter and more heavily built than the OCC male.

Ctenoid fleshy papillae normally are well developed on the rays of the anal fin, and "wrinkled" ctenoid scales are present on the head. The chromosome conditions are given under OCC.

TRO: *Adiniops troemneri* Myers 1926 **(=ORT)**

Myers (1926: 91) based this species on aquarium-kept individuals of unknown origin. Later on (1933: 184) he considered *Adiniops* to represent a synonym for *Nothobranchius* and TRO to represent a synonym for ORT. He also said that TRO came from Beira, Mozambique.

UNC: *Panchax unicolor* Ahl 1924 **(=BIV?)**

Ahl (1924: 309) had a single specimen from the aquarist Matte of Berlin, who gave "West Africa" as its origin. Ahl placed UNC closest to DEC and EXI. This individual has not, however, been redescribed by other zoologists, and for that reason, we do not know whether the meristic data given by Ahl for his UNC are correct; it seems that Ahl very frequently was not able to count fin rays and scales and to measure proportions.

The caudal fin of UNC was pointed, longer than the head, indicating UNC to be an *Epiplatys*. The caudal peduncle was 1.5 times as long as deep, indicating a position in *Aphyosemion*, although this ratio also develops in some species of *Epiplatys*. D.9, A.12, Sq-l. 26. Uniform yellowish-white (the usual colors of a dead aquarium fish) scales darkly edged. The fins were colorless or very faint reddish. By an analysis of meristic characters UNC was not separated from 60 of the about 120 nominal rivuline species of Atlantic Africa, and on the basis of the published information it is impossible to determine to which previously described species Ahl's UNC belongs. It may represent a badly preserved or decomposing specimen of BIV, as it is known that Matte at that time supplied Ahl with some dead BIV individuals from his aquaria. See BIT and RUS.

UNS: *Aphyosemion (Fundulopanchax) unistrigatus* Ahl 1935 (=BIV)

Ahl (1935: 123) had eight specimens collected by Zenger at Bipindi in southern Cameroon. This locality is situated in the lowlands at the 63 meters level, within the Lokundje drainage.

The first anal fin ray is below the corresponding dorsal fin ray. The number of fin rays and of scales is low, thus strongly indicating a species of the BIV complex. This idea is fully supported by the color description. Yellowish above, silvery below. From the eye there is a dark brown longitudinal band as broad as the pupil. Dorsal of male with rows of small red spots, dorsal and anal narrowly edged with reddish. These fins are posteriorly pointed in the male.

UNS cannot be separated from BIV by meristic characters or by color patterns, and as BIV is common in the Lower Lokundje River, I place UNS in BIV without hesitation. Ahl placed UNS closest to PAP, another synonym for BIV.

Adult hybrid male of GAR-Owo/WAL crossing.

Young (above) and old (below) male of WAL from southeast Ivory Coast. Arnoult's strain.

WAL: *Fundulus walkeri* Boulenger 1911 **(Valid sp.)**

Boulenger (1911: 262) had three specimens, probably females, caught by R. B. N. Walker at the Bokitsa Mine (Wasa) of Ghana. They were badly preserved and were brownish all over.

Under PET and CHA I have already mentioned one specimen of PET and several specimens of CHA which also were taken by Walker at the Bokitsa Mine. PET is a species of the humid forest of Southwest Ghana and Southeast Ivory Coast, and thus it seems that the Bokitsa Mine belongs to this zoogeographical area.

Under THI it has been mentioned that individuals of THI have been referred to WAL. Under SPU it was said that SPU and WAL cannot be separated by meristic characters, by colors nor by geographical distribution, and thus SPU was considered to represent a synonym for WAL.

In 1953 Sheljuzhko (1954: 313) collected live fishes for Werner of Munich in southeast Ivory Coast. About 85 kms north of Abidjan, towards Agboville, he caught specimens of a species of *Aphyosemion*. Live members

Fundulus walkeri.

Type.

The type for WAL as figured by Boulenger 1915.

of this population were sent to Germany, and on this basis an aquarium strain was established. This strain was identified as GAR by Meinken (1953: 86–89), referring to Dieterich's ARN (see FIL) of 1915, which Meinken said had been identified as GAR by Boulenger. The GAR name was generally accepted for this strain by aquarists, but in 1963 the present author (Scheel 1963: 166) pointed out that this strain was even closer to SPU and that for zoogeographical reasons the name GAR should be replaced by the SPU name. This point of view also was generally accepted by aquarists. Sheljuzhko's strain soon became extinct, although it was reproduced by several aquarists in Europe and in the United States (I have color-slides sent by F. Werner of Detroit).

About 1963, Arnoult of the Paris Museum brought to France live specimens of a strain of WAL which he caught in Southeast Ivory Coast. I had eggs from this strain. At the same time, Bruce Turner of the United States had live specimens from a friend in Ghana. These two strains of WAL were rather similar, although in the Ghana male there were fewer red markings on the sides and a strong tendency for the development of thin red crossbars or lines posteriorly on the sides. Other Ghana strains might have developed from the fish exported by Kluge. See THI. These strains have been described by Franz (1966: 22–24), and Hesse (1966: 196–200).

It seems that the unpaired fins of the male WAL generally develop a strong yellowish-orange color at the tips. This color was described by Boulenger for the live types for SPU, by Meinken (1953) for Sheljuzhko's strain, and this color also is present in Werner's color slides. Klee and Turner (1964: 590), however, said that the edging of unpaired fins might be white (rarely), and Senfft's color photo in Meder (1953: 221) and in *Die Aquarien Fische* seems to picture a male without yellow fin color.

Ctenoidy often is strongly developed in the adult WAL male, which has numerous spines on the scales of median rows. Short fleshy ctenoid fin ray papillae may occur in the anal fin. Pectoral fin ray papillae generally are present. The frontal scalation develops the G-pattern, very rarely with H-scales.

In some males the central posterior part of the throat may develop a faint dark marking rather similar to the marking which characterizes the GUL male. The red pattern on the sides of the male WAL usually is very irregular; the spots often are vertically oblong and tend to develop vertical red lines posteriorly on the side. No (distinct) submarginal red bands develop in unpaired fins, which are broadly edged with very dark red.

WAL appears to be more "annual" than the closely related GAR superspecies, but less than the more distantly related GUL species. Eggs often do not develop at once after the spawning, not even when kept in clean water with free entrance of air.

The closest relatives of WAL are the different forms of the GAR superspecies, but unlike these forms, WAL is a form of the humid forest. WAL also might be rather closely related to ARN and GUL, and it may represent a rather ancient form, now isolated in the humid forests just west of the Dahomey Gap.

WER: *Aplocheilus werneri* Meinken 1966 **(Valid sp.)**

Meinken (1966: 76–78) based his description of WER on aquarium-kept specimens which were imported in 1964 by Werner of Munich. The origin of this strain is unknown, but Meinken suggested that it came from Ceylon. The first specimens were imported together with specimens of BLO and of DAY. Meinken placed WER closest to DAY, but in his figure showing the dorsal part of the head, it seems that the two snout-neuromasts are widely separated, whereas in DAY these two neuromasts develop the closed pattern. Also the number of scales in a longitudinal series (35 scales) places WER closer to LIN than to DAY.

Live individuals are yellowish-olive to greenish-grey-olive, the belly being more brownish to reddish brown. The lower part of the caudal peduncle is very dark, almost black. The scales of the back are golden. Those of the median part of the sides are silvery-greenish, and those of the lower part of the sides are greenish-golden in both sexes. There are some red spots on the median part of the sides. Unpaired fins are green to bluish-green, with dark red spots and streaks. The lips are red. Sexual dimorphism is very faint, but in fully grown males the anal fin is more pointed posteriorly than in females.

The color pattern of WER differs so markedly from other known species of *Aplocheilus* (and *Epiplatys*) that WER probably represents a distinct species. I have not seen any material of WER.

VEX: *Panchax vexillifer* Meinken 1929 **(=CAL)**

Meinken (1929: 225) probably had one male and one female for the

The types for VEX redrawn after Meinken's drawing of 1929.

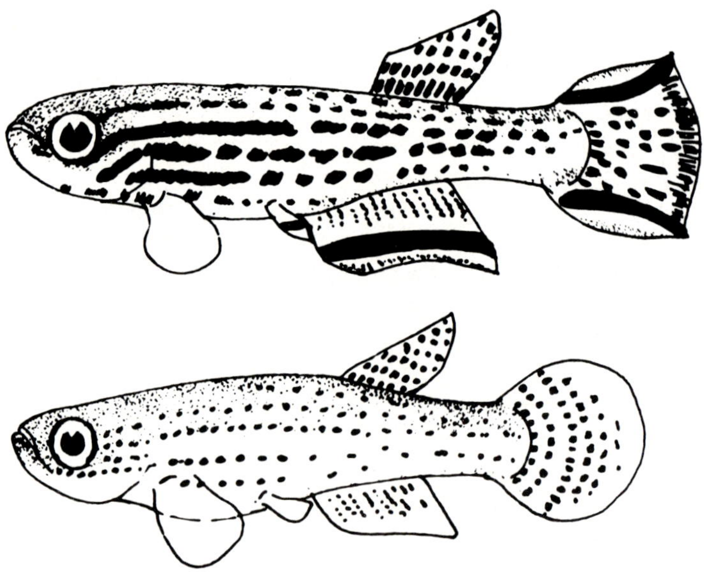

description. These individuals were part of a large shipment of aquarium fishes from Africa, imported in 1929 by the "Platy-Tischrunde" of Hamburg. Meinken considered VEX to be very close or even identical with OES. See also OES.

On the basis of meristic characters, VEX cannot be separated from CAL and 67 other nominal species of rivulins from Atlantic Africa. The color pattern corresponds well to that of CAL, and I do not hesitate to consider VEX to represent a simple synonym for CAL.

The VEX name has frequently been used for CAL in recent years. See Selvey (1964: 3) and Wilhelm (1966: 116–18).

VOS: *Nothobranchius vosseleri* Ahl 1924 **(Invalid sp. ??)**

Ahl (1924b: 144–45) had a single specimen, probably a female, from Mombo of the former German East Africa (Tanganyika). The modern Mombo is situated near the Pangani River about 90 kilometers from the coast (at Tanga). He placed his VOS closest to WAL and to ORT.

The number of scales along the side is 28, thus placing VOS as intermediate to PAL and GUE which both are likely to occur at the type locality for VOS. On the basis of the published information VOS cannot be identified, but undoubtedly VOS does not represent a valid species.

The colors of VOS were described: dark greyish-olive, brownish beneath. Fins yellowish grey.

ZEN: *Panchax zenkeri* Ahl 1928

Ahl (1928b: 117) had three specimens caught by Zenker at Bipindihof in southern Cameroon. He did not compare ZEN to previously described species. Holly (1930: 195–281) inspected the types and did not change the data given by Ahl (!!). Myers (1933: 182) placed ZEN in *Epiplatys*.

The caudal fin is rounded and as long as the head (about 30% of standard length) and the caudal peduncle is as long as deep, indicating that ZEN is a species of *Epiplatys*. The dorsal and anal are pointed in the male. Yellowish-white, with three or four faint crossbars, a narrow whitish stripe from the upper part of the gillcover to the root of caudal. The colors of the individuals indicate a bad state of preservation.

The type for ZEN as figured by Ahl 1928.

ZEN cannot be separated from 42 nominal species of rivulins from Altantic Africa, including such important forms as ORN, MAC and SEX. ZEN might represent a valid distinct species, but on the basis of the very incomplete description, it cannot be defined at the present time. Bipindihof might represent the same locality as Bipindi. If this is true, SEX and perhaps GRA occur in this locality, situated in the lowlands of coastal Cameroon, on the Lokundje River.

ZIM : *Fundulus zimmeri* Ahl 1924 **(Valid sp.??)**

Ahl (1924: 139) had two specimens from the aquarium trade in Germany. He supposed that the material originated from the mouth of the Niger, but he did not say on what evidence he based this idea. Meinken (1930: 221) said that he had not been able to disclose when and from where Ahl's types for ZIM were imported. He said that the types were from the Berlin Aquarium and that ZIM resembled BIT, but differed markedly by colors and the long streamers in dorsal, anal and caudal. Ahl placed ZIM closest to LOE, another synonym for BIV. He thought that ZIM differed from LOE by the more posterior position of the dorsal, the broader interorbital space, the smaller scales and the coloration. He also placed ZIM close to GAR and said that ZIM took an intermediate position between LOE and GAR. If he was right in this, ZIM certainly represented a most interesting species.

D.11–12, A.13–14, with the dorsal exactly above the anal, indeed indicate that ZIM is a synonym for BIV, but the caudal fin was described as being trilobate. Brownish above, yellowish below, with 10–12 broad vertical bands of dark color on the sides and the caudal peduncle. Dorsal and anal blackish, caudal blackish in upper and lower third, lightly colored in central part. The largest individual measured 70 mm.

Radda, who examined the types a few years ago, informed me that the broad dark crossbars were not really broad and also that they were not true crossbars of the *Epiplatys* type.

On the basis of published information, ZIM cannot be considered to be similar to any known rivuline species from Atlantic Africa, and thus ZIM might represent a distinct species unknown to us from the aquarium.

XVI

THE SPECIATION IN RIVULINAE

Where tropical forests have grown for very many years, the soil becomes highly leached, leaving very little mineral matter to dissolve into the waters of the forest. The composition of the forest waters are then influenced not by the soil but by the vegetation (decomposition of vegetable matter) and by the composition of the rain. Within these old forests the climate is very stable, with only minor changes in temperature and humidity. Such biotopes are not likely to encourage speciation (multiplication of species) within a rather flat terrain having no important geographical barriers such as mountain ranges. In spite of this, the speciation (both morphologically and biologically) within rivulins of Atlantic Africa and probably also of East Africa has produced a high number of distinct bio-species and morpho-species.

One of the reasons for this abundance of species might be the result of the major changes in climate which took place within the Pleistocene epoch, which began about 1,000,000 years ago and which was characterized by four glacial periods separated by interglacial periods. It appears to be generally accepted that these glacial and interglacial periods greatly influenced the climates of tropical Africa. During the peak of the glacial periods the temperature was about 5°C. lower than it is at the present time, and the level of the sea was about 100 meters below the present level. There is not, however, any generally accepted theory for the interactions of the climatic zones of the northern hemisphere and those of tropical Africa. Some experts assume that the major climatic zones of Africa, north of Equator, move southward during glacial periods, whereas others believe in just the reverse interactions of these climatic zones, and still others have developed more complicated theories, with maximum or minimum rainfalls at the first or last phases of a glacial period. There is, however, an abundance of evidence that the tropical forests both north and south of the Equator varied in their northern and southern limits during the Pleistocene.

According to Butzer (1957) and others, the West African tropical forest was much reduced during some epochs of long duration during the Pleistocene, the coastal forest belt being almost eliminated. In some places forest refuges survived, and in these isolated forest fragments the rivuline species of the forest also were able to survive until the climatic belts again moved northward and the forests again were able to expand and the fragments were able to reunite.

These isolated forest fragments thus represented a number of "Noah's arks" within which living animals and plants were stored during the long dry periods.

The best known of these forest fragments is situated in extreme southwest Ghana and southeast Ivory Coast. See Leneuf and Mangenot (1960) and Ahn (1961). Other similar fragments undoubtedly existed near Freetown in Sierra Leone, in Liberia, in southern Nigeria, in West Cameroon and south of the Sanaga River of East Cameroon. See also Boot (1958a and 1958b) and Moreau (1963), who studied the distribution of primates and of birds. Schiötz (1967) in his extensive study of West African tree frogs also dealt with this theme.

During the isolation within the forest fragment, the different major populations of certain rivuline species were not able to exchange genes; during an isolation of long duration, the genetic and karyotypic composition might have been changed by gene- and chromosome-mutations to the extent that even after the isolation ended, these populations were not able to produce fully viable and fully fertile offspring in crossings that resulted when individuals of the different populations expanded within the reuniting forests. The results of such encounters cannot be predicted. If the reproductive isolation is not complete, the two populations may fuse, perhaps producing a more variable phenotype. A species border may develop for some time, the two populations being restricted to their geographical areas. What is more likely, one of the new "species" may eliminate the other form by competition, or one of the forms may be able to shift into a new ecological niche, the two forms becoming sympatric.

It seems to me that when such speciation occurs within isolated forest fragments the phenotype probably will remain rather unchanged, because within the different forest fragments (if they are sufficiently large) the forces of natural selection probably will not be different, and as the phenotype is the thing on which natural selection acts, it is likely that it remains constant independent of the genetic and karyotypic alterations. Where genetic mutations produce alterations of the phenotype, this may be restored by other mutations. The apparent large number of sibling (morphologically similar, but genetically different) species within the tropical forest of Atlantic Africa might be the result of a speciation of this nature.

This theory does not, however, explain why, as it appears, the chromosome number within the *Roloffia-Aphyosemion* group decreases when one moves from the western part of West Africa towards the east and then southwards to the Congo drainages.

Some aspects of the population structure of these rivuline species might partly explain this problem.

The *Aphyosemion* and *Roloffia* species appear to be composed of a large number of very small populations or micropopulations, each having a very limited number of individuals which may or may not be much reduced or eliminated during the dry season, the eggs surviving in the soil. The exchange of genes between these micropopulations probably is much more restricted than between other populations of fish.

The local micropopulation living at the species-border (the extreme limit of the range of the species) probably is not so well adapted to the local conditions as are the populations living near the center of distribution, and gene-mutations which improve the adaptation of the border-population thus probably is much favored by natural selection. The border population is not, however, able to produce stable alterations of its genotype using these mutations, because the constant influx of genes carried by the migrants from the more central micropopulations, where a higher concentration of individuals is more likely to occur, will restore the original genotype of the species. The border-population can succeed in building up a better-adapted genotype only where it is able to eliminate or greatly reduce the influx of genes from other populations. Chromosome mutations of some types are able to produce varied grades of sterility in individuals which are heterozygous in respect to the chromosome mutation. Depending on where the breaks in the different chromosomes occur and how the broken parts reunite, complete sterility or full fertility may result. Thus it seems that a high degree of sterility, which is not complete, may occur as the result of some recombinations of chromosome material. If an individual in which such mutation occurred is able to produce offspring which survive and reproduce, the mutation may become established in the border-population, and in individuals which become homozygous for the new karyotype the fertility is restored. At the same time, this new bio-species is almost completely reproductively isolated from the populations of the species from which it arose. Now the bio-species is able to reconstruct its genotype and to become fully adapted to the conditions at the species-border, and in time it may expand beyond this border and conquer new land. But as we have seen in the section dealing with chromosome conditions, it seems that such alterations of the karyotype generally will result in a reduction of the chromosome number.

The new bio-species again probably will establish a species border, beyond which the new genotype is not sufficiently adapted, and again some border-population may by chromosome-translocations become reproductively isolated and in time expand the range of the superspecies. In this way a chain of very closely related species may develop, the chain being characterized by a steady reduction of the chromosome number when we move out from the original center of the superspecies. See Darlington (1963: 75–93) regarding chromosome-geography within plants and Mayr (1963) concerning the border-populations in animals. I have been unable to point out any other theory by which the differences in chromosome numbers between the Nigerian and the Congo areas might be explained.

It is difficult to understand why the rivuline fauna of the Central Congo area is poor compared to that of southern Nigeria and West Cameroon, because the fauna of other sorts of fishes appear to be very rich. Poll and Gosse (1963) reported 408 different species of fishes from the Central Congo. Of *Aphyosemion* there seem to be only one or two morphospecies living within the whole central basin. See CHR and LUJ, which again are rather similar.

According to de Heinzelin (1952) the whole central basin lost its forest at some time during the Pleistocene, and the area became so dry that the wind was able to carry sand from the Kalahari Desert right up to the northern part of the basin (Ubangui River) producing sand dunes. He assumed that forest survived this dry epoch of long duration only in the mountains to the east, west and north of the basin. This desertification of the Central Congo might greatly have influenced the fish-fauna of this geographical area, but certainly it does not explain why the rivuline fauna of this area is so poor and that of other fishes apparently so rich.

Males of BIV from a brook 20 miles east of Bata, Rio Muni have n = 13, 2n = 26. This finding supports the idea of a certain reduction of the chromosome numbers when species are expanding from north to south in western Central Africa.

Fernando Poo males of BIV have n = 19, 2n = 38.

Males of CAM from a brook 20 miles east of Bata, Rio Muni have n = 17, 2n = 34 and the karyotype seems to be identical to that of OBS from Yaunde. These two forms undoubtedly are very closely related.

The species of *Aphyosemion* figured on page 338 under OES has n = 20, 2n = 40 and the karyotype differs rather markedly from that of OBS and CAM. Superacrocentrics are longer.

XVII

THE GENERA AND SUBGENERA

The different nominal species of *Rivulinae* of the Old World have been referred to a rather large number of genera. Some of these are of no importance at the present time:

Esox L.

Poecilia Bl. Scheider 1801. (*P. viviparae* Bl. Sch. 1801)

Cyprinodon Lacepede 1803. (*C. variegatus* Lacepede 1803)

Fundulus Lacepede 1803. (*F. heteroclitus* L.1766)

whereas others are still used (more or less):

Aplocheilus McClelland 1839. (*A. panchax* (Hamilton 1822))

Epiplatys Gill 1862. (*E. sexfasciatus* Gill 1862)

Nothobranchius Peters 1868. (*N. orthonotus* (Peters 1844))

Aphyosemion Myers 1924. (*A. castaneum* Myers 1924)

Pachypanchax Myers 1933. (*P. playfairi* (Günther 1866))

Fundulosoma Ahl 1924. (*F. thierryi* Ahl 1924).

Recently, some more generic names have been created, but it is not known to which extent these new names will be accepted:

Roloffia Clausen 1966. (*R. occidentalis* (Clausen 1966))

Pseudepiplatys Clausen 1967. (*P. annulatus* (Blgr. 1915))

Aphyoplatys Clausen 1967. (*A. duboisi* (Poll 1953))

Other genera are considered to represent synonyms for some of the genera mentioned above, but where different type-species are pointed out for the synonym generic names, these names might be useful in a future division of genera into subgenera:

Lycocyprinus Peters 1868 with *L. sexfasciatus* (Peters 1863) = *Epiplatys dageti monroviae* Daget and Arnoult 1964 as type-species is considered to represent a synonym for *Epiplatys* Gill (see SEF, SEX, DAG and MON).

Adiniops Myers 1924 with *A. guentheri* (Pfeffer 1893), as the type-species first was created as a subgenus of *Aphyosemion* Myers, but in 1926 he used it as a generic name for *A. troemneri* Myers 1926. In 1933 (: 184) Myers, however, considered *Adiniops* to represent a synonym for *Nothobranchius* Peters.

Fundulopanchax Myers 1924. (*A.* (*F.*) *coeruleum* (Blgr. 1915))
Callopanchax Myers 1933. (*A.* (*C.*) *sjöstedti* (Lnbgr 1895))
represent subgenera of *Aphyosemion* Myers, but the name first mentioned frequently was used as a generic name by Meinken during the 1930's. Clausen (1966: 331–41), however, pointed out that the type-species for *Fundulopanchax* represented a synonym for the type-species for *Callopanchax* (COE = SJO), and thus it seems that the latter name represents a synonym for the former name. It is, however, very likely that Myers (1933: 180–81) had *F. sjöstedti* Arnold 1911 (: 137) and not *F. sjöstedti* Lönnberg 1895 in mind, because the true identity of the latter was not recognized in 1933, because Lönnberg's species was known under the *A. coeruleum* (Blgr. 1915) name at that time, whereas the Sierra Leone species, which in 1966 became the name *A. occidentale* Clausen, was considered to represent Lönnberg's species in 1933. If the point of view last mentioned is true, *Callopanchax* Myers is an invalid name because it was attached to an unnamed species.

Quite recently a new subgeneric name in *Epiplatys* Gill has been created:
Parepiplatys Clausen 1967 (*E.* (*P.*) *grahami* (Blgr. 1911))

The generic names *Panchax* Cuvier and Valenciennes 1846 and *Haplochilus* Günther 1866 appear to represent simple synonyms for *Aplocheilus* McClelland 1839, all with *A. panchax* (Hamilton-Buchanan 1822) as type-species.

APLOCHEILUS AND EPIPLATYS

There seems to be no good reason to separate *Epiplatys* and *Aplocheilus*, except for zoogeographical reasons. I have been unable to disclose important differences in external morphology, color patterns, behavior, ecology, egg-type, hemoglobin patterns, scale patterns or cytology. Indeed, females and juveniles of *Aplocheilus* (except for WER) develop a dark marking in basal part of dorsal, this marking being absent in all known species of *Epiplatys*, but a similar marking also develops in *Pachypanchax* (in PLA, not in HOM). Also in some species of *Aplocheilus* (LIN and DAY) the ventrals of both sexes are rather long and pointed, whereas in *Epiplatys* long and pointed ventrals develop in males only (LAM, LON and SEX). Such minor morphological differences which do not even characterize the genus should not possibly be useful for a generic separation. Unfortunately, Sethi (1960: 123–39) did not study the osteology of any individual of *Epiplatys*, and I have been unable to find any further osteological study which deals with *Epiplatys* and *Aplocheilus*. Myers (1933: 180) separated *Panchax* (=*Aplocheilus*) from *Epiplatys* first by the degree of the freedom of the maxillary (a character very difficult to study in most preserved material) and also by the presence or absence of expanded haemal arches, these arches being expanded in *Aplocheilus*, but not in *Epiplatys*. Clausen (1967: 29), however, found expanded haemal arches in 12 of the 13 species of *Epiplatys* he studied. At the present time, there seems to be no reason to maintain *Epiplatys* even as a subgenus of *Aplocheilus*.

APLOCHEILUS AND PACHYPANCHAX

Myers (1933: 180) separated *Panchax* and *Epiplatys* on the one hand from all other genera of African and Asian Rivulinae first by the freedom of the maxillary, the snout of the group last mentioned being less sharp.

Pachypanchax also was separated from the rather similar *Aplocheilus-Epiplatys* group by the scalation of the caudal fin, which in the former reaches beyond the middle of that fin, whereas in the latter it is more reduced. These differences are present in my specimens of PLA, in which the scalation may cover the basal two-thirds of the caudal, whereas in my individuals of HOM the scalation does not reach beyond the basal fourth or third of that fin.

Pachypanchax might, however, be separated from the *Aplocheilus-Epiplatys* group by other characters. In the former, the central rays of the caudal fin do not extend, developing a distinct lobe, this lobe being present in almost all (see ACU, SEN and SPI and the closely related FAS) nominal species of the *Aplocheilus-Epiplatys* group. Furthermore, the frontal scalation is different. *Pachypanchax* develop the rare E-pattern and prominent H-scales in all individuals of PLA and HOM studied, whereas in *Aplocheilus* (except for some populations of BLO) the G-pattern without H-scales is present.

Aplocheilus and *Pachypanchax* are, however, very close, but probably the latter does not represent a link between the former and *Epiplatys*, but rather a link between the former and *Roloffia*.

PACHYPANCHAX AND ROLOFFIA

Pachypanchax differs from *Roloffia* first of all by the frontal scalation. All species of *Roloffia* develop the G-pattern (except for GUI) whereas in *Pachypanchax* the rare E-pattern develops. GUI appears to represent a real problem, because it seems to be much closer to *Pachypanchax* than all other species of *Roloffla*, and also GUI appears not to represent a close relative of OCC (the type-species for *Roloffia*) or of any other nominal species of *Roloffia*. It may indeed represent a certain link between these two groups, although the chromosome conditions indicate that GUI is a specialized form. "Wrinkled" (ctenoid) scales so frequently present in adult males of *Roloffia* seem to be absent in *Pachypanchax*, whereas in *Aplocheilus* and *Epiplatys* some of the scales of large males resemble the "wrinkled" scales of *Roloffia* (and *Cynopoecilus* among South American killifishes), but ctenoid fin papillae have not been found in these large males. It has been mentioned, under PLA and HOM, that these two species of *Pachypanchax* are rather different when minor details of the head are studied.

APHYOSEMION AND ROLOFFIA

Individuals of these two genera are very similar in all respects except for the pattern of lateral line organs of the dorsal part of the snout. Clausen (1966: 331–41 and 1967: 15–18) used these differences for the separation of *Roloffia* from *Aphyosemion.* They are indeed remarkably constant and a tendency for the development of intermediate patterns occurs in only one species (MAE). Also, the species of *Roloffia* generally appear to have higher haploid chromosome numbers (n = 20–23, when GUI is not considered), whereas in *Aphyosemion* slightly lower or much lower numbers occur (n = 9–20), the species having 20 haploid chromosomes apparently representing a very divergent and important group. The chromosome conditions indicate that *Aphyosemion* derived from *Roloffia* or some group closely related to *Roloffia.* Also, the two genera inhabit different geographical areas in Atlantic Africa, with *Roloffia* being restricted to the western part of West Africa and *Aphyosemion* generally living east of the "Dahomey Gap." Species of the two genera are sympatric in Ghana and in the Ivory Coast. See PET and WAL.

The type of ctenoidy in males is different. In *Roloffia* no ctenoid spines develop in scales; instead, the so-called "wrinkled" scales develop particularly on the head and near the head. The supporting ctenoid element of the anal fin ray fleshy papillae is a rounded bony body (as in *Cynopoecilus* and *Cynolebias whitei* Myers of South Anerica) and not a bony spine as in *Aphyosemion* (not all species) and *Nothobranchius.*

Some minor differences in color patterns exist. Whereas in *Aphyosemion* males frequently develop a golden edge on the anal fin, such golden edge is absent in all species of *Roloffia* in which the caudal fin is edged with golden above and below. Generally, also, the color pattern of the throat of both sexes is different. Males of *Roloffia* (except for GER) develop no red pattern on the throat, which in all activated males develop a deep black color. The "*Roloffia*-pattern" occurs, however, also in AHL, AUS and CIN of *Aphyosemion.* In the female *Roloffia* (except for GER) a rounded dark marking appears anteriorly on the throat in pre-mating display and in both sexes when they are afraid. A similar marking occurs in females of most species of *Aplocheilus, Epiplatys* and *Pachypanchax* and in AHL, AUS, CAL and OBS in *Aphyosemion.*

Thus the separation of *Roloffia* from *Aphyosemion* appears to be justified.

APHYOSEMION AND FUNDULOSOMA

Ahl (1924a: 52) separated *Fundulosoma* from *Fundulus* on the basis of supposed differences in the development and pattern of teeth of the lower jaw. It seems that his new genus was not generally accepted.

Myers (1933: 180) said that the tooth structure described as diagnostic was to be seen in many partly dried or shriveled specimens of other genera.

Under THI it has, however, been indicated that THI apparently represents a form rather intermediate between *Nothobranchius* and *Aphyosemion* (*Fundulopanchax*). The eggs from THI are very similar to those of *Nothobranchius* and different from those of *Aphyosemion*. The eggs resemble, however, those of ARN by being non-sticky and by the evenly distributed short and rather stiff filaments. They differ markedly from those of ARN by the absence of a hexagonal reticulated pattern and by shape; before they decompose (un-fertilized eggs), oil-drops are released as in *Nothobranchius*, but not in *Aphyosemion*. THI and *Nothobranchius* are species of the savanna, whereas *Aphyosemion* generally is restricted to the forest. The development of shallow open grooves for the neuromasts on the sides of the head of THI resembles the condition in *Nothobranchius* and differs from the closed canals with pores found in *Aphyosemion* and *Roloffia*.

The color pattern, the development of the caudal fin, the frontal scalation and the general appearance place THI very close to *Aphyosemion* and to ARN in particular, and are different from *Nothobranchius*. Also the geographical distribution of *Fundulosoma* (the savannas of West Africa) place *Fundulosoma* intermediate to *Aphyosemion* and *Nothobranchius*.

If *Aphyosemion* and *Nothobranchius* should not be joined into a single genus (*Nothobranchius*), which I think they should not at the present time, it appears to be best to keep THI and its allies separated from both genera in a genus of their own: *Fundulosoma*.

APHYOSEMION AND NOTHOBRANCHIUS

As indicated by Clausen (1967: 25), *Nothobranchius* differs from *Aphyosemion* and *Roloffia* by the development of the lateral line on the sides of the head. In *Nothobranchius* (and *Fundulosoma*) the neuromasts are housed in open rather shallow grooves, whereas in *Aphyosemion* and *Roloffia* closed canals with pores occur.

Also, the lateral line pattern of the dorsal part of the head is somewhat different, and in *Nothobranchius* the supraorbital and the upper postorbital depressions or grooves have fused (but see PAL) as in ANN, whereas in *Aphyosemion*, *Roloffia* and *Fundulosoma* these two depressions are well separated by the F-scale bridge, which is absent in *Nothobranchius*. The frontal scalation also is different. In *Nothobranchius* the whole frontal scalation has restrained from the anterior part of the front and usually there are no scales in front of a line through the centers (or even the posterior part) of the eyes; the frontal scales also are deeply embedded in the spongy skin, not very different from the situation in the OCC male.

The frontal scalation of *Nothobranchius* does not resemble the frontal scalation of other rivuline genera in that the frontal scales are very small and often so irregularly arranged that no distinct pattern develops. Generally, however, the scalation has reversed completely, the foremost scales resting on the subsequent scales. "H-scales" very rarely occur.

Thus *Nothobranchius* appears to be sufficiently separated from *Aphyosemion* even in the presence of *Fundulosoma*, but undoubtedly these three genera represent an evolutionary unit of their own, separated from other rivuline genera and characterized by their low chromosome numbers which appear not to exceed n = 20, and by their superacrocentrics.

APHYOPLATYS, APHYOSEMION AND EPIPLATYS

As it has already indicated during the present short review of rivuline genera of Africa and Asia, *Epiplatys* and *Aphyosemion* do not appear to be closely related genera. Clausen (1967: 32) indicated his new genus (*Aphyoplatys*) to be intermediate to *Epiplatys* and *Aphyosemion* and based this point of view on a study of lateral line patterns, color patterns, shape of caudal fin and general appearance. Under DUB, the type species for *Aphyoplatys*, these and other characters are discussed. The result of this discussion is that *Aphyoplatys* appears to be very close to *Epiplatys*, the resemblance with *Aphyosemion* probably being a matter of convergence, but also *Aphyoplatys* appears to be so aberrant that it should be separated from *Epiplatys*; now it has been separated from that genus.

PSEUDEPIPLATYS AND EPIPLATYS

Clausen (1967: 30–31) considered ANN to be so different from *Epiplatys* that it should be removed from that genus and that a separate genus, named *Pseudepiplatys*, should be created for ANN. Clausen separated ANN from *Epiplatys* on the basis of several characters including biology. It appears to me that most of these distinguishing characters are not sufficiently distinct to permit a separation of ANN from *Epiplatys* on the generic level, but maybe these characters will prove to be sufficient for a separation on the subgeneric level. I believe that ANN is a typical member of the *Aplocheilus-Epiplatys* group, representing perhaps a form rather close to the ancient form which some or all species of this genus originated. See also ANN.

ADINIOPS AND NOTHOBRANCHIUS

Although the author of *Adiniops* (Myers) considered this genus to represent a synonym for *Nothobranchius*, these two genera have different type-species, GUE and ORT respectively. Pfeffer (1895: 47), however, considered GUE and ORT to be so different (patterns of the teeth) that he considered these two forms to belong to different genera (*Fundulus* and *Nothobranchius* respectively). In my opinion, it is highly unlikely that GUE and ORT will ever be separated, even subgenerically, and thus *Adiniops* probably will not be a useful name in the future.

FUNDULOPANCHAX AND APHYOSEMION

Myers (1924: 6) separated *Fundulopanchax* from *Aphyosemion s.s.* by the juxtaposition of the dorsal and anal fin, the former being inserted forward, above or very slightly behind the anal origin and by the subequality of these two fins, the size being usually considerably larger than in *Aphyosemion s.s.*

He also said that *Fundulopanchax* graded directly into *Aphyosemion s.s.* through LIB and GAR. At the present time we know more species which grade rather smoothly between these two subgenera, which thus are not well separated, and never were and probably do not need to be separated. To the aquarists, the term "Fundulopanchax" is still very useful to point out a certain major phenotype in *Aphyosemion* which is characterized by its robustness, the large dorsal fin, the annual or semi-annual reproduction and the ctenoid spines on scales of males. When the latter character is used, the subgenus *Fundulopanchax* becomes rather well separated from *Aphyosemion,* although the CAL superspecies (see AHL, AUS and CAL) may represent a certain problem. Crossing experiments indicate, however, that probably these two subgenera are rather closely related, and for that reason it is not likely that *Fundulopanchax* will become a generic name in the future.

PAREPIPLATYS AND EPIPLATYS

Clausen (1967: 27–29) separated the species which develop the "open" pattern of lateral line organs on the dorsal part of the snout (SEX and LON plus two more supposed species, one of these being ESE) from all other species of *Epiplatys* (except for ANN and DUB) which develop the "closed" pattern. He also considered the group of species last mentioned to be more lightly built generally, a point of view which cannot be accepted. Thus the separation of SEX and its allies from other species of *Epiplatys* seems to be based on the small difference of the lateral line pattern. Under LAM it has been said that in this species some individuals develop a lateral line pattern somewhat intermediate to that of Clausen's two subgenera of *Epiplatys.* Also, in *Aplocheilus* both patterns occur in different species, with the open pattern in LIN and PAN and the closed in DAY, the BLO species being variable.

Unfortunately the separation of one group of species in *Epiplatys* from another group places SEX in one subgenus and CHA in the other. Undoubtedly these two species are very closely related and may represent the two allopatric components of a superspecies. See CHA, SHE and SEX. The genus *Epiplatys* (with or without *Aplocheilus*) indeed appears to be heterogeneous, but I think that a future division into different subgenera should not be based on a single small detail in morphology but on a combination of many biological and morphological characters.

Clausen made GRA the type-species of the new subgenus, named *Parepiplatys.* This choice of type-species undoubtedly was well-chosen, because any future division of the genus *Epiplatys* (or *Aplocheilus*) probably will separate GRA (plus, ORN, MAC and other species) from SEX and its allies. As it has been mentioned under SEF, the name *Lycocyprinus* Peters 1868 has DAG (the MON subspecies of DAG) as type-species, and because DAG has a "closed" pattern of lateral line organs, it belongs to *Parepiplatys*

Clausen. DAG appears to be a close relative of GRA, the type-species for the subgenus last mentioned. Thus it seems that *Parepiplatys* is a synonym for *Lycocyprinus*, and it is not likely that GRA and DAG will be separated by a future subdivision of *Epiplatys*.

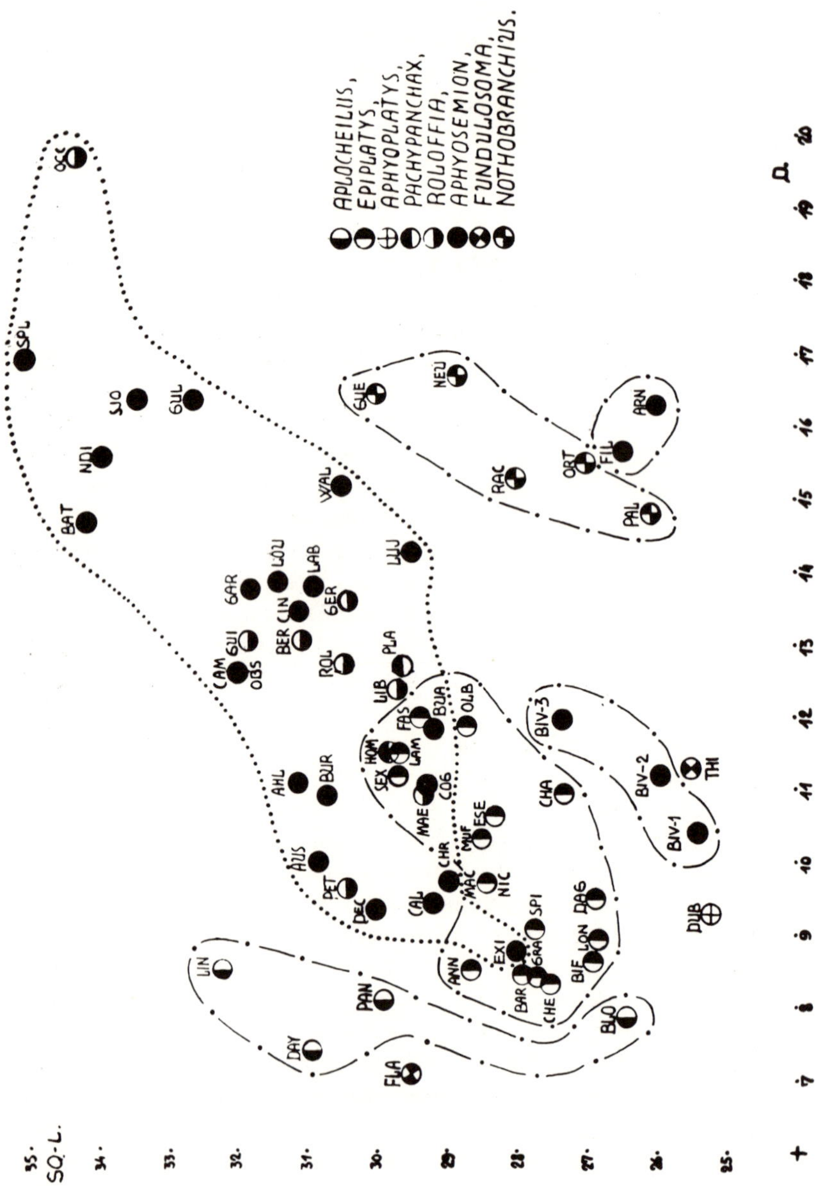

Among the many meristic characters used in the orthodox definition of the rivuline species, the number of dorsal fin rays and the number of scales in a longitudinal series probably are the most important. The *mean* figures for these two characters have been calculated for 66 species of *Rivulinae* of the Old World on the basis of my own counts of preserved material. These figures are plotted on an "D/Sq-l" diagram showing an interesting distribution of these species.

The 16 species of *Epiplatys* produce a rather narrow group which grades into the combined *Aphyosemion-Roloffia* group through SEX, LAM and FAS. The total variation of these two characters is rather limited in *Epiplatys*. The four species of *Aplocheilus* do separate to some extent from *Epiplatys* and exhibit a rather constant number of dorsal fin rays, whereas the variation in number of scales is considerable, and this character seems to separate BLO from the other three species. The five species of *Nothobranchius* (limited material only) also produce a rather narrowly limited group separated from *Aphyosemion* except for the ARN superspecies (ARN + FIL). It should be kept in mind, that when egg characteristics are considered, the ARN superspecies is much closer to *Nothobranchius* than any other species of *Aphyosemion* and *Roloffia*. FLA and DUB probably represent two monotypic genera separate somewhat from other species groups, but DUB is close to BIV and when juveniles of these two species are studied, those of DUB are so close to those of BIV that only the absence of the dark longitudinal bands separates these two forms when seen in the aquarium.

The eight species of nominal *Roloffia* do not separate from *Aphyosemion* and it is seen that the type species (OCC) is well separated from all other species of these two nominal genera except for a certain group in *Fundulopanchax* (BAT, GUL, NDI, SJO, SPL). The EXI species separates from the main lot of species of *Aphyosemion* (including *Roloffia*) and falls among the species in *Epiplatys*. The temporary dark lateral band of EXI correspond to that of many species of *Epiplatys*. The three groups of populations in BIV separate rather distinctly from other species in *Aphyosemion*.

Thus, it appears that on the basis of these two important characters the genera *Epiplatys*, *Aplocheilus* and *Nothobranchius* may represent natural taxonomic units, whereas the combined *Aphyosemion-Roloffia* group needs further study.

For the two major genera, *Epiplatys* and *Aphyosemion* (including *Roloffia*) there seems to be a connection between the number of dorsal fin rays and the number of scales: Sq-l = 20.6 + 0.82.D, more or less.

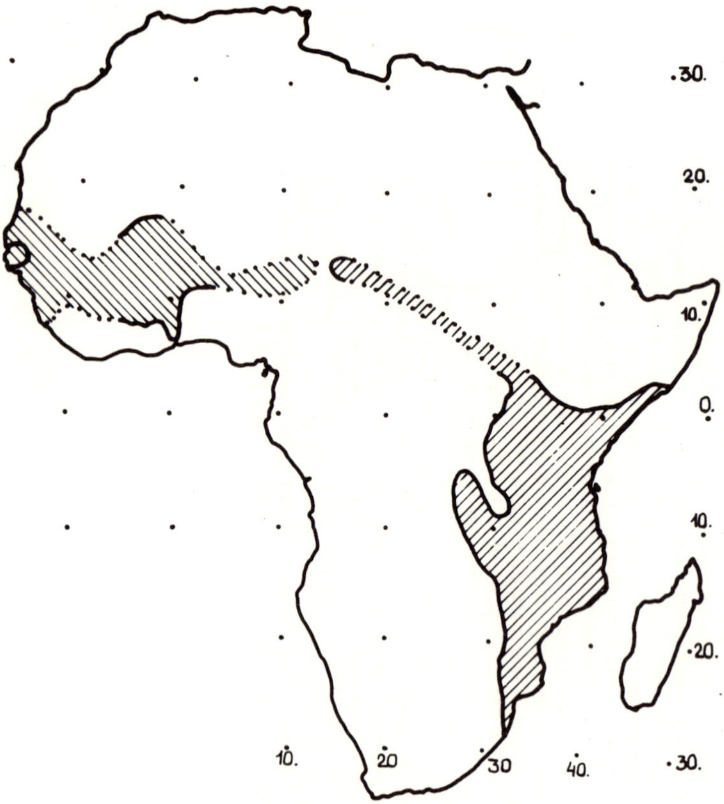

Geographical distribution of *Nothobranchius* (area lined //////) and *Fundulosoma* (area lined \\\\\\).

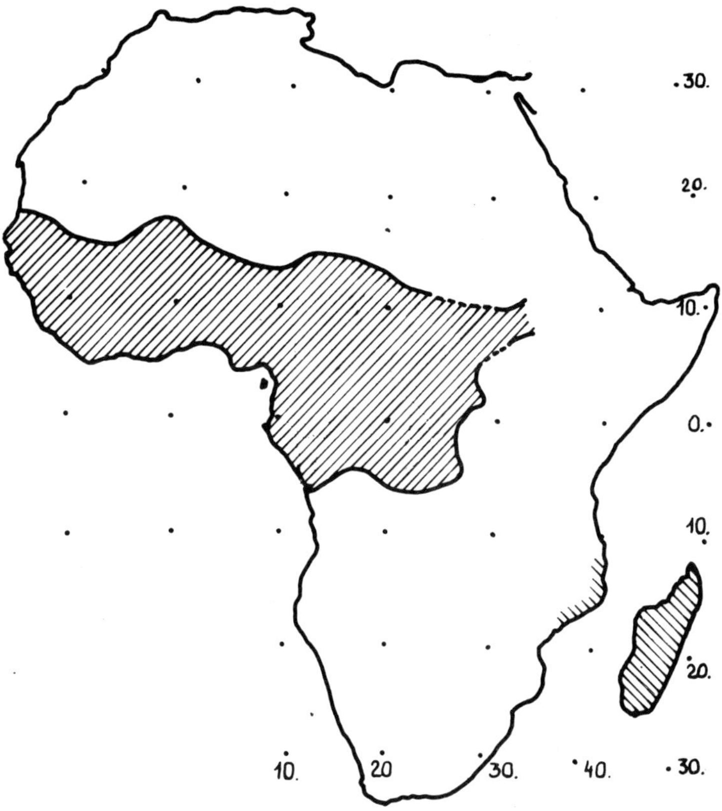

Geographical distribution of *Epiplatys* (area lined//////) and *Pachypanchax* (area lined\\\\ ).

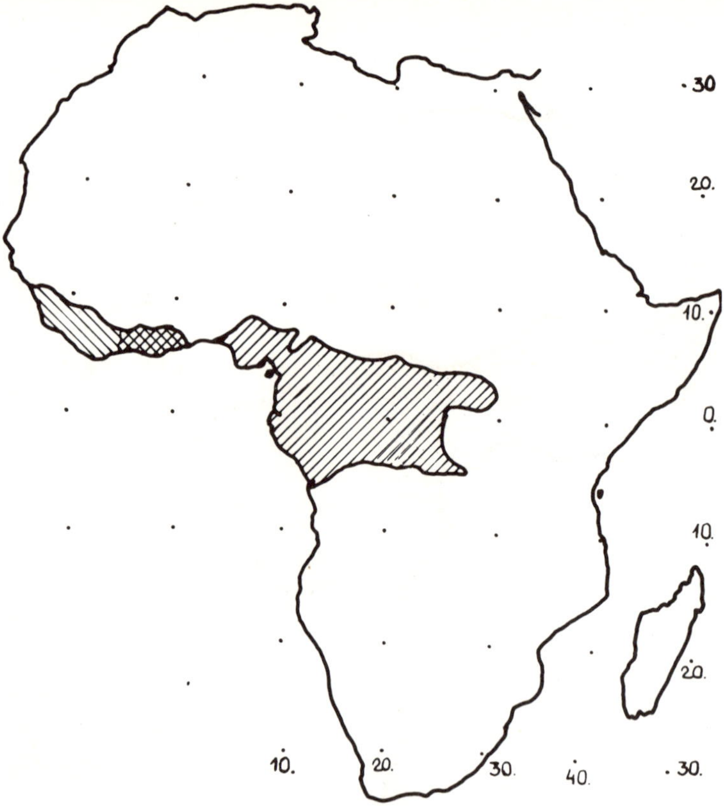

Geographical distribution of *Aphyosemion* (area lined/////) and *Roloffia* (area lined \\\\\).

Geographical distribution of *Aplocheilus* (area marked/////).

XVIII

THE SYSTEMATIC STATUS OF THE RIVULINE SPECIES

Myers (1955) and Sethi (1960) both considered the rivulins to represent the most generalized forms within the family Cyprinodontidae and based this point of view on osteological features (details of the skeleton). Hoedeman (1961: 82) developed a similar point of view from his studies of the frontal scalation, but if the point of views of the present author are used, the scalation of the rivuline species of the Old World seems to represent the final step in the evolution of the scalation of the head, and this character thus does not point out these species as the most generalized cyprinodonts. Primitive and highly advanced scalations of the head occur within two subfamilies at least (*Fundulus chrysotus* versus *Lucania goodei* in Fundulinae and *Lamprichthys tanganicanus* versus *Aplocheilichthys normani* in Procatopodinae) and thus it seems that the frontal scalation is not a useful character in the phylogeny of cyprinodonts.

No generally accepted subdivision of the family Cyprinodontidae is available at the present time, but probably the subsequent division is not too far from the point of view of other authors:

Cyprinodontinae (the pursy minnows): 17 species of the New World: *Cyprinodon*, *Floridichthys*, *Garmanella*, *Jordanella* and *Cualac*.

Fundulinae (the fundulins): 40 species of the New World: *Profundulus*, *Fundulus*, *Oxyzygonectes*, *Adinia*, *Leptolucania*, *Lucania*, *Cubanichthys*, *Empetrichthys* and *Crenichthys*.

Orestiatinae (the orestias): 18 species of the New World: *Orestias* (Andes Mts.).

Aphaniinae (the aphanins): 13 species of the Old World: *Aphanius*, *Kosswigichthys* and *Valencia* .

Rivulinae (the rivulins): 250 species of the New World and the Old World, see below.

Procatopodinae (the lampeyes): 80 species of Africa: *Lamprichthys*, *Plataplochilus*, *Hylopanchax*, *Procatopus*, and *Aplocheilichthys*.

Pantonodontinae (the toothless lampeyes): 2 species in Africa and Madagascar: *Pantonodon*.

Oryziatinae (the medakas): 10 "species" in Asia: *Oryzias*.

I am aware that the Cyprinodontinae and Fundulinae often are considered to represent a single subfamily, because *Cualac* appears to be intermediate, and that the Oryziatinae have been removed from Cyprinodontidae by Rosen (1964) and have been considered to represent a distinct family.

On the basis of present evidence, the rivuline species of the Old World cannot be separated from those of the New World; these two geographical groups of genera appear to grade into each other through the *Cynolebias-Cynopoecilus* group (Cynolebiatidi of Hoedeman 1961: 87) and in particular through *Cynopoecilus* to *Roloffia*.

Hoedeman (*op. cit*) considered three tribes of rivuline species and he placed the New World species in Rivulidi (except for *Cynolebias* and *Cynopoecilus* = Cynolebiatidi and *Austrofundulus*), whereas all species of the Old World were placed in Aplocheilidi. This division corresponds well to the opinion of the present author.

He considered *Austrofundulus* to represent a specialized outshoot from the Cyprinodontinae and not to represent a rivuline genus. I have seen only a very limited number of nominal *Austrofundulus*, but it seems that the eggs of *A. dolichopterus* separate this species distinctly from all other species of Rivulidi and place it close to *Nothobranchius* in Aplocheilidi. I have seen the eggs and preserved adult individuals of *Rachovia brevis* and found this species to be in agreement with other Rivulidi species. The *Austrofundulus-Rachovia* group, which by some authors (Myers 1963: 8, also Weitzman and Wourms 1967: 12) are considered to represent a single genus, appear to be an unnatural assemblage of species.

Both Myers and Sethi considered *Aplocheilus* to represent the most generalized form within the Rivulinae. The chromosome conditions support this idea and place *Aplocheilus* (with or without *Epiplatys* included) and *Pachypanchax* among the ancient cyprinodont species. Ctenoid spines, which occur in all or almost all subfamilies of Cyprinodontidae are, however, completely absent in these two (three) genera and the perfect or almost perfect scalation of the head also indicates that in respect to these two characters *Aplocheilus* and *Pachypanchax* are not generalized. The membrane pattern of the eggs of *Aplocheilus* also represents a certain problem, because in some species of this genus a perfect hexagonal reticulation occurs. Such modification of the egg indicates that the species belongs to or derived from an annual rivuline form, but *Aplocheilus* is certainly not an annual rivuline, although it might have originated from an ancient annual rivuline species.

Appendix 1

***Epiplatys barmoiensis* sp. nov.**

Holotype: Adult male (standard length 46 mm, total length 64 mm) caught by E. Roloff 1965 at Rokupr of western Sierra Leone. No. P. 35.2582 in the Copenhagen Zoological Museum and *allotype:* Adult female (standard length 40 mm, total length 55 mm) same collector, same locality, No. P. 35.2583.

Data of the holotype and the allotype (in parentheses). Measurements in percentages of standard length:

Greatest depth of body: $20\frac{1}{2}\%$ ($22\frac{1}{2}\%$), least depth of caudal peduncle: 13% (13%). Length of head: 29% (29%), of snout: 11% (10%), of eye: 8.7% (8.8%) and of interorbital width: 13% (13%). Distance from end of snout to origin of dorsal: 76% (78%), of anal: 61% (62%) and of ventrals: 50% (52%).

There are 9 (8) dorsal fin rays; the anterior-most ray is placed above the 14th (13th) anal fin ray. The anal fin has 17 (16) rays and is posteriorly pointed in the male.

There are 27 (27) scales in a longitudinal series, 21 (19) predorsal scales and $9\frac{1}{2}$ ($9\frac{1}{2}$) scales in an oblique row from root of ventrals to the back. The caudal peduncle is 1.1 (1.2) times as long as deep.

The pattern of frontal neuromasts is closed. The frontal scalation is of the usual G-type, the basic frontal scales being very large and their edges curved. There is one row of large scales on the preopercle.

Colors of the male: There are four rather regular rows of vertically oblong or (anteriorly) crescent-shaped red spots along the sides. The spots of the median row are largest; those of the lower row are broadly edged with black, producing an indistinct and irregular lateral band. Other spots are narrowly edged with black. The red markings of fins are much darker, sometimes almost black. The dorsal has an irregular pattern of spots and three larger spots at base. The anal is narrowly edged with black and has a submarginal row of rather large vertically oblong spots. The upper and lower parts of the caudal have rows of dark spots between fin rays, and the lower part has a thin red submarginal band. The rather pointed pectorals have a number of very dark spots in the upper half, tending to produce an evenly distributed black color posteriorly. The lobes around the lateral line organs of the dorsal part of the head are brilliant red. The lower lip has a broad red band, and there are very small red spots anteriorly on the throat, but no regular pattern is produced. In front of a line between the corners of the mouth there are evenly distributed dark pigments, but behind that line there are almost no dark pigments.

Colors of the female: The color pattern of the body is very similar to that of the male, but the red spots are much smaller or even absent; the dark longitudinal band produced by the lower row of red spots is more marked, but also in this individual this band is rather indistinct. The throat pattern corresponds to that of the male, but behind the line between the corners of the mouth there are no dark pigments. The fins have an irregular pattern of dark spots in particular in the dorsal and in the upper part of the caudal. There is a row of dark spots below the root of anal, as in CHA, DAG, GRA, MAC and other species of the genus.

There is no color pattern on the belly in either sex.

The holotype has been cytologically studied (Scheel 1966a: 134 and Pl. IX. fig. 7a (showing the live holotype) and 7b (showing the primary meiotic divisions)) and 17 haploid chromosomes were found.

The allotype was used for hemoglobin tests and the usual four-line rivuline pattern developed. Its egg measured 1.3–1.4 mm, and the membrane had a hexagonal reticulation which was more or less regular.

Epiplatys esekanus sp. nov.

Holotype: Small male (standard length 32½ mm, total length 43½ mm) caught by the author in a brook 15 miles north of Eseka, near the Eseka-Jaunde-Edea road-crossing in East Cameroon. No. P. 35.2584 in the collections of the Copenhagen Zoological Museum. *Allotype:* small female (standard length 24 mm, total length 32½ mm) same collector and locality. No. P. 35.2585.

Data for the holotype and the allotype (in parentheses). Measurements in percentages of standard length:

Greatest depth of body: 22% (23%), least depth of caudal peduncle: 12% (12%). Length of head: 32% (33%), of snout: 9% (10%), of eye: 9% (8%) and of interorbital width: 15% (15%). Distance from end of snout to origin of dorsal 74% (83%), of anal: 62% (67%) and of ventrals: 49% (54%).

There are 10 (10) dorsal fin rays; the anteriormost ray is placed above the 10th (11th) ray of anal, which has 16 (16) finrays and is posteriorly pointed in the male.

There are 29 (28) scales in a longitudinal series, 19 (21) predorsal scales and 10 (11) scales in an oblique series from the root of ventrals to the back. The caudal peduncle is 1.3 (1.2) times as long as deep.

The colors of the male (see also color picture of live holotype on page 185 under ESE). Almost every scale of the sides has a very large red spot, these spots being crescent-shaped in upper rows and more rounded in lower rows, but no red spot is completely rounded. There are about eight rather indistinct, oblique, irregular dark cross bars on the sides, these rather narrow bars being of the *E. grahami* and not the *E. sexfasciatus* type. There is a dark marbled pattern on the belly, tending to produce thin dark lines across the belly posteriorly. The lower parts of the head have a rather uniform dark pigmentation with very little irregularly scattered red pigments. No pattern develops. The fleshy lobe around the neuromasts of the dorsal part of the head are brown, not red. The markings of fins are very dark, almost black. The dorsal fin has an irregular pattern of dark spots. The anal fin has a few similar spots and is narrowly edged with black. The caudal fin has rather large dark spots between fin rays, and the lower edge is black. The color pattern of the female is similar, but the red spots on the sides are few and small. The black bars are more marked but very irregular. The fins have irregularly distributed small dark spots. The lobes of the dorsal part of the head are brilliant red. The anterior part of the throat is darker than the posterior part, but no distinct pattern develops. There is an irregular red line just behind the lower lip.

The pattern of frontal neuromasts is open (as in LON and SEX); the frontal scalation is normal for the genus (G-type, no H-scales), and there is a single row of scales on the preopercle below the eyes.

The male has been cytologically studied and 21 haploid chromosomes were found. Further details on the karyotype are given under ESE.

Aphyosemion ndianum **sp. nov.**

Holotype: Adult male (standard length 49 mm, total length 59 mm) from an aquarium strain based on individuals caught by H. S. Clausen in 1959 near Osomba of East Nigeria, near the southern part of the Mamfe-Eyomojok-Calabar road. No. P. 35.2586 in the collections of the Copenhagen Zoological Museum. *Allotype:* Adult female (standard length 44 mm, total length 52 mm) same strain No. P. 35.2587.

Data of the holotype and the allotype (in parentheses). Measurements in percentages of standard length:

Greatest depth of body: 20% (18%), least depth of caudal peduncle: 12% (9%). Length of head: $26\frac{1}{2}$% (27%), of snout: 6% ($4\frac{1}{2}$%), of eye: 6% (8%) and of interorbital width: 10% (9%). Distance between end of snout and origin of dorsal: 61% (61%), of anal: 57% (59%) and of ventrals: 45% (50%).

There are 15 (15) dorsal fin rays; the anteriormost ray is above the 4th (6th) ray of anal which has 16 (17) fin ray.

There are 33 (34) scales in a longitudinal series, 18 (18) predorsal scales and 13 (12) in an oblique row from the root of ventrals to the back. The caudal peduncle is 1.5 (1.5) times as long as deep.

The pattern of frontal neuromasts is open. The frontal scalation is of the usual G-type, without H-scales. There are two rows of scales on the preopercle below the center of the eyes. In the male ctenoid sensory papillae occur on the upper ray of pectorals, these organs being small and low in number. No similar papillae occur in other fins. Ctenoid spines occur on some of the scales of the opercle and on median rows above the anal, but in this particular male the spines are short and few in number. In the female there is a small scaled "pouch" just in front of the origin of the anal fin.

The color pattern of this species is highly variable and has been described on page 311–14 .The general features of this pattern are close to those described for *Aphyosemion cameronense* (Blgr.); from this the new species differs by geographical distribution, by the higher number of dorsal fin rays and of scales in a longitudinal series, by the color pattern of the throat. and by the presence of ctenoid spines in sexually matured males. No other species of African rivulins develop similar color patterns.

The holotype was cytologically studied and twenty haploid chromosomes were found (Scheel 1966a: 135 and Pl. XIII fig. 20a). The details of the karyotype are given on page 311.

Appendix 2

REVIEW OF MORPHOLOGICAL CHARACTERS

Table 1: In this table 17 different distinguishing characters are published for 65 different morpho-species of Rivulinae of the Old World. Except for the hemoglobin pattern figures (prepared by Dr. K. Sick) all figures have been based on the observations of the author.

CODES USED IN TABLE 1

1. The haploid chromosome number (numbers)
2. The hemoglobin pattern (see also fig. 6)
3. The lateral line pattern (O = Open, C = Closed, V = Variable)
4. The frontal scalation (V = Variable) (see also fig. 3)
5. The presence of H-scales (Y = Yes, N = No)
6. Ctenoidy in scales or fin rays (S = Spines, W = Wrinkled type, N = No)
7. Egg-diameter in millimeters.
8. The pattern of the egg-membrane (R = Reticulated, S = Streaks, D = Dots and P = Plain)
9. "Do eggs adhere to mud?" (Y = Yes, N = No)
10. Dark crossbars on the sides of body? (Y = Yes, N = No)
11. Dark (premating) lateral band in females? (Y = Yes, N = No)
12. Dark premating throat-color in males? (Y = Yes, N = No)
13. Dark temporary throat-pattern in females? (Y = Yes, N = No)
14. The red pattern on the sides of males (R = Reticulation, S = Spots, N = No)
15. Number of dorsal fin rays observed by the author
16. Number of anal fin rays observed by the author
17. Number of scales in a longitudinal series observed by the author
18. Number of individuals on which the figures under 15–17 are based.

TABLE 1.

Spc.	1	2	3	4	5	6	7	8	9	10	11	12	13	14	15	16	17	18
AHL	18	?	O	G	(Y)	N	1.1–1.2	R	Y	N	N	Y	Y	S	10–12	14–16	30–33	11
ANN	25	A	C	V	(Y)	N	0.9–1.0	P	Y	Y	Y	N	N	N	8–10	14–18	26–29	32
ARN	19	A	O	G	N	S	1.0	R	N	N	N	N	N	S	15–17	15–18	24–29	12
AUS	15	A	O	G	N	N	1.2	S	Y	N	N	N	N	S	9–11	14–16	29–32	15
BAR	17	A	C	G	N	N	1.3–1.4	R/S	Y	N	Y	N	Y	S	8–9	15–18	26–29	18
BAT	?	?	O	G	N	N	?	?	?	N	N	?	?	V	14–16	14–17	33–35	4
BER	21	A	C	G	N	W	1.1	R	Y	N	N	Y	Y	S	11–14	15–18	29–33	38
BIF	20	C	C	G	N	N	1.2–1.3	P	Y	N	Y	N	Y	R	7–10	14–19	25–29	141
BIV	13–20	A	O	G	N	N	1.2–1.5	D	Y	N	Y	N	(N)	S	9–14	11–15	24–28	244
BLO	24	H	C	(E)	(Y)	N	1.5	P	Y	N	N	N	N	(S)	6–9	14–16	24–29	23
BUA	19, 20	A	O	G	(Y)	(W)	1.3–1.5	D	Y	(Y)	N	(Y)	(Y)	S	10–12	15–17	28–31	45
CAL	16	A	O	G	N	N	1.2–1.4	S	Y	N	N	N	Y	S	8–10	12–15	29–32	39
CAM	17	?	O	G	N	N	?	?	?	N	N	N	N	S	11–13	15–17	31–33	9
CHA	25	B	C	G	N	N	1.1–1.3	R/S	Y	Y	Y	Y	Y	S	9–12	14–19	25–30	153
CHE	?	B	C	G	N	N	?	?	?	N	Y	N	(Y)	S	8–9	13–15	26–29	3
CHR	9, 15	A	O	G	N	N	1.2	S	Y	N	N	N	N	S	9–11	14–16	27–30	20
CIN	20	A	O	G	N	S	1.5	R	Y	N	N	Y	N	N	13–14	16–17	30–33	20
COG	?	A	O	G	N	N	1.4–1.5	D	Y	N	N	N	N	S	11	15–16	29–30	2
DAG	25	A	C	G	N	N	1.0–1.1	R/P	Y	Y	Y	N	N	R	8–11	14–17	25–29	113
DAY	24	G	C	G	N	N	1.6	R/S	Y	Y	Y	N	N	S	7–8	15–18	29–32	12
DEC	?	?	O	?	?	N	?	?	?	N	N	?	?	S	9–10	14–15	28–32	3
DUB	24	A	O	G	N	N	1.1	D	Y	N	Y	N	Y	S	9–10	15–17	24–26	8
ESE	21	B	O	G	N	N	1.4	D	Y	Y	Y	N	Y	S	10–12	15–17	27–30	9
EXI	18	A	O	G	N	N	1.3–1.4	D	Y	(Y)	(Y)	N	(Y)	S	8–9	14	28	5
FAS	19	A	C	G	N	N	1.1–1.5	R/S	Y	Y	Y	Y	Y	R/S	10–15	15–20	27–31	112
FIL	18	A	O	G	N	S	1.0–1.3	R	N	N	N	N	N	S	14–17	14–17	24–28	62
FLA	?	?	O	G	N	N	?	?	?	N	N	N	N	N	6–8	12–15	28–31	38
GAR	18, 20	A	O	G	(Y)	S	1.0–1.5	R/S	Y	N	N	N	N	S	12–16	14–18	29–34	262
GER	20	A	C	G	N	W	1.1	R	Y	N	Y	N	N	S	12–15	15–18	28–31	18
GRA	23	A	C	G	N	N	1.0	S	Y	Y	N	N	Y	S	7–11	15–17	26–30	65
GUE	19	C	O	V	V	S	1.3	P	N	N	N	N	N	R	15–18	16–19	29–32	8
GUI	19	A	C	E	N	W	1.6	R	Y	N	N	Y	Y	N	11–15	14–18	29–34	61
GUL	16	A	O	G	Y	S	1.5–1.7	R/S	Y	N	N	N	N	S	15–18	17–19	29–35	24

TABLE 1 (*continued*).

Spc.	1	2	3	4	5	6	7	8	9	10	11	12	13	14	15	16	17	18
HOM	?	?	V	E	Y	N	1.5	?	Y	N	N	?	N	N	11–12	15–19	29–31	4
LAB	14	A	O	G	N	N	1.6–1.8	D	Y	N	N	N	N	S	13–14	16–17	30–31	5
LAM	?	?	V	G	N	N	?	?	?	N	N	?	?	S	10–12	14–17	28–31	13
LIB	?	A	C	G	(Y)	W	1.1–1.2	R	Y	(N)	(N)	Y	Y	S	11–13	15–16	28–31	9
LIN	25	E	O	G	N	N	1.6–1.9	R/P	Y	Y	Y	N	(Y)	S	8–9	17–20	31–34	15
Spc.	1	2	3	4	5	6	7	8	9	10	11	12	13	14	15	16	17	18
LON	?	B	O	G	N	N	1.4	D	Y	Y	Y	N	Y	(S)	7–10	15–18	25–29	44
LOU	10	A	O	G	N	N	1.6	S/D	Y	N	N	N	N	S	12–14	15–16	31–32	7
LUJ	20	A	O	G	N	N	1.3–1.4	P	Y	N	N	N	N	S	13–15	15–17	29–31	4
MAC	?	A	C	G	N	N	1.1	P	Y	N	N	N	Y	S	08–10	15–17	26–29	39
MAE	?	?	(C)	G	N	N	?	?	?	N	N	?	(Y)	(S)	09–14	14–18	27–34	476
MUF	?	?	C	G	(N)	N	?	?	?	Y	N	?	?	S	09–11	14–16	26–31	23
NDI	20	A	O	G	N	S	1.3	R	Y	N	N	N	N	S	14–16	15–18	31–35	18
NEU	?	?	O	V	V	S	?	?	?	N	N	N	N	R	16–18	18	27–31	4
NIC	?	?	C	G	N	N	?	?	?	N	Y	?	?	S	9–10	15–16	28–29	5
OBS	17	?	O	G	N	N	1.3	R	Y	N	N	N	Y	S	12–13	15–18	31–34	10
OCC	23	A	C	G	N	W	1.4–1.5	R	Y	(N)	Y	Y	Y	R	17–23	17–20	32–37	62
OLB	?	?	C	G	N	N	?	?	?	Y	Y	?	?	S	10–13	15–18	28–30	21
ORT	18	?	O	V	N	S	?	P	N	N	N	N	N	(R)	15–18	16–19	26–28	10
PAL	18	?	O	V	N	S	0.9–1.0	P	N	N	N	N	N	R	14–15	15–17	25–27	6
PAN	18	F	O	G	N	N	1.6–1.8	R	Y	N	N	N	N	(S)	06–09	15–19	27–33	88
PET	?	?	C	G	N	N	1.3	P	Y	Y	N	Y	Y	(S)	08–11	15–16	29–32	21
PLA	24	D	(C)	E	Y	W	?	?	Y	N	N	N	Y	S	12–13	17–19	29–30	5
RAC	9	A	O	G	N	S	0.8–1.1	P	N	N	N	N	N	R	15	16–17	28	3
ROL	21	A	C	G	N	W	1.1–1.2	R	Y	(N)	N	N	Y	S	11–14	15–18	29–32	33
SAN	24	?	C	G	N	N	1.7	R	Y	(N)	Y	N	Y	S	11	16	30	1
SEX	24	B	O	G	N	N	1.4–1.6	D	Y	Y	Y	Y	Y	S	9–13	14–19	27–32	802
SJO	20	A	O	G	Y	S	1.3–1.5	R/S	Y	(N)	N	N	N	S	14–18	16–19	31–37	28
SPI	17	A	C	G	N	N	1.1–1.3	R/S	Y	Y	Y	Y	Y	R	7–10	15–19	26–29	71
SPL	?	?	O	G	N	S	?	?	?	N	N	?	?	V	17	17–19	35–39	9
THI	?	?	O	G	N	N	1.0	P	N	N	N	N	N	S	12–13	15–17	24–26	3
WAL	18	A	O	G	(Y)	S	1.4–1.5	R/S	Y	N	N	N	N	S	14–16	16–17	29–32	12
"BUR"	20	A	O	G	N	(N)	?	?	Y	N	N	N	N	S	10–12	13–14	31–32	8

Table 2: In this table the most important meristic characteristics for 156 nominal species of Rivulinae of the Old World are presented. These data have been taken from the descriptions.

CODES USED IN TABLE 2

For each nominal species (name) the figures published in the original description or in subsequent descriptions, based on the type material (the names for these species are marked +) only, are given.

D: The number of dorsal finrays.

A: The number of anal finrays.

D/A: The juxtaposition of the dorsal and anal fin. The figure D/A = 8A indicates that the anteriormost dorsal fin ray is inserted above the eighth anal fin ray. D/A = 12S indicates that this fin ray is above the 12th scale of the longitudinal series, counted from the head (root of pectorals) and D/A = >2D indicates that the anteriormost anal fin ray is in front of the 2nd dorsal fin ray.

Sq-l: The number of scales in a longitudinal series.

Sq-t: The number of scales in a vertical (or oblique) row (figures below 14) or round the body in front of ventrals (larger figures).

C.P.R.: The length of the caudal peduncle divided by the least depth.

C/Hd: The length of the (central rays of) the caudal fin compared to the length of the head. C/Hd = >H indicates that the caudal fin is longer than the head.

SL/Ht.: The depth of the body, indicated by the number of times which this figure is contained in the standard length.

SL/Hd.: The length of the head, indicated by the number of times which this figure is contained in the standard length.

Hd/Eye: The length of the head divided by the diameter of the eye.

Int/Eye: The width of the interorbital space divided by the diameter of the eye.

mm: The length in millimeters of the largest individual. The figure "32+11" indicates that the standard length is 32 mm and the length of the caudal fin is 11 mm. Most of the figures given are referable to the standard length (length of individual without caudal fin), but some figures might indeed refer to the total length of the fish.

TABLE 2.

spc.	D	A	D/A	Sq-l	Sq-t	C.P.R..	C/Hd	SL/Ht	SL/Hd	Hd/Eye	INT/Eye	mm
ACU	9–10	15–16	10A	26–29	9½–10½	1.0	=H	4.7–5.0	3.0–3.7	3.0–4.0	?	32+11
AND	11	?	?	?	?	?	?	?	?	?	?	?
ANN	7	13	9–10A	28–29	?	1.7	>H	5.0	3.5	2.7	1.3	16
ANS	11	16	8A	30–31	18	1.0	>=H	4.0	3.3–3.5	3.7–4.0	2.0	70
ARN	15–16	15–17	>10	25–27	20–22	1.0	?	4.5–5.5	3.0–3.3	3.5–4.0	1.3	55
BAR	8–9	16–17	13–14A	27	9½	1.1–1.2	>H	4.4–4.9	3.5	3.3	1.5	46
BAT	14–15	14–15	1A	33–35	24–26	1.7	?	4.5	3.5	3.7	1.5	66
BAU	8	16	11A	25	18	1.0	>H	4.3	3.5	3.5	1.5	40+12
BEA	15	16	2–3A	36	24	1.5	=H	5.0	3.5	3.3	1.3	32
BEL	11–12	14–15	5A	30–34	22–24	1.8–2.0	>=H	5.0–5.3	3.7–4.0	3.3	1.5–1.6	42
BER	13–14	16–17	7A	31–32	?	?	?	?	?	?	?	40
BIF	6–7	15	?	28	7½	?	>H	4.5–5.0	3.3–3.4	>4.0	?	?
BIT	9–10	12	2–3A	27–28	24	1.5	?	4.5–5.0	3.7–4.0	<3.0	1.2	36
BIV	12	14	13S	26	7	?	<H	4.7	3.5	3.6	1.8	30
BOU	9–11	14–17	10–12A	28–29	20–22	>=1.0	=H	3.8–4.3	3.0–4.5	3.0–3.5	1.0–1.5	44+11
BRI	15–16	15–18	1A	28–31	23–24	1.0–1.5	?	2.8–4.0	?	3.5–4.7	1.3–2.0	58
BRU	12–13	14–15	4–5A	30–34	24–26	1.5	=H	4.0–4.5	3.0–3.5	3.3–3.5	1.6–1.7	32
BUA	11	16	5A	32	22	?	<H	5.0	3.5	3.7	1.5	32
CAB	9–11	14–15	5A	30	20–22	1.7–2.0	=H	4.5–5.0	4.5	3.5	2.0	42
CAL	9–10	12–13	6–7A	29–30	20–22	1.3–1.5	?	3.5–4.3	3.3–3.7	3.7–4.0	2.0	50
CAM	11–12	14–15	18–19S	30–33	10–11	<2.0	=H	4.5–5.5	4.0	3.3–3.5	1.6–1.8	55
CAP	13	13	6D	36	11	?	?	?	?	2.5	>1.0	26
CAR	11	16	8A	30	24	1.7	=H	4.5	4.0	3.5	2.0	36
CAS	8	15	7–8A	30	7	?	>=H	5.0	3.7	3.5	1.5	34
CHA	7	15	18S	25	?	?	<=H	>5.0	>4.0	3.5	1.7	70
CHE	7–8	13–14	19–20S	27–28	8	1.0	>H	4.0–4.5	3.5	3.0	1.5	48
CHI	8	14	11A	27	20	1.5	>H	4.8	3.3	3.5	1.5	40
CHR	10–11	14–15	7–8A	32–34	22–24	1.7–2.0	>=H	4.0–4.7	3.8–4.0	2.7–3.0	1.5	50
CIN	13–14	16–17	?	30–31	9–10	2.0	?	4.0–5.0	3.3–3.9	3.1	1.4	40
COG	8–9	13–14	17S	30–31	22	1.3–1.4	?	4.4–4.8	3.8–3.9	3.6–3.7	1.8–1.9	60
CON	9–10	13–15	9–10A	29–31	24	1.8–2.0	<H	4.8–5.0	3.8–4.0	3.0–3.3	1.5	39
DAG	9	14–16	9–11A	26	19–20	?	>=H	4.6–5.0	3.4–3.5	3.6–3.7	1.4–1.5	44
DAY+	6–7	15	?	29–30	?	?	?	5.0	3.5–4.3	?	?	90
DEC	8–9	13–14	16–17S	25–28	11–12	1.5	<=H	3.7–4.0	3.5	<4.0	1.5	40
DEF	7–8	16	?	28–29	20	1.0	>=H	4.0–4.5	2.8–3.3	2.8–3.0	1.3–1.5	27+8
DOM	8	15	?	30	?	?	=H	6.0	4.1	3.5	?	40
DOR	14	18	7A	28	22	1.0	>H	4.0	3.7	3.3	2.0	77
DUB	9–10	14–16	7–9A	25–26	18	?	>=H	5.0–5.5	3.3–4.0	2.7–3.3	1.0	30
EIM	15	14	>2D	26	26	2.0	?	3.5	3.5	4.7	?	55

TABLE 2 (*continued*).

spc.	D	A	D/A	Sq-l	Sq-t	C.P.R.	C/Hd	SL/Ht	SL/Hd	Hd/Eye	INT/Eye	mm
ELB	11–12	16–17	8–9A	30–31	24	1.8	?	4.5–5.0	3.5	3.5	1.5	43
ELE	8	14–15	7–8A	30–32	20	<=2.0	?	4.5–5.3	3.5–3.7	3.0–3.3	1.5	40
ESC	11–12	15–17	5A	30–32	22–24	1.8–2.0	?	4.5–5.0	3.5–4.0	3.3–3.5	1.5–2.0	50
ESE	10	16	10–11A	28–29	10–11	1.2–1.3	>H	4.4–4.6	3.0–3.1	3.5–4.0	1.7–1.8	32
EXI	8	12	6A	28	22	1.5	<=H	4.0	3.3	3.5	1.5	32
FAL	13–14	15–16	1A	34	28–30	>1.0	?	4.8–5.0	3.7–3.8	3.0–3.3	1.5	70
FAS	11	18	17S	28	9–10	?	?	4.0	<4.0	<4.0	2.0	75
FER	10	13–14	6–7A	30–32	20	1.7	=H	6.0–6.5	3.7–4.0	3.5	1.7	50
FIL	13	14	11S	28	20–22	1.5	?	4.0–4.5	3.3–3.6	3.5	1.7–2.0	55
FLA	6–7	13–14	20S	30–31	20	2.2	=H	4.2–5.0	3.6–3.8	2.7	1.1–1.2	27
FLV	7	14	7A	29	22	1.5	=H	5.0	4.3	3.3	1.5	40
GAM	15	15	5D	30	24	1.5	?	4.0	3.5	4.5	2.0	47
GAR	12–13	14–16	2A	28–32	22–26	>1.0	?	4.0–4.5	3.0–3.5	3.3–4.0	1.6–2.0	60
GER	13–15	16–17	4A	31–34	20–22	1.5–1.8	=H	5.4–6.8	4.2–5.2	3.0–4.0	1.2–2.0	30
GRA	7	15–16	11–12A	?	?	1.0	>H	4.0–4.5	3.0–3.3	3.0–3.3	1.5	38
GUE	15–17	15–16	1A	30–32	9½–14	?	?	<3.5	3.5	4.5	?	?
GUI	12–14	14–17	5–6A	28–32	26–28	1.0–1.4	?	4.3–5.3	3.4–4.0	3.9–4.5	2.0–2.4	70
GUL	15–16	16–18	1A	30–31	12–13	>1.0	?	3.5–4.0	3.5–4.0	3.8–4.0	2.0	63
GUS	15–16	16–17	1A	36	24	1.7	?	4.5–5.0	3.3	2.7–2.8	1.3	32
HJE	11–12	14	7A	34–35	22–23	1.3	?	5.0	4.0	3.0	1.7	—
HOL	12	14	2A	28–29	24	1.6–1.7	<=H	4.1–4.7	3.4–3.9	3.0–3.5	1.3	31
HOM+	11	14	7A	28	?	<1.0	?	5.0	3.5–4.0	3.5–4.0	?	95
INF	11	14	19S	28	9	1.7	?	4.5	3.7	3.3	?	62
JAC	6	11–12	8–9A	27–28	20	1.7	<H	4.0–4.3	3.5–3.8	2.8–3.0	1.5	31
JAU	9	14	7A	30	24	1.7	=H	4.0	4.0	3.3	1.5	35
KIY	13–15	14–15	>1D	26	24	1.0	<H	3.8–4.3	3.3	3.5–3.7	1.5	40
KUH	14–16	13–15	2D	30	26	>1.5	?	3.8–4.0	3.0	3.5	1.3	44
LAB	12–14	14–17	5–8A	30–32	22–24	?	?	4.4–5.0	3.9–4.3	3.4–4.0	?	51
LAM	11–13	16–17	7–8A	28–32	22	1.0–1.2	=H	4.0–4.8	3.1–3.5	3.4–3.8	1.4–1.8	54+16
LIB	11	15–16	4–5A	32	20	2.0	<=H	5.5–6.0	4.0	3.5	1.6–1.7	36
LIN+	7	17	?	32	8	?	=H	5.0–5.3	4.0–4.3	3.5	1.5	100
LOB	8	15	7–8A	27	20	1.5	<H	4.5	3.5	3.0	1.3	29
LOE	11–12	12–13	1A	27	10	<2.0	?	4.5	3.7–4.0	3.5	1.7	50
LOL	8–9	14–15	7–8A	24–27	20–22	1.5	<H	4.5–5.0	3.5	3.5–3.7	1.5	34
LON	9	15–16	11–12A	26–27	20	1.0	>H	4.3–5.0	3.0–3.3	3.5–3.7	1.7–1.8	55
LOU	11	14	?	31	?	?	?	?	?	?	?	31+9
LUJ	10–11	14–15	7–8A	29–32	20–22	1.7–2.0	=H	4.0–5.0	3.5–4.0	3.5–3.7	1.5–1.7	45
MAC	9	16	11A	28	18	1.0	>H	4.3	3.5	3.5	1.7	40
MAE	11	15	7–8A	30	20–22	1.7	=H	5.0	3.8	3.7–4.0	1.6	37

TABLE 2 (*continued*).

spc.	D	A	D/A	Sq-l	Sq-t	C.P.R.	C/Hd	SL/Ht	SL/Hd	Hd/Eye	INT/Eye	mm
MAG	10–11	14–15	9–10A	28–29	6	?	?	5.8–6.7	3.8–4.3	3.2–3.5	1.3–2.0	44
MAR	8	14–15	?	27	$8\frac{1}{2}$	?	?	4.5–5.4	3.3–3.5	3.0–3.3	1.5–1.6	?
MAT	8–9	17	?	25	8	?	?	4.5–5.0	3.0–3.1	3.0–3.5	?	68
MAY	16	17	?	32	40	1.5	?	3.6	>3.0	4.0	2.0	70
MEI	12	14	14–19S	34	20	2.0	<H	5.0	4.0	3.3	1.7	48
MEL	7	10	22S	27–28	7–8	?	?	4.3–5.3	3.5–4.0	3.0–3.1	2.0	25
MEP+	14	16	1A	30–31	22–24	1.4–1.5	?	3.7	3.0–3.3	4.0	1.8–2.0	53
MIC	11	15	7–8A	31–32	22–24	1.8	>H	4.8–5.3	>4.0	3.3	1.7	35
MIL	16–17	17–18	1A	40–42	26–28	?	?	3.0–3.3	3.0–3.3	5.8–6.0	2.4–2.5	65
MKU												
MUC	10	13	13S	26	24	=1.5	<=H	4.0–4.5	3.3–3.6	3.7–4.1	1.7	33
MUF	8–9	14–15	11A	29–30	20	1.0	>=H	5.0–5.3	3.3–3.5	3.3	1.5	45
NDE	10	18	19S	26–27	7	?	>H	4.3–5.1	2.9–3.0	2.8–3.0	1.8–2.0	25
NDI	15	16–17	4–6A	33–34	12–13	1.5	<H	5.0–5.6	3.7–3.8	3.4–4.4	1.1–1.7	49
NEU	16	16–17	2D	33	15	=1.0	?	3.1	3.1	?	?	?
NIG	14–15	15–17	?	31–32	10	1.7	?	4.0–4.4	3.2–3.6	2.2	1.6	38
NII	11	15	2–3A	30	26	1.0	>H	4.8	3.3	3.5	1.7	42
NIC	9	15	10–11A	28–29	20	=1.0	>H	4.5–4.7	3.3–3.5	3.5	1.5	47
NIM	9	15–16	13A	29–30	20	1.5	>H	4.0–4.3	3.8–4.0	3.5	1.5–1.7	40
NIS	12	10–11	6D	36	12	?	?	3.7	4.0	4.0	2.0	100
NOR	12	17	8–9A	30	20	=2.0	<=H	4.5	3.5	3.3	1.5	45
NUC	10	13	6–7A	30	?	<1.0	<H	4.0	3.5	3.5	1.5	64
NYO+	8	16	8A	28	20	1.8	>=H	4.3	3.7	3.5	1.7	55
OBS	11	15	7–8A	31	22	1.7	=H	4.5	3.7	3.3	1.7	28
OCC	17–23	17–20	3–6D	32–36	?	1.0–1.5	?	3.6–4.5	3.6–4.3	3.7	?	?
OGO	10–11	14–15	<7A	29–31	22–24	?	?	4.3–5.0	3.0–4.0	3.0–3.7	?	40+10
OLB	11–12	15–16	8A	28–30	20–22	1.0	>=H	4.3–4.8	3.3–3.6	3.5	1.7	60
ORN	9	16	14A	29	20	1.3	>H	4.8	3.8	3.3	1.8	60
ORT+	15	15	?	28–30	24	1.0	?	3.0–3.5	2.7–3.0	4.0	1.7–1.8	70
PAL	16	15	>2D	27–28	10	>1.0	?	2.8–3.0	3.5	4.0	1.7	?
PAN+	7–8	15–16	?	30–33	8–9	?	?	4.5	3.5	3.5	>1.0	80
PAP	12	14	1A	26	22	1.5	?	4.5	4.0	3.0	1.5	31
PAR	7–8	14–15	?	26–27	7	?	?	4.0–4.5	3.3–3.5	3.4–3.5	2.0	42
PAS	11–12	14	7A	30–31	22	1.5	>H	4.5–5.0	4.0	3.3–3.4	1.5–1.7	47
PET	7	14	21S	32	?	?	=H	6.7	5.5	=3.0	=1.5	48
PLA	12	18	18S	32	9	?	?	4.0	3.3	2.5	<2.0	90
PAT	16	15	>2D	25–26	24–26	?	?	3.0–3.8	3.0–3.5	2.7–4.4	1.5–1.6	34
POL	10–11	?	?	31–33	20–22	1.5–1.7	?	4.5–5.0	4.3–4.7	3.0–3.3	2.0	43
PRE	11	16	5–6A	30	20	1.7	<H	4.8	4.0	3.3	1.7	34

TABLE 2 (*continued*).

spc.	D	A	D/A	Sq-l	Sq-t	C.P.R.	C/Hd	SL/Ht	SL/Hd	Hd/Eye	INT/Eye	mm
RAC	15	15–16	1A	25–26	22	1.0	?	3.0	3.0	3.3–3.5	1.7	60
RIG	13–15	13–14	1A	28–29	20–22	1.3	?	4.0–4.3	3.5	3.3	1.5–1.7	43
ROB	17	17	>2D	33	24	>1.0	?	3.5	3.5	4.0	2.0	52
ROL	12–13	14–15	9–10A	28–30	22	1.3–1.5	>H	5.0	3.5–4.0	=4.0	2.0	35
RUF	10–12	15–17	?	27–29	9–10	1.5	?	4.0–4.4	3.2–3.6	4.1	1.6	35
RUR	16–20	16–20	>2D	29–34	24–30	1.1–1.7	?	2.6–4.0	3.0–3.6	2.6–4.4	1.1–2.8	48
RUS	10–11	10–12	2A	26–28	24	1.3	?	4.3–4.5	3.5	3.0	1.2	38
SAK	12	17–19	11–12A	32–34	24–26	< 1.0	<H	3.6–4.2	2.8–3.5	3.1–4.0	1.4–2.0	71
SAN	10	17	10–11A	31	20	1.8	<H	4.5	3.8	3.7	1.7	53
SCH	9	14–15	?	28–30	20	1.7–2.0	>H	4.0–4.3	4.0–4.3	3.3–3.5	1.5	42
SEN	8	15	?	28	10	?	>H	5.2–5.7	3.4	>4.0	2.0	?
SEF	7	14	?	25	7	?	?	?	?	?	1.5	30
SEX+	10–11	15	?	28	8	?	=H	3.9–4.5	3.0–3.3	3.5–4.0	1.5–1.6	79
SHE	11	16	9–10A	28–30	21–22	1.0	=H	4.4–4.8	3.4–3.5	4.2–4.3	1.8–1.9	59
SEY	14–15	17–18	1D	30–31	22–24	=1.0	?	3.3–3.5	3.0	4.0	1.7–1.8	55
SIA	8–9	16–17	21–22S	31	8½–9	?	>H	4.8–4.9	3.8	3.3	1.5	64
SIN	8	14	14A	27–28	20	1.0	?	4.5	3.7	3.5	1.5	45
SJO	17–18	17	1A	35	11	?	?	?	3.6–3.7	4.0–4.2	1.7	78
SPI+	10	14–15	?	26–27	22	1.0	?	4.5	3.0–3.3	3.0	1.5	45
SPL	15–17	15–17	1–2A	33–36	28–34	1.5–2.0	?	4.0–5.0	3.3–3.5	3.7	1.8–2.0	70+30
SPM	10	16	8A	27	18	?	?	4.6–4.8	3.4	3.6	?	40+7
SPP	11	13	13S	26–27	22	1.5–1.7	=H	4.0–4.3	3.4–3.6	3.4	1.7	30
SPU	13–14	15–16	>2A	29–31	24–26	>1.0	?	4.0–4.3	3.3–3.8	3.5	1.7	42
STE	8–9	15–18	10–12A	26–27	20–22	1.0	>H	4.5–5.5	3.5–4.0	3.0–3.5	1.5–2.0	43
STR	9–10	13	6–7A	30–31	20–22	1.3	?	4.0–4.3	3.5–3.7	3.3	1.5	40
SUP	7	14	12A	27	22	1.5	>=H	4.5	4.0	3.5	1.5	56
TAE	8	16	12A	27	20	1.0	?	5.0	3.3	3.6	1.4	40
TAP+	17	17–18	>2D	30–31	22–24	1.3	?	3.8	3.0–3.5	3.5	1.5	52
TES	11	13	5A	30	22–24	1.5	?	4.0	3.5–3.7	3.3–3.5	1.5	46
THI	11–12	14–16	1A	26–27	20–22	1.0	<H	3.3–3.5	3.0–3.2	3.0–3.5	1.0–1.3	30
TOD	20	19	6D	32–33	?	1.0	?	3.7	?	3.1	?	?
UNC	9	12	6A	26	20	1.5	>H	5.0	3.7	3.5	=2.0	39
UNS	10–11	13	1A	25–27	20	1.5	?	4.0–4.5	3.5	3.5	1.5	30
WAL	14	15–16	>2A	28–30	22–24	1.0	?	3.7–4.0	3.0–3.3	3.5	< 2.0	42
WER	7	15–16	26–27S	35	8–8½	2.0	<H	4.3–4.7	4.1	< 3.0	1.5–1.6	57
VEX	9	14	16S	32	24	?	?	4.3–4.8	4.0–4.5	3.5–3.8	1.5	65
VOS	15	14	1A	28	22	1.3	?	4.3	3.3	3.0	1.5	33
ZEN	9	15–16	10–11A	29–30	22	1.0	?	4.0–4.5	3.4–3.6	3.3–3.5	2.0	49
ZIM	11–12	13–14	1A	30–32	22–24	< 2.0	?	4.2–5.0	3.0–3.3	4.0	1.8–2.0	70

LITERATURE CITED

Ahl, E. 1924a: Zool. Anz. 60.
— 1924b: ibid 61.
— 1926: Blätter Aquar. Terr. 37.
— 1928a: Ann. Mag. Nat. Hist. (10) 2.
— 1928b: Zool. Anz. 79.
— 1928c: Das Aquarium 2.
— 1935a: Zool. Anz, 110.
— 1935b: Sitzber. Ges. Nat. Freunde. Berlin.
— 1935c: Zool. Anz. 112.
— 1937: Mitt. Zool. Mus. Berlin 22.
— 1938: Zool. Anz. 124.
Ahn, P. M. 1961: UNESCO BULL. Unit. Nat. (1957).
Arnold, J. P. 1906: Wochenschr. Aquar. Terr. 3.
— 1908a: ibid. 5.
— 1908b: Blätter Aquar. Terr. 19.
— 1911: Wochenschr. Aquar. Terr. 8.
— 1912: ibid. 9.
— 1913: ibid. 10.
— 1914: ibid. 11.
— 1925: ibid. 22.
— 1933: ibid. 30.
— 1934: ibid. 31.
Arnold, J. P. and Ahl, E. 1936: "Fremdländische Süsswasserfische". Wenzel.
Arnoult, J. 1960: Bull. Mus. Nat. Hist. (2).
Aslaiani, R. 1966: The Aquarium (Innes') 35. (2).
Axelrod, H. and Schultz, L. 1955: Handb. Trop. Aquar. Fish. (McGraw-Hill.)
Basch, H. 1965: Aquar. Terr. Zeitschr. (DATZ) 18.
Benl, G. and Foersch, W. 1966: ibid. 19.
Beyer, L. 1930a: Wochenschr. Aquar. Terr. 27.
— 1930b: Blätter Aquar. Terr. 41.
— 1931: ibid. 42.
— 1932: ibid. 43.
Blanche and Miton 1960: Bull. Mus. Nat. Hist. Nat.
Bleeker, P. 1850: Verh. Batav. Genootsch. 23.
Boot, A. 1958a: Evolution 12: 46–62.
— 1958b: Bull. I.F.A.N. 20: 587–622.
Boulenger, G. A. 1899: Ann. Mus. Congo. Belge. Zool. 1.
— 1901: Proc. Zool. Soc. London 2.
— 1902: ibid. 3.
— 1903: Ann. Mag. Nat. Hist. (7). 12.
— 1907: "Fish. Nile".
— 1908: Ann. Mag. Nat. Hist. (8). 2.
— 1910: Bull. Soc. Nat. Luxembourg.
— 1911: Ann. Mag. Nat. Hist. (8). 8.
— 1912: Bull. Soc. Nat. Luxembourg.
— 1912b: Ann. Mus. Congo Belge Zool. (1). 2.
— 1913a: Bull. Soc. Nat. Luxembourg.
— 1913b: Rev. Zool. Afr. 2.
— 1913c: Ann. Mag. Nat. Hist. 1.

Boulenger, G. A. — 1913d: Blätter Aquar. Terr. 24.
— 1915: Cat. Freshw. Fish. Africa. 3.
— 1915b: Ann. Mag. Nat. Hist. 15.
— 1918: ibid. (9). 2.
— 1920: Ann. Mus. Congo. Belge 2. Zool. 4.
Bozkurt, B. — 1945: Rev. Fac. Sci. Univ. Istanbul. B.10.
Brieger, Th. — 1913: Wochenschr. Aquar. Terr. 10.
Brüning, Chr. — 1911: ibid. 8.
— 1929: ibid. 26.
— 1930: Aquarienfische & Naturfreunde (5).
Burkhart, — 1912: Blätter Aquar. Terr. 23.
Butzer, K. W. — 1957: Georg. An. Stockholm 39.
Cahalan, T. L. — 1957: Aquarium Journ. 28.
— 1966: The Aquarium (Innes') 35.
Clausen, H. S. — 1963: Vidensk. Medd. Dansk. naturh. Foren. 125.
— 1964: ibid. 126.
— 1966: Rev. Zool. Bot. Afr. 74.
— 1967: "Tropical Old World Cyprinodonts". Akademisk Forlag. Copenhagen.
Clausen, H. S. and Scheel, J. — 1966: Ichthyologica 37.
Cope, E. D. — 1877: Proc. Amer. Philos. Soc. 11.
Corte, R. la. — 1961a: Aquarium Journ. 32.
— 1961b: The Aquarium (Innes') 30. (2).
Cuvier and Valenciennes — 1846: Hist. Nat. Poiss. 18.
Daget, J. — 1948: Cat. I.F.A.N. 3.
— 1951: Bull. I.F.A.N. 13.
— 1954: Mem. I.F.A.N. 36.
— 1957: Bull. I.F.A.N. 22.
— 1960: Bull. Mus. Nat. Hist. Nat. (2). 32.
— 1962: Mem. I.F.A.N. 65.
Daget, J. and Arnoult, J. — 1965: Bull. Mus. Nat. Hist. Nat. 36.
Daget, J. and Iltis, A. — 1965: Mem. I.F.A.N. 74.
Darlington, C. D. — 1963: "Chromosome Botany etc." London.
Dieterich, W. von. — 1915: Wochenschr. Aquar. Terr. 12.
Dumeril, A. — 1861: Arch. Mus. Hist. Nat. 10.
Dvoskin, A. — 1956: The Aquarium (Innes') 25.
Elek, V. — 1965: Aquar. Terr. 12.
Fischer, — 1911: Wochenschr. Aquar. Terr. 8.
Foersch, W. — 1956: Mitt. Con. Europ. Vivar. 2. (1).
— 1957: Aquar. Terr. Zeitschr. (DATZ) 10.
— 1958: ibid. 11.
— 1959: Aquaria (Switzerland) 6.
— 1961: Aquar. Terr. Zeitschr. (DATZ) 14.
— 1966: ibid. 19.
— 1967: ibid. 20.
Fowler, H. W. — 1930: Proc. Acad. Nat. Sci. Philadelphia 82.
— 1934: Ann. Natal. Mus. 7.
— 1936: Proc. Acad. Nat. Sci. Philadelphia 88.
— 1949: ibid. 101.
— 1950: Notulae Naturae 225.

Franz, J.	1958: Aquar. Terr. 5.
—	1959: ibid. 6.
—	1960: Trop. Fish. Hobbyist. 9. (4).
—	1966a: ibid. 14. (9).
—	1966b: Aquar. Terr. 13.
Freund, L.	1913: Bibl. Aquar. Terr. Kunde. 24.
Garman, S.	1895: Mem. Mus. Comp. Zool. 19 (Harvard).
Gellner, F.	1911: Wochenschr. Aquar. Terr. 8.
Gerlach, G.	1912: Blätter Aquar. Terr. 23.
Gill, T.	1862: Proc. Acad. Nat. Sci. Philadelphia.
Goldstein, R. J.	1966: The Aquarium (Innes') 35. (9).
Gosline, W. A.	1949: Occ. Pap. Mus. Zool. Univ. Michigan (519).
Gosse, J. P.	1963: Ann. Mus. Afr. Centr. Zool. 116.
Graichen, H.	1909: Wochenschr. Aquar. Terr. 6.
Gras, R.	1961: Bull. I.F.A.N. 23.
Grote, J.	1910: Wochenschr. Aquar. Terr. 7.
Guichen,	1866: Mem. Soc. Sci. Nat. Cherbourg. 12.
Günther, A.	1866: Cat. Fish. 6.
—	1902: Proc. Zool. Soc. London. 2.
Günther, A. and Playfair, R. L.	1867: "The Fishes of Zanzibar".
Haas, F. and Schwarz, E.	1913: Geol. Rundschou 4.
Haas, R.	1965: World Health Or. WHO/EBL/39.65 and WHO/EBL/66.48.
Häfelin, W.	1965: Aquar. Terr. Zeitschr. (DATZ) 18.
Hajek, K.	1967: Trop. Fish. Hobbyist. 16. (6).
Hamilton-Buchanan, F.	1822: "Fishes Ganges River".
Hansen, H.	1957: The Aquarium (Innes'). 26.
Heger, R.	1953: Aquar. Terr. Zeitsch. (DATZ), 6.
Henzelmann, E.	1937: Wochenschr. Aquar. Terr. 34.
Hesse, U.	1966: Aquar. Terr. Zeitschr. (DATZ), 19.
—	1967: ibid. 20.
Hilgendorf, F. M.	1888: Sitzber. ges. natf. Fr. Berlin.
—	1905: Zool. Jahrb. Syst. Thiere, 22.
Hiller, W. A.	1961: Aquar. Terr. 8.
Hjerresen, G.	1954: Dansk Akvarieblad (8).
Hoedeman, J. J.	1956: Aquar. Terr. 3.
—	1957: ibid, 4.
—	1958: Bull. Aquar. Biol. 1.
—	1961: ibid. 2.
Holly, M.	1927: Sitzber. Akad. Wiss. Wien. 1.
—	1927b: Zool. Anz. 73.
—	1928: ibid. 76.
—	1930: Sitzber. Akad. Wiss. Wien. 139.
Hsu, T. S. and Pomerate, C.	1953: Journ. Heredity 44.
Isis, München	1963: Aquar. Terr. Zeitschr. (DATZ), 16.
Job, T. J.	1940: Rec. Indian Mus. 42.
Johnels, A. G.	1954: Arkiv Zool. 6.
Jerdon,	1849: Madras Journ. Lit. & Sci. 15.
Kaden, D.	1967: Aquar. Terr. 14.
Karbe, L.	1961: Mitt. Zool. Mus. Inst. Hamburg, 59.
Keilhack, L.	1910: Mitt. Zool. Mus. Berlin, 5.

Klausewitz, W.	1957: Senk. Biol. 38.
Klee, A. and Turner, B.	1962: Aquar. Journal. 33.
— —	1964: ibid. 35.
Kluge, K.	1961: Aquar. Terr. Zeitschr. (DATZ), 14.
Köhler, W.	1906: Blätter Aquar. Terr. 17.
—	1907: ibid. 18.
Kosswig, C.	1917: Wochenschr. Aquar. Terr. 14.
Kramp, E.	1913: Blätter Aquar. Terr. 24.
—	1914: ibid. 25.
Krüger, B.	1913: Wochenschr. Aquar. Terr. 10.
Ladiges, W.	1951: "Der Fisch in der Landschaft", Wenzel.
Lambert, J.	1956: L'Aquarium & les Poissons, 6.
—	1957: Rev. Zool. Bot. Afr. 57.
—	1958: ibid. 57.
—	1961: Ann. Mus. Centr. Afr. (8). 93.
Laursen, K. K.	1921: Stuekultur 18.
Leipzig, Aquar. Soc.	1937: Wochenschr. Aquar. Terr. 34.
Leneuf, N. and Mangenot, G.	1960: Soc. Bot. France (1959).
Lönnberg, E.	1895: Oefv. Kungl. Sv. Vet-Akad. Förh. 52.
—	1903: Ann. Mag. Nat. Hist. (7). 12.
—	1907: Y. Sjöstedt's Kilimandjaro-Meru Exp. 1. (5).
Looze, A. J. de	1964: Aquarium Journ. 32.
Mackessak, T.	1958: The Aquarium (Innes'), 27. (6).
Magis, N.	1963: Bull. Soc. Sci. Liege, 32.
Makino, S. and Nishimura	1952: Stain Tech. 27.
Matthes, H.	1964: Ann. Mus. Afr. Centr. Zòol. 126.
Mayer, F.	1936: Wochenschr. Aquar. Terr. 33.
—	1937: ibid. 34.
Mayr, E.	1963: "Animal Species and Evolution". Harvard.
McClelland, J.	1839: Jour. Asian Researches 19.
Meinken, H.	1929: Blätter Aquar. Terr. 40.
—	1930a: Wochenschr. Aquar. Terr. 27.
—	1930b: Blätter Aquar. Terr. 41.
—	1932a: Kalender Aquar. Terr. Fr.
—	1932b: Blätter Aquar. Terr. 43.
—	1933: ibid. 44.
—	1951: Aquar. Terr. Zeitschr. (DATZ), 4.
—	1952: ibid. 5.
—	1953: ibid. 6.
—	1960: ibid. 13.
—	1964: ibid. 17.
—	1966: ibid. 19.
Meder, E.	1953: ibid. 6.
Memmler, M.	1912: Wochenschr. Aquar. Terr. 7.
Milewski, A.	1913: Wochenschr. Aquar. Terr. 10.
—	1915: ibid. 12.
Monard, A.	1940: Arq. Mus. Bocage, 11.
Moreau, R. E.	1963: Proc. Zool. Soc. London, 141 (2).
Myers, G. S.	1924: Amer. Mus. Novt. 116.
—	1926: The Fish Culturist 6.
—	1933: Copeia

Myers, G. S.	1952: Aquarium Journ. 23.
—	1954: ibid. 25.
—	1955: The Trop. Fish. Mag.
—	1963: AKA. Journ. 2. (2).
Nichols and Griscom	1917: Bull. Amer. Mus. Nat. Hist. 37.
Nichols and La Monte	1933: Amer. Mus. Novt. 626.
Nieuwenhuizen, A.v.d.	1958: Aquar. Terr. Zeitschr. (DATZ), 11.
—	1960: ibid. 13.
—	1960b: Het Aquarium 30.
—	1963a: Aquarium Journ. 34.
—	1963b: Het Aquarium 33.
—	1964: Aquar. Terr. Zeitschr. (DATZ) 17.
Nymphea, Aquar. Soc.	1913: Wochenschr. Aquar. Terr. 10.
Oeser, R.	1928: Blätter Aquar. Terr. 39.
Ohlendorf, H. v.	1917: Wochenschr. Aquar. Terr. 17.
Pappenheim,	1924: Sitzber. Ges. naturf. Fr. Berlin.
—	1926: ibid.
Pellegrin, J.	1904: Bull. Mus. Hist. Nat. 10.
—	1907: ibid. 13.
—	1911: Bull. Soc. Zool. France. 36.
—	1913: ibid. 38.
—	1914: "Les Poissons du Bassin du Tchad".
—	1921: Bull. Mus. Nat. Hist. Nat. 27.
—	1922: Bull. Soc. Zool. France. 47.
—	1924: ibid. 49.
—	1924b: "Poissons Afr. Occ."
—	1925: Bull. Soc. Zool. France. 50.
—	1926: Rev. Zool. Afr. 14.
—	1928: Bull. Soc. Zool. France, 53.
—	1929: ibid. 54.
—	1930: ibid. 55.
—	1931: ibid. 56.
—	1934: Arch. Hydrobiol. 26.
Peters, N.	1963a: Intern. Rev. Hydrobiol. 48. (2).
—	1963b: Aquar. Terr. Zeitschr. (DATZ) 16.
—	1964a: Mitt. Hamburg Zool. Mus. Inst. (Kosswig F).
—	1964b: Aquar. Terr. Zeitschr. (DATZ) 17.
—	1965: Roux' Arch. Entwick. Mechanik 156.
—	1966: Mitt. Hamburg Zool. Mus. Inst. 63.
Peters, W.	1844: Monatschr. Akad. Wiss. Berlin.
—	1863: ibid.
—	1868: ibid.
Pfaff, J. R.	1933: Vidensk. Medd. Dansk. naturh. Foren. 94.
Pfeffer, G.	1893: Jahrb. Hamburg. Anst. 10.
—	1895: "Die Thierwelt Ost-Afrikas". 3.
Poll, M.	1938: Rev. Zool. Bot. Afr. 30.
—	1939: Ann. Mus. Congo. Belge. Zool. (1). 4. (1).
—	1941a: Bull. Mus. Hist. Nat. Belge. 17. (50).
—	1941b: Rev. Zool. Bot. Afr. 34.

Poll, M. 1952: ibid. 45.
— 1952b: ibid. 46.
— 1953: ibid. 48.
— 1957: "Les Genres des Poissons d'Eau Douce de l'Afrique".
— 1959: Ann. Mus. Congo Belge, Zool. (8). 71.
Poll, M. and Gosse, J. 1963: Ann. Mus. Afr. Centr. Zool. 116.
Post, A. 1964: Zeitschr. Zool. Syst. Evol. 3.
Pugh, E. 1914: Wochenschr. Aquar. Terr. 11.
Rachow, A. 1911: Blätter Aquar. Terr. 22.
— 1914: ibid. 25.
— 1920: ibid. 31.
— 1921: Bibl. Aquar. Terr. Fr. (16).
— 1924: Wochenschr. Aquar. Terr. 21.
— 1925: ibid. 22.
Radda, A. 1963: Aquar. Terr. Zeitschr. (DATZ). 16.
Roberts, M. F. 1954: The Aquarium (Innes'). 23. (7).
Rochebrune, 1883: Faune de Sénégambié, Poissons.
Roloff, E. 1936: Wochenschr. Aquar. Terr. 33.
— 1957: The Aquarium (Innes'). 26. (9).
— 1959: Trop. Fish. Hobbyist. 7. (10).
— 1965a: ibid. 14. (7).
— 1965b: Aquar. Terr. Zeitschr. (DATZ) 18.
— 1967: Trop. Fish. Hobbyist.
Rosen, D. E. 1964: Bull. Mus. Nat. Hist. New York. 127. (5).
Sachs, W. 1917: Wochenschr. Aquar. Terr. 14.
— 1919: Blätter Aquar. Terr. 30.
Sauvage, H. E. 1882: Bull. Soc. Zool. France. 7.
Scheel, J. J. 1960: Aquar. Journ. 31.
— 1962a: ibid. 33.
— 1962b: Tropicals 7. (2).
— 1963a: Aquar. Journ. 34.
— 1963b: The Aquarium 32. (6).
— 1963c: ibid. 32. (9).
— 1964: Aquar. Journ. 35.
— 1965: The Aquarium (Innes'). 34. (11).
— 1966a: Vidensk. Medd. Dansk. natur. Foren. 128.
— 1966b: Ichthyologica 37.
— 1966c: AKA. Killie Notes. 3. (3).
— 1967: ibid. 4.
Schiötz, A. 1967: Spolia Zool. Mus. Copenhagen. 25.
Schmidt, W. 1901: Blätter Aquar. Terr. 12.
Schreitmüller, W. 1919: ibid. 30.
— 1933: Wochenschr. Aquar. Terr. 30.
— 1935: ibid. 32.
Schultz, L. P. 1942: Proc. U.S. Nat. Mus. 92.
Schultze, O. 1937: Wochenschr. Aquar. Terr. 34.
Seleuthner, J. 1936: ibid. 33.
Selvey, H. M. 1964: The Aquarium (Innes'). 33. (11).

Sethi, R. P.	1960: Xerographed Dissertation. Mic. 60–1909. Ann. Arbor.
Sheljuzhko, L.	1954: Aquar. Terr. Zeitschr. (DATZ). 7.
Sick, K.	1961: Nature 192.
—	1965: Hereditas 54.
Siegal, H.	1963: Aquar. Journ. 34.
Sjöstedt, Y.	1904: "I Västafrikas Urskogar". Stockholm.
Smith, H. M.	1945: Smithsonian Inst. Bull. 188.
Soc. Essen-Ruhr	1912: Blätter Aquar. Terr. 23.
Soc. Plauen	1936: Wochenschr. Aquar. Terr. 33.
Soc. Wiesbaden	1936: ibid. 33.
Stansch, K.	1908: Wochenschr. Aquar. Terr. 5.
—	1910: ibid. 7.
—	1911: ibid. 8.
Steindachner, F.	1867: Sitzber. Akad. Wiss. Wien. 56.
—	1870: ibid. 61.
—	1881: ibid. 83.
—	1892: Denkschr. Akad. Wiss. Wien. 59.
—	1895: Notes Leyden Mus. 16.
Stolk, A.	1957: Het Aquar. 28.
—	1958: ibid. 29.
—	1960: ibid. 30 & 31.
Sundaja Raj, B. M. A.	1916: Rec. Indian Mus. 12.
Svensson, G. S. O.	1934: Kungl. Sv. Vet. Handl. (3). 12.
Vanderplanck, F. L.	1941: East Afr. med. Journ. 17.
Witzman, S. H. and Wourms	1967: AKA Journ. 4. (1).
Velke, O.	1930: Wochenschr. Aquar. Terr 27
Werner, F.	1919: Denkschr. Akad. Wiss. Wien 96.
—	1929: Wochenschr. Aquar. Terr. 26.
Whitehead, P. J. P.	1962: Bull. Brit. Mus. (Nat. Hist.). Zool. 9. (3).
Wickler, H.	1938: Wochenschr. Aquar. Terr. 35.
Wilhelm, H.	1966: Aquar. Terr. 13.
Vinciguerra, D.	1897: Ann. Mus. Genova (2). 17.
—	1927: ibid. 52.
Wolterstorff, W.	1913: Blätter Aquar. Terr. 24.
Träber, G.	1911: ibid. 22.
—	1912: ibid. 23.
Trewavas, E.	1962: Bonn. Zool. Beitr.
Zimmermann,	1908: Wochenschr. Aquar. Terr. 5.
Zindler, A.	1922: ibid. 19.

GLOSSARY

Acrocentric chromosome: A chromosome with the centromere (center of the moving chromosome) near one end of the chromosome.

Alkaline water: A type of water with pH above 7.0. Such water is just the reverse of being acid. Usually a water containing much dissolved lime.

Allopatric populations: Populations living in separate geographical areas.

Anaerobic condition: A condition without oxygen available.

Annual species: A species of which the free-swimming individuals live for less than one year, the fertile eggs hibernating in the soil during the dry period.

Bio-species: A zoological species defined by its biological characters. A bio-species is reproductively isolated from all other bio-species.

Bivalent chromosome: The ordinary (somatic) cell nucleus contains a coupled set of chromosomes (rivulins). One set originated from the sperm of the father; the other set from the egg of the mother. During the first phases of the meiotic cell division the similar (homologous) chromosomes pair up producing bivalents.

Conspecific populations: Populations belonging to the same species.

Ctenoid structures: Usually spine-like structures developing on edges of scales or on the ends of fin ray segments.

Cytology: The study of cells. Usually the study of cell nuclei, and in particular, of the chromosomes.

Ecological conditions: All external factors influencing the life of the population in a certain place. Usually climactic conditions, soil type, water type, forest, savannah, etc.

Embryo: The juvenile fish when still in the egg.

F-scales: A series of frontal scales which usually bridge the grooves for the lateral line organs of the front, dividing these grooves into a supra-orbital and a post-orbital part. The scales just above the eyes are also part of this series.

Gonatropine hormones: Hormones produced by the pituitary glands and governing the development of the sexual organs (gonads).

H-scales: Scales, usually rather small, in front of the G-scale of the front and situated between the supraorbital grooves.

Haploid: Cells having only a single set of chromosomes.

Hemoglobin: Chemical compounds producing the red color of the blood.

Hibernation: A phase of life during which the metabolism of the individual is highly reduced.

Holotype: A preserved individual to which a certain zoological name has been attached. Often the holotype is simply called the "type."

Hybrid: Individual produced by crossing of individuals originating from different populations, usually originating from different bio-species or morpho-species.

Karyotype: A schematic figuration of the haploid chromosome complement.

Lateral line: In Rivulinae the lateral line sense organs are generally restricted to the head, situated in open grooves, depressions or in canals with pores.

Maxillary: The upper jaw.

Meiosis: A certain sort of cell division during which sperm or eggs are produced.

Meristic character: A characteristic which can be measured (proportions) or counted (fin rays, scales).

Metacentric chromosome: A chromosome with the centromere (center of the chromosome during movement) situated near the middle of the chromosome.

Mitosis: The normal cell division.

Morpho-species: A zoological species defined by the external morphology, usually including colors.

Neuromasts: The sense organs of the lateral line system.

Opercle: Gill cover.

Osteology: The study of the skeleton.

Phenotype: The sum of visible characters of the individual.

Pre-opercle: The disc-like bone situated just below the eye of the rivuline species and in front of the opercle.

Pre-orbital: The (narrow) space between the mouth and the anterior rim of the eye, very narrow in most rivulins.

Sibling species: Morphological similar or identical populations which are reproductively isolated (Mayr 1963: 671).

Super-acrocentric chromosomes: A term introduced for acrocentric chromosomes originating from the fusion of two small acrocentric chromosomes.

Super-species: A group of closely related, usually strictly allopatric, species which have originated from a common ancestor.

Sympatric: Living in the same geographical area.

Synonym: One of two or more different names for the same species.

Syntypes: In the case where no holotype has been designated for a certain species the individuals which were used for the preparation of the original description are syntypes, unless otherwise stated.

Taxonomy: Classification of organisms.

Univalent: When homologous chromosomes fail to pair (see: *bivalent chromosome*) during the first meiotic division two (or more) univalents are produced.

GENERAL INDEX

Bold numbers indicate main reference.

INDEX OF SCIENTIFIC TERMS

Species-names are referred to the genera to which they are presently considered to belong. **Bold** references are major references.

Aph. = *Aphyosemion* Myers 1924
Apl. = *Aplocheilus* McClelland 1839
Apy. = *Aphyoplatys* Clausen 1967
E. = *Epiplatys* Gill 1862
F. = *Fundulosoma* Ahl 1924
N. = *Nothobranchius* Peters 1868
P. = *Pachypanchax* Myers 1933
R. = *Roloffia* Clausen 1966

GEOGRAPHICAL INDEX

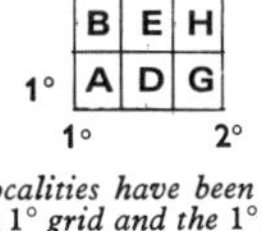

Most localities have been referred to the 1° × 1° grid and the 1° × 1° square has been divided into nine equal squares. Thus the individual locality can be plotted on any map accurate to twenty miles or less.